RADIO RECOMBINATION LINES

ASTROPHYSICS AND SPACE SCIENCE LIBRARY

A SERIES OF BOOKS ON THE RECENT DEVELOPMENTS
OF SPACE SCIENCE AND OF GENERAL GEOPHYSICS AND ASTROPHYSICS
PUBLISHED IN CONNECTION WITH THE JOURNAL
SPACE SCIENCE REVIEWS

VOLUME 80
PROCEEDINGS

RADIO RECOMBINATION LINES

PROCEEDINGS OF A WORKSHOP
HELD IN OTTAWA, ONTARIO, CANADA, AUGUST 24–25, 1979

Edited by

P. A. SHAVER
European Southern Observatory, Geneva, Switzerland

D. REIDEL PUBLISHING COMPANY
DORDRECHT : HOLLAND / BOSTON : U.S.A.
LONDON : ENGLAND

Library of Congress Cataloging in Publication Data

CIP

Radio Recombination Lines.
(Astrophysics and space science library ; v. 80)
Includes indexes.
1. Radio recombination lines–Congresses.
I. Shaver, Peter Albert, 1943– II. Series
QB462.5.R32 522'.682 80-1253
ISBN-13: 978-94-009-9026-5 e-ISBN-13: 978-94-009-9024-1
DOI: 10.1007/978-94-009-9024-1

Published by D. Reidel Publishing Company,
P.O. Box 17, 3300 AA Dordrecht, Holland.

Sold and distributed in the U.S.A. and Canada
by Kluwer Boston Inc., Lincoln Building,
160 Old Derby Street, Hingham, MA 02043, U.S.A.

In all other countries, sold and distributed
by Kluwer Academic Publishers Group,
P.O. Box 322, 3300 AH Dordrecht, Holland.

D. Reidel Publishing Company is a member of the Kluwer Group.

Softcover reprint of the hardcover 1st edition 1980

TABLE OF CONTENTS

FOREWORD

Since their first detection 15 years ago, radio recombination lines from several elements have been observed in a wide variety of objects including HII regions, planetary nebulae, molecular clouds, the diffuse interstellar medium, and recently, other galaxies. The observations span almost the entire range from 0.1 to 100 GHz, and employ both single-dish and aperture synthesis techniques.

The theory of radio recombination lines has also advanced strongly, to the point where it is perhaps one of the best-understood in astrophysics. In a parallel development, it has become possible over the last decade to study these same highly-excited atoms in the laboratory; this work provides further confirmation of the theoretical framework. However there has been continuing controversy over the astrophysical interpretation of radio recombination line observations, especially regarding the role of stimulated emission. A workshop was held in Ottawa on 24-25 August, 1979, bringing together many of the active scientists to review the field and discuss these questions of interpretation.

A broad concensus has emerged: the subtleties of the line-formation process are understood, and the conditions under which reliable information can easily be extracted from the line measurements are known. It thus appears likely that the emphasis will shift increasingly from the study of the line phenomenon itself to further application in other areas of astrophysics, ranging from physical processes in plasmas (temperatures, densities, ionization structure), to the large-scale properties of our galaxy (abundances, kinematics, structure), and studies of extragalactic systems.

This volume contains the 22 papers presented at the Ottawa workshop. It is the first book on radio recombination lines, and provides a comprehensive review of research in this field; it should prove useful both to active scientists in the field and to graduate students of astrophysics.

I would like especially to thank Dr. L.A. Higgs, my collaborator in organising this workshop. We are grateful to Drs. J.L. Locke and N.W. Broten of the Herzberg Institute of Astrophysics for making available the facilities of the institute, and providing financial support. The papers were prepared by the authors themselves and submitted in camera-ready form; we thank the D. Reidel Publishing Co. for their cooperation in producing this volume.

December 1979 P.A. Shaver
Geneva

P.A. Shaver (ed.), Radio Recombination Lines, vii.

FOREWORD

Since their first detection 15 years ago, radio recombination lines from several elements have been observed in a wide variety of objects including HII regions, planetary nebulae, molecular clouds, the diffuse interstellar medium, and recently, other galaxies. The observations span almost the entire range from 0.1 to 100 GHz, and employ both single-dish and aperture synthesis techniques.

The theory of radio recombination lines has also advanced strongly, to the point where it is perhaps one of the best-understood in astrophysics. In a parallel development, it has become possible over the last decade to study these same highly-excited atoms in the laboratory; this work provides further confirmation of the theoretical framework. However there has been continuing controversy over the astrophysical interpretation of radio recombination line observations, especially regarding the role of stimulated emission. A workshop was held in Ottawa on 24-25 August, 1979, bringing together many of the active scientists to review the field and discuss these questions of interpretation.

A broad concensus has emerged: the subtleties of the line-formation process are understood, and the conditions under which reliable information can easily be extracted from the line measurements are known. It thus appears likely that the emphasis will shift increasingly from the study of the line phenomenon itself to further application in other areas of astrophysics, ranging from physical processes in plasmas (temperatures, densities, ionization structure), to the large-scale properties of our galaxy (abundances, kinematics, structure), and studies of extragalactic systems.

This volume contains the 22 papers presented at the Ottawa workshop. It is the first book on radio recombination lines, and provides a comprehensive review of research in this field; it should prove useful both to active scientists in the field and to graduate students of astrophysics.

I would like especially to thank Dr. L.A. Higgs, my collaborator in organising this workshop. We are grateful to Drs. J.L. Locke and N.W. Broten of the Herzberg Institute of Astrophysics for making available the facilities of the institute, and providing financial support. The papers were prepared by the authors themselves and submitted in camera-ready form; we thank the D. Reidel Publishing Co. for their cooperation in producing this volume.

December 1979 P.A. Shaver
Geneva

P.A. Shaver (ed.), Radio Recombination Lines, vii.

LIST OF PARTICIPANTS

BELL, M.B.	Herzberg Institute of Astrophysics, National Research Council of Canada, Ottawa, ON K1A 0R6, Canada
BROWN, R.L.	National Radio Astronomy Observatory, P.O.Box 2, Green Bank, WV 24944, USA
CHURCHWELL, E.B.	Washburn Observatory, Univ. of Wisconsin, 475 N. Charter St., Madison, WI 53706, USA
DAVIES, R.D.	Nuffield Radio Astronomy Laboratories, Jodrell Bank, Macclesfield, GB-Cheshire SK11 9DL, England
DOHERTY, L.H.	Herzberg Institute of Astrophysics, National Research Council of Canada, Ottawa, ON K1A 0R6, Canada
FALGARONE, E.	DERAD, Observatoire de Paris, F-92190 Meudon, France
GOLDBERG, L.	Kitt Peak National Observatory, P.O.Box 26732, Tucson, AZ 85726, USA
GORKOM, J.H. van	Kapteyn Astronomical Institute, Postbus 800, NL-9700 AV Groningen, The Netherlands
HIGGS, L.A.	Herzberg Institute of Astrophysics, National Research Council of Canada, Ottawa, ON K1A 0R6, Canada
KESTEVEN, M.J.	Astronomy Group, Dept. of Physics, Queen's University, Kingston, ON K7L 3N6, Canada
KLEPPNER, D.	Research Laboratory of Electronics, Massachusetts Institute of Technology, Cambridge, MA 02139, USA
LOCKMAN, F.J.	National Radio Astronomy Observatory, Edgemont Rd. Charlottesville, VA 22901, USA

P.A. Shaver (ed.), Radio Recombination Lines, ix.

MACLEOD, J.M.	Herzberg Institute of Astrophysics, National Research Council of Canada, Ottawa, ON K1A 0R6, Canada
MATHIS, J.S.	Washburn Observatory, Univ. of Wisconsin, 475 N. Charter St., Madison, WI 53706, USA
MEZGER, P.G.	Max-Planck-Institut für Radioastronomie, Auf dem Hügel 69, D-5300 Bonn 1, West Germany
PANAGIA, N.	Laboratorio di Radioastronomia CNR, Via Irnerio 46, I-40126 Bologna, Italy
PANKONIN, V.	Max-Planck-Institut für Radioastronomie, Auf dem Hügel 69, D-5300 Bonn 1, West Germany
PAULS, T.A.	Max-Planck-Institut für Radioastronomie, Auf dem Hügel 69, D-5300 Bonn 1, West Germany
RADHAKRISHNAN, V.	Raman Research Institute , Hebbal P.O., Bangalore 560006, India
SCHMID-BURGK, J.	Max-Planck-Institut für Radioastronomie, Auf dem Hügel 69, D-5300 Bonn 1, West Germany
SEATON, M.J.	Dept. of Physics and Astronomy, University College London, Gower St., GB-London WC1E 6BT, England
SHAVER, P.A.	European Southern Observatory, c/o CERN, CH-1211 Geneva 23, Switzerland
TERZIAN, Y.	National Astronomy and Ionosphere Center, Space Sciences Building, Cornell University, Ithaca, NY 14853, USA
VINER, M.R.	Astronomy Group, Dept of Physics, Queen's University, Kingston, ON K7L 3N6, Canada
WALMSLEY, C.M.	Max-Planck-Institut für Radioastronomie, Auf dem Hügel 69, D-5300 Bonn 1, West Germany
WILSON, T.L.	Max-Planck-Institut für Radioastronomie, Auf dem Hügel 69, D-5300 Bonn 1, West Germany

PART I

PHYSICS OF RADIO RECOMBINATION LINES

"The uncertainty is in the astronomy,
not in the physics"

— M.J. Seaton

THEORY OF RECOMBINATION LINES

M. J. Seaton*
Joint Institute for Laboratory Astrophysics,
University of Colorado and National Bureau of Standards,
Boulder, Colorado 80309 U.S.A.

1. INTRODUCTION

In a recent review, Brown, Lockman and Knapp (1978, to be referred to as BLK) say that radio recombination lines have been observed at frequencies from 0.242 GHz to 86 GHz. In the present paper I discuss the radio lines and also the recombination lines observed in the infrared, optical and ultraviolet spectral regions; this takes us out to He^+ 2α at 1.83 MGHz (λ = 1640 Å). I will attempt to provide some historical perspective and to present a simple conceptual approach to the basic ideas of the theory. I will not give all of the detailed formulae which can be found in the literature (see, for example, Brocklehurst and Seaton (1972, to be referred to as BS)); nor will I attempt any detailed discussion of the radio observations, since these problems have been discussed by BLK and will be discussed further by other speakers at the present meeting.

2. SOME EARLIER HISTORY

In some of the earliest attempts to interpret observed frequencies of spectrum lines, it was assumed that atoms might have internal structures and that perturbations of the atom could give rise to characteristic atomic frequencies ν(atom). Maxwell's electromagnetic theory suggested that the frequency of the emitted radiation, ν(radiation), would be equal to ν(atom). Since, however, the internal motions of the perturbed atom might not correspond to those of a simple harmonic oscillator, overtone frequencies could also be expected, giving

$$\nu(\text{radiation}) = m \times \nu(\text{atom}) \qquad (1)$$

*Permanent address: Department of Physics and Astronomy, University College London, Gower St., London WC1E 6 BT, ENGLAND.

P.A. Shaver (ed.), Radio Recombination Lines, 3–22.

with m=1,2,3,... . In the early work, such series of spectral lines, equally spaced in frequency, were searched for but not found. In most cases the observed frequencies seemed to be distributed almost at random but Huggins (1880) observed 14 lines in stellar spectra which formed a regular series. The first few members of the series had also been observed in hydrogen discharges and the entire series was therefore attributed to hydrogen. It was shown by Balmer (1885) that the observed frequencies could be fitted to the formula

$$\nu_n = C \times \left(\frac{1}{2^2} - \frac{1}{n^2}\right) \tag{2}$$

with n=3,4,5,... and C a constant. Later, Paschen (1908) discovered a series in the near infrared, with frequencies given by

$$\nu_n = C \times \left(\frac{1}{3^2} - \frac{1}{n^2}\right) \quad . \tag{3}$$

Development of the theory by Bohr (1913) became possible following the discovery of the electron by Thomson (1897), the introduction by Einstein (1905) of quantum conditions in connection with the interactions between matter and radiation, and the discovery of the nuclear nature of the atom by Rutherford (1911).

Bohr assumed that hydrogenic atoms contain one electron and a nucleus of charge Z. He took the angular momentum to be quantized,

$$L = \left(\frac{h}{2\pi}\right) \tau \tag{4}$$

where τ is an integer (I use Bohr's original notation; it is of interest to note that quantization of angular momentum had been introduced earlier by Nicholson (1912) in an attempt to interpret the observed frequencies of what are now known as the "forbidden lines" of gaseous nebulae and of the solar corona). Using (4) and ordinary classical mechanics, Bohr derived expressions for the radii r_τ of circular orbits and for the corresponding energies E_τ.

Bohr further assumed that the frequency of the emitted radiation is not, in general, given by relations of the form of equation (1) but by the quantum condition

$$h\nu = E_{\tau_1} - E_{\tau_2} \quad . \tag{5}$$

For the frequency of the observed radiation he then obtained

$$\nu = C \times Z^2 \times \left(\frac{1}{\tau_2^2} - \frac{1}{\tau_1^2}\right) \tag{6}$$

with

$$C = \frac{2\pi^2 me^4}{h^3} \quad . \tag{7}$$

Taking $Z = 1$ for hydrogen, $\tau_2 = 2$ gives the Balmer formula (2) and $\tau_2 = 3$ the Paschen formula (3); and the constant C is given in terms of fundamental atomic constants. Bohr also noted that a series of lines observed by Pickering (1896) in ζ Puppis could be attributed to He^+, the observed frequencies being obtained on taking $Z=2$ and $\tau_2=4$ in (6). This identification was confirmed by subsequent laboratory work, but precision measurements showed that some further refinement was required. According to (6) the Balmer line with τ_1 has the same frequency as the Pickering line with $\tau_1' = 2\tau_1$. Precision measurements revealed small differences in these frequencies, and Bohr showed that they could be explained on introducing the refinement of replacing the electron mass m in (7) by the reduced mass

$$\mu = \frac{m}{1 + m/M} \tag{8}$$

where M is the mass of the nucleus. It is this reduced mass correction which gives the separation between the radio lines of hydrogen and helium.

I give three quotations from Bohr's paper which are of direct relevance to the topics we are discussing at the present meeting.

Quotation 1. Having explained the Balmer and Paschen series, Bohr predicted the existence of other series:

> "If we put $\tau_2 = 1$ and $\tau_2 = 4,5,\ldots$, we get series respectively in the extreme ultraviolet and the extreme ultrared, which are not observed, but the existence of which may be expected."

The series with $\tau_2 = 1$ was later named after Lyman. Our meeting is concerned with "the extreme ultrared."

Quotation 2. At the time when Bohr's paper was written, 12 Balmer lines had been observed in the laboratory and 33 in spectra of celestial bodies. Bohr calculated numerical values for the radii of orbits, proportional to τ^2, and compared them with typical mean free paths in laboratory discharges and in celestial bodies. He concluded that the differences in the number of lines observed was due to differences in density:

> "According to the theory the necessary condition for the appearance of a great number of lines is therefore a very small density of the gas; for simultaneously to obtain an intensity sufficient for observation the space filled with the gas must be very great."

Quotation 3. Bohr paid particular attention to the case of transitions between highly excited states, in order to establish the correspondence between his theory and ordinary classical theory:

"If we consider the passing of the system between two states corresponding to $\tau = N$ and $\tau = N-n$ where n is small compared with N, we get $\nu = n\omega$ ... if ω is the frequency of revolution of the electron."

Let us use modern notation. In (6), put $\tau_2 = n$ and $\tau_1 = n+m$ and assume $n >> m$. We then have

$$\nu \simeq m \times \nu_n(\text{atom}) \tag{9}$$

where

$$\nu_n(\text{atom}) = \frac{2cZ^2}{n^3} \tag{10}$$

is Bohr's ω, the frequency of revolution of the electron. Putting $m = 1,2,3,...$ in (9) we have the frequencies of the $n\alpha$, $n\beta$, $n\gamma$,... lines. Thus at radio frequencies we have reached the semiclassical limit of equation (1); we observe the fundamental frequencies and the harmonics for which the early spectroscopists had searched in vain.

3. PHYSICAL PROCESSES OF IMPORTANCE FOR THE INTERPRETATION OF RECOMBINATION SPECTRA

3.1 The Correspondence Between Classical and Quantal Theories

In considering the interpretation of radio recombination lines it is necessary to consider the correspondence between classical and quantal theories, in the limit of large quantum numbers, and also the existence of certain correspondence identities (Percival, 1968).

3.2 Rutherford Scattering

The problem of Rutherford scattering provides an example of a correspondence identity; for scattering of electrons by nuclei, and considering only Coulomb forces, classical and quantal theories give identical results for the differential cross section $\sigma(\theta)$.

This cross section becomes large for small values of θ, which in classical theory occur for large values of the impact parameter p (see Fig. 1). Using the Rutherford formula, one can calculate the cross sections for energy transfer in electron-electron collisions (it is necessary to introduce a cutoff in p, related to the Debye radius). These cross sections are very large and electron-electron collisions are therefore very efficient in tending to establish a Maxwell distribution of electron velocities, characterized by a temperature T_e. In ionized hydrogen regions, processes of energy gain (injection of photoelectrons) and energy loss (mainly excitation of forbidden lines) tend to perturb the Maxwell distribution but such perturbations are negligible compared with the electron-electron collisions. The electron velocity distribution is therefore always very close to

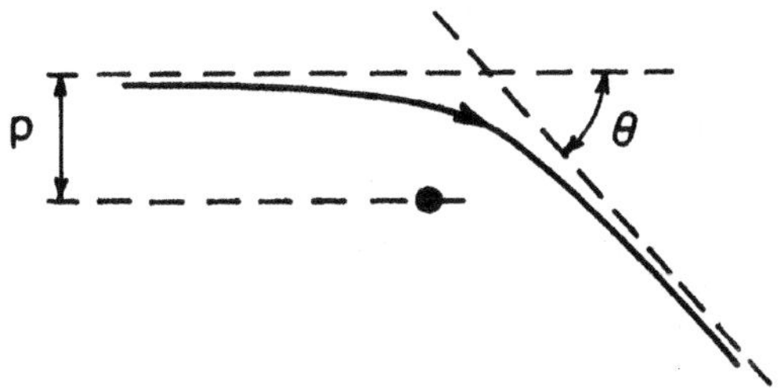

Figure 1. Rutherford scattering.

Maxwellian. The actual numerical value of T_e is, of course, determined by the gain and loss processes.

3.3 Free-Free Transitions

When scattered by a nucleus, an electron is accelerated and emits electromagnetic radiation. Let $P(E,p,\nu)d\nu$ be the probability for emission of radiation in the range $\nu,\nu+d\nu$, for an electron of energy E and impact parameter p. This is the process which gives the free-free radio continuum. At thermal energies and radio wavelengths, classical and quantal theories give results for $P(E,p,\nu)$ in very close agreement (Oster, 1961).

The total cross section is

$$\sigma(E,\nu)d\nu = \int [P(E,p,\nu)d\nu] \times 2\pi p dp \qquad (11)$$

(where it is, again, necessary to introduce a cutoff in p).

3.4 Bound-Free Transitions

The energy lost by the colliding electron is neglected in the classical theory of free-free transition. If a quantum of frequency ν is emitted, the final energy of the electron is $E' = E-h\nu$. For ν sufficiently large this energy can be negative, corresponding to radiative capture to a bound state. According to quantum theory the bound state energies are given by $E_n = -hC/n^2$ and the separation between adjacent levels is $dE_n = 2hC/n^3$. For large n the capture cross section $\sigma_n(E)$ could be calculated on putting $dE_n = hd\nu$ giving $d\nu = 2C/n^3$ and

$$\sigma_n(E) = \int \left[P(E,p,\nu) \frac{2C}{n^3}\right] 2\pi p dp \quad . \qquad (12)$$

In practice, formulae obtained directly from quantum theory are usually employed.

3.5 Bound-Bound Radiative Transitions

We have seen that, for transitions $n+m \rightarrow n$ with $n \gg m$, the frequencies of emitted radiation correspond to the m'th harmonic of the frequency of the orbital motion of the electron. The amount of radiation in these harmonics can be calculated using classical theory (see, for example, Landau and Lifshitz, 1962) and radiative transition probabilities can be deduced on introducing the appropriate quantum numbers (Hjellming, Mezger, Schraml and Burke; see Mezger, 1968). The expression obtained for the oscillator strength is

$$f(n,n+m) = \frac{4n}{3m^2} J_m(m)[J_m(m) - J_{m+1}(m)] \tag{13}$$

where J_m is a Bessel function. The exact quantum mechanical expression for f involves hypergeometric functions. On re-arranging the expansions for these functions, Menzel (1968,1969) obtained (13) for $m \ll n$, and also gave more accurate formulae.

3.6 Inelastic Collisions Between Electrons and Hydrogen Atoms

We have noted that, for conditions in nebulae, electron-electron collisions are very effective in re-distributing energy. It may be expected that, in the limit of large n, electron collisions with hydrogen atoms will also be important. Inelastic collisions, and collisional ionization and the inverse process of three-body capture, will tend to give a thermodynamic (Saha) distribution for the highly excited states.

In considering electron collisions with hydrogen atoms we are faced with a more difficult three-body problem. Direct use of quantum mechanical methods does not provide practical techniques for calculating the cross sections in the energy range of interest. In classical mechanics there are no known analytical solutions for the three-body problem, but for a given impact parameter and specified initial state of the target atom it is possible to carry out numerical integrations of particle trajectories. The difficulty here is that many parameters are required to specify the initial conditions and, until the integrations have been completed, one does not know what the final state of the atom will be. In practice, Monte Carlo techniques are used, involving essentially random samplings of initial conditions.

The classical problem can also be solved using a variety of approximations with well-defined regions of validity (Percival, 1978). The ultimate test of their validity is provided by the Monte Carlo calculations. Having obtained solutions of the classical problem, cross sections for transitions between quantized energy levels are obtained using correspondence principles discussed by Percival and Richards (1975). Results, and further references to the literature, are given by Burgess and Percival (1968), Percival and Richards (1975) and Percival (1978). It is estimated that errors in the recommended cross sections are unlikely to exceed 15% and that a typical error is 6%.

3.7 Redistribution in Angular Momentum

The hydrogen atom has quantum numbers n,ℓ,m with $\ell = 0,1,\ldots(n-1)$ and $m = (-\ell),(-\ell+1),\ldots(+\ell)$. If all quantum states are equally populated for a given value of n, we have

$$N(n\ell) = \frac{(2\ell+1)}{n^2} N(n) \tag{14}$$

where $N(n\ell)$ is the number of atoms with quantum numbers n,ℓ and $N(n)$ the total number in energy level n.

If the level populations are determined by purely radiative processes, (14) will not be valid. We must therefore consider collisions which give a redistribution in ℓ. The most important are collisions with protons,

$$H(n,\ell) + H^+ \rightarrow H(n,\ell\pm 1) + H^+ \quad . \tag{15}$$

Very large impact parameters contribute (once again, a cutoff must be used) and accurate calculations can be made using a semiclassical perturbation treatment (Pengelly and Seaton, 1964). Comparing the rates for (15) with rates for radiative de-excitation, we obtain a critical value N_c of $N(H^+)$ such that (14) is valid for $N(H^+) \gg N_c$. Figure 2 shows $\log(N_c)$ against $\log(n)$. There is a dramatic difference between the radio and optical regions: for $n = 100$, $N_c \sim 10^{-3}$ while for $n = 10$, $N_c \sim 10^{+6}$. The He^+ 3α line at λ = 4686 Å provides an example of the errors which can arise if (14) is assumed at densities for which it is not valid; the line intensity can be underestimated by a factor of 1.8.

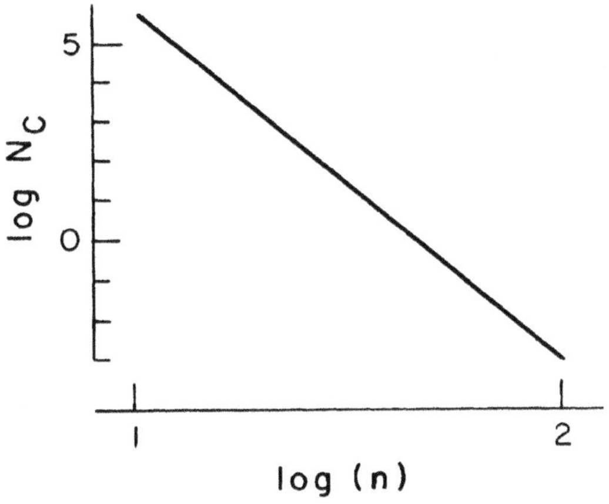

Figure 2. N_c, such that (14) is valid for $N(H^+) \gg N_c$.

The relation (14) has been assumed in deriving expressions such as (13). The possibilities of departures from (14) should be borne in mind as work continues on the exploration of new ranges of frequencies and different types of objects.

3.8 Pressure Broadening

The term "pressure broadening" is used to describe all mechanisms which lead to broadening of spectral lines with increasing pressure (or density).

One possible mechanism is the Stark effect. Its particular importance for hydrogen may be understood from the following simple semiclassical argument. According to classical theory, an electron in a bound state of a hydrogen atom describes a closed elliptical orbit. Let $\vec{r}(t)$ be the positive vector of the electron in a coordinate system with the proton at the origin and let $T = 1/\nu(\text{atom})$ be the time taken to complete the orbit. Such an orbit has a time averaged dipole moment

$$\vec{P} = \frac{-e}{T}\int_0^T \vec{r}(t)dt \quad . \tag{16}$$

The value of $\vec{P}$ depends on the eccentricity of the orbit and vanishes for circular orbits. For a hydrogen atom in a constant electric field $\vec{F}$ there is a first-order interaction energy

$$\Delta E = \vec{P}\cdot\vec{F} \quad . \tag{17}$$

This linear dependence on F is characteristic of hydrogenic systems. For other atoms the orbits precess, and there is no time-averaged dipole moment in the absence of a field; the effect of a field is to induce a dipole moment and hence to give an energy shift quadratic in F. (From the standpoint of quantum perturbation theory, the linear Stark effect for hydrogenic systems is a consequence of the energy degeneracy of states having the same value of n but different values of ℓ; and the quadratic effect for non-hydrogenic systems a consequence of the lack of such degeneracy.)

At any point in a plasma one has fluctuating electric fields $\vec{F}(t)$ due to neighbouring particles. The broadening of the spectrum lines can be described in terms of the linear Stark effect so long as $\vec{F}(t)$ varies sufficiently slowly (the quasi-static approximation). It is essentially this effect -- but developed in a much more sophisticated way -- which explains the pressure broadening of hydrogen lines in stellar atmospheres.

For large values of n, the orbit times T become very large and the quasi-static Stark approximation is no longer valid. This is discussed by Griem (1967), who shows that the broadening is mainly determined by electron impacts. We have already noted that, for highly excited states, such impacts are very effective in redistributing energy.

Consider a damped harmonic oscillator with angular frequency ω_0. The electric vector F(t) of the emitted radiation is proportional to $\exp\{i\omega_0 t - (1/2)\gamma t\}$, and the power radiated to $\exp(-\gamma t)$, so that $(1/\gamma)$ is a characteristic decay time. The Fourier transform of F(t) is proportional to $\{i(\omega_0-\omega) - (1/2)\gamma\}^{-1}$ and the radiation intensity to $\{(\omega-\omega_0)^2 + [(1/2)\gamma]^2\}^{-1}$. Now consider a hydrogen atom in level n. Inelastic collisions with electrons give a decay time $(1/\gamma)$ where $\gamma = N_e\langle vQ_n\rangle$, $\langle\ldots\rangle$ indicates an average over the velocity distribution v, and

$$Q_n = \sum_{n''\neq n} Q(n\rightarrow n'') \tag{18}$$

is the total inelastic collision cross section. For radiation in the transition $n\rightarrow n'$ the relevant parameter is $\langle v(Q_n + Q_{n'})\rangle$ and when the difference between n and n' is small we obtain

$$\gamma = 2N_e\langle vQ_n\rangle \tag{19}$$

for the damping constant. The conclusion of Griem's paper is that this mechanism of impact broadening is the dominant pressure broadening mechanism for the radio lines. Griem does not use the notation of inelastic cross sections and his computations of γ involve some approximations. In the expansion for the electron-electron interaction,

$$\frac{1}{r_{12}} = \sum_{\lambda=0}^{\infty} \frac{r_1^\lambda}{r_2^{\lambda+1}} P_\lambda(\hat{r}_1\cdot\hat{r}_2) \quad \text{for} \quad r_2 > r_1 \quad , \tag{20}$$

he retains only the dominant dipole term, $\lambda = 1$. This introduces an error of a factor of about 1/2, since the contribution from all terms with $\lambda \neq 1$ is comparable to the contribution from $\lambda = 1$. However, Griem calculates the contribution from $\lambda = 1$ using a perturbation theory which gives an overestimate by a factor of about 2. It is seen that the two sources of error tend to cancel.

A calculation using (19) and best available estimates for the cross sections was made by Brocklehurst and Leeman (1971), and the validity of the theory is further discussed by Peach (1972).

4. CALCULATION OF RECOMBINATION LINE INTENSITIES

4.1 Level Populations

Let N_i be the number density of atoms in states i. We take i = n if (14) can be assumed, i = (n,ℓ) otherwise. We put

$$N_i = b_i \times N_i\,(\mathrm{LTE}) \tag{21}$$

where "LTE" denotes "local thermodynamic equilibrium." In full thermodynamic equilibrium ("TE") at temperature T the intensity of radiation is given by the Planck function $B_\nu(T)$ and there are no spectrum lines. For LTE we assume only that the material particles satisfy TE laws, such as the Saha law which gives N_i (LTE) in terms of ion and electron densities.

The equations for the determination of N_i are obtained on considering the processes: radiative capture and cascade; redistribution in energy due to electron impacts; redistribution in ℓ due to ion impacts (if $i = (n,\ell)$); and in some special cases absorption of radiation (usually negligible).

For $i = n$, b_n depends on the energy variable $\varepsilon_n = -1/n^2$ and varies quite slowly as a function of this variable. In the matrix condensation method of Burgess and Summers (1969), a very large system of equations for the determination of the b_n coefficients is replaced by a much smaller system for the calculation of b_n at selected representative values of ε_n. Calculations for the case of $i = n$ have been made by Brocklehurst (1970), who tabulates b_n and $d(\ell n\ b_n)/dn$ for T_e and N_e in the ranges 2500 to 20,000 K and 10 to 10^5 cm^{-3}. A computer program for the calculations of these quantities has been published by Brocklehurst and Salem (1977). The best available estimates of cross sections are used, and computations can be made for lower values of T_e.

Figure 3 shows typical results for b(E) as a function of E. For $E < 0$ we have $b_n = b(E_n)$ and for $E > 0$ we have $b(E) = 1$ corresponding to the Maxwell distribution. The departures from LTE are such as to give $b_n < 1$; this is because downward radiative transitions are not compensated by the radiative absorptions which would occur with a full TE radiation field.

Results for smaller values of n (using $i = (n,\ell)$) are given by Brocklehurst (1971) and further results for H Brackett lines ($n \rightarrow 4$) are given by Giles (1977) and for He^+ UV lines by Seaton (1978).

4.2 Intensity of Radiation

The intensity of radiation in the vicinity of a line is given by the equation of radiative transfer

$$\frac{d}{ds} I_\nu = -\kappa_\nu I_\nu + j_\nu \tag{22}$$

where j_ν and κ_ν are emission and absorption coefficients for the line and for the continuum. Detailed formulae are given in many papers (see, for example, BS) and need not be repeated here. For the transition $n' \rightarrow n$ the line coefficients are given by

$$j_\nu^L = h\nu \times N_{n'} \times \frac{A_{n'n}}{4\pi} \times \phi(\nu) \tag{23}$$

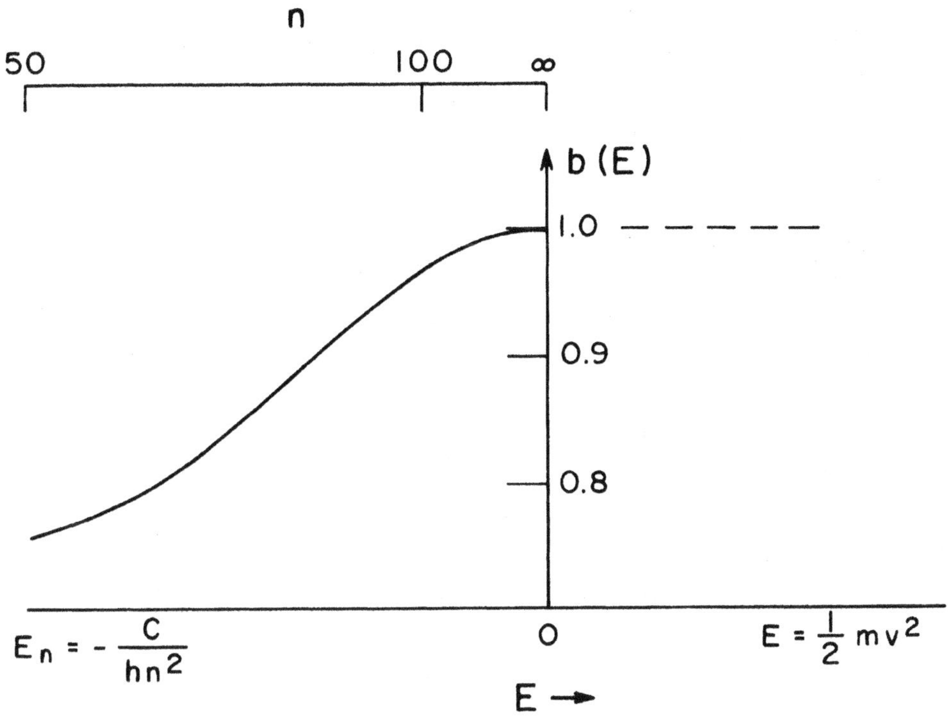

Figure 3. $b_n(E)$ for $T_e = 10^4$ K, $N_e = 10^3$ cm^{-3}.

$$\kappa_\nu^L = \frac{c^2}{2\nu^2} \times \left[\frac{N_n}{\omega_n} - \frac{N_{n'}}{\omega_{n'}}\right] \times \frac{\omega_{n'} A_{n'n}}{4\pi} \times \phi(\nu) \tag{24}$$

where $\omega_n = 2n^2$ and $\phi(\nu)$ is a profile factor normalized to

$$\int \phi(\nu) d\nu = 1 \quad . \tag{25}$$

These equations may be written

$$j_\nu^L = b \times j_\nu^L \text{ (LTE)} \tag{26}$$

$$\kappa_\nu^L = b\beta \times \kappa_\nu^L \text{ (LTE)} \tag{27}$$

where

$$\beta = 1 - kT_e \frac{d}{dE} \ell n(b) \quad . \tag{28}$$

In practice one usually finds that β is negative; the background free-free radiation can then produce enhanced line intensities due to stimulated emission, that is to say, maser action (Goldberg, 1966).

The solution of the transfer equation (22) is discussed by BS.

5. RECOMBINATION LINES NOT IN THE RADIO REGION

Two complications arise when we leave the radio region: equation (14) can no longer be assumed; and we must correct for interstellar absorption. The relation between the observed and emitted intensities is

$$I(\lambda,\mathrm{obs}) = I(\lambda,\mathrm{emm}) \times 10^{-0.4\, A_\lambda} \tag{29}$$

where A_λ is the extinction in magnitudes. We can put

$$A_\lambda = K \times X_\lambda \tag{30}$$

where X_λ, shown in Figure 4, gives the known λ-dependence of the extinction and K depends on the amount of extinction for a particular object. Assuming T_e to be known, $I(\lambda,\mathrm{emm})$ can be calculated using the intensity in the optically thin radio continuum. The observed line intensity $I(\lambda,\mathrm{obs})$ can then be used to calculate A_λ from (29).

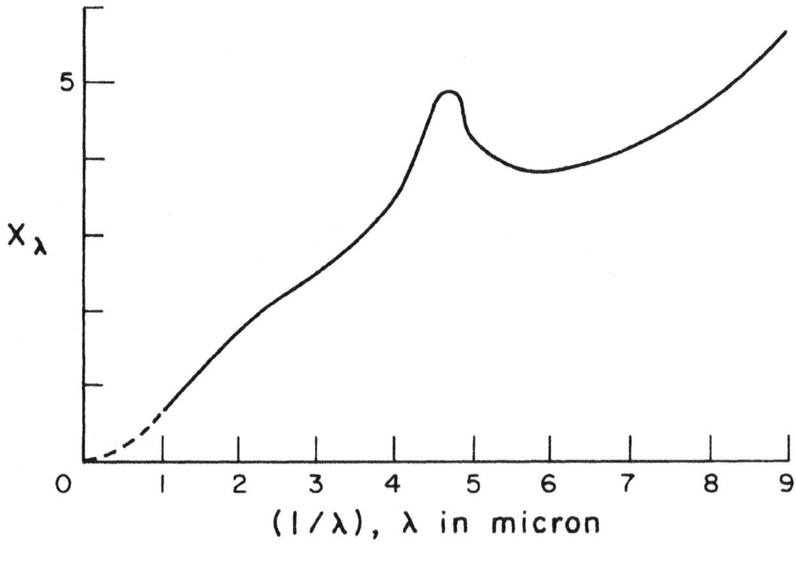

Figure 4. The wavelength dependence of interstellar extinction, X_λ against $(1/\lambda)$.

The observational data are particularly rich for the planetary nebula NGC 7027. Figure 5 (Seaton, 1979) shows A_λ deduced from (29) against X_λ. The observed lines extend from H 4γ at λ = 2.17 μ to He^+ 2α at λ = 0.164 μ. Most points come very close to a straight line, in accordance with (30), but there are some discrepancies for the largest values of λ, probably due to obscuration local to the object.

Many other checks on the theory have been made. For example, Miller (1971) has considered the high Balmer lines, out to n = 18. Such checks are important for radio astronomers in that they confirm the accuracy of the calculated departure coefficients in the limit of smaller values of n.

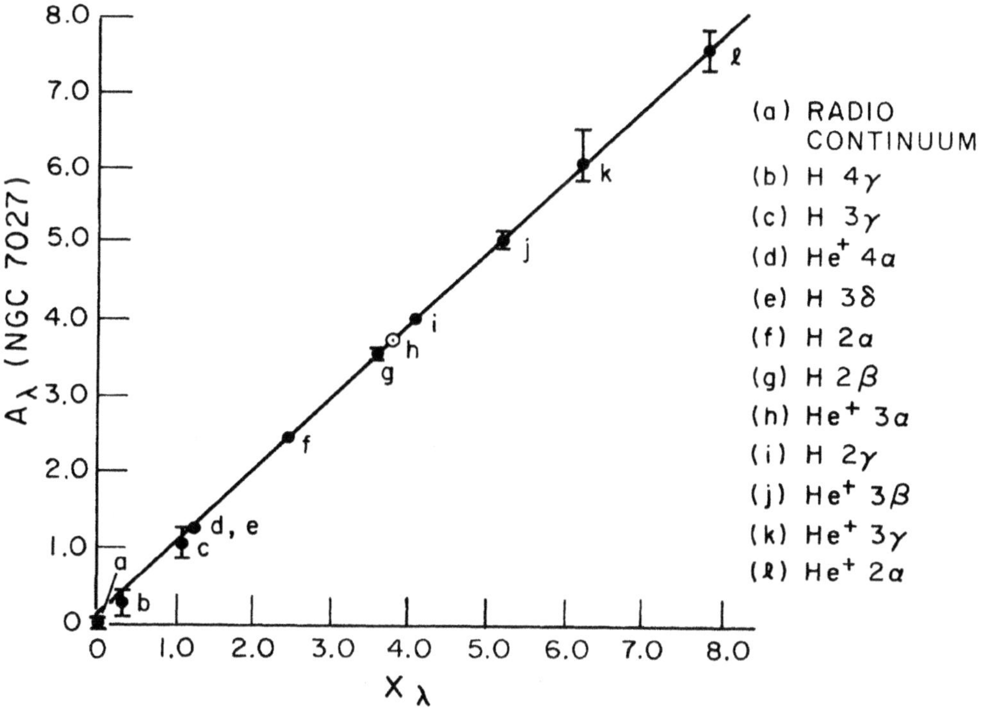

Figure 5. Extinction for NGC 7027 deduced from recombination lines, against X_λ.

6. THE RADIO LINES

6.1 Profiles

I consider first the question of line profiles, since it is important in judging how one should best attempt to interpret the observations.

Let I_ν be the total intensity, I_c the continuum intensity. The total intensity in a line is

$$I_L = \int_{\text{Line}} (I_\nu - I_c)d\nu \quad . \tag{31}$$

We can put

$$(I_\nu - I_c) = I_L\ \theta\ (\nu) \tag{32}$$

which defines the line profile factor $\theta(\nu)$, normalised to

$$\int \theta(\nu)d\nu = 1 \quad . \tag{33}$$

For a homogeneous nebula, $\theta(\nu) = \phi(\nu)$ where $\phi(\nu)$ is the profile factor in the expressions (23), (24) for the emission and absorption coefficients. Thermal Doppler effects give a Gauss profile,

$$\phi_G(\nu) = \frac{\alpha}{\pi^{1/2}\nu_o} \exp\{-[\alpha(\nu-\nu_o)/\nu_o]^2\} \tag{34}$$

with width (FWHM) $\Delta_G = 2(\ell n\ 2)^{1/2}\ \nu_o/\alpha$. The observed widths are almost invariably larger than thermal Doppler widths and it is generally assumed that non-thermal motions contribute to the broadening and that the combined Doppler profile has the Gauss form (34) and can be characterised by a Doppler temperature T_D. Impact broadening gives a profile close to the Lorentz form

$$\phi_L(\nu) = \frac{\delta}{\pi}\left[(\nu-\nu_o)^2 + \delta^2\right]^{-1} \tag{35}$$

with $\delta = \gamma/4\pi$ and width $\Delta_L = 2\delta$. Convolution of (34) and (35) gives the Voigt function $\phi_V(\nu)$ which, in general, has a narrow Doppler core and broad damping wings. For an inhomogeneous nebula the line profile $\theta(\nu)$ is a superposition of Voigt profiles.

The quantities usually measured are the width $\Delta\nu$ of the line (FWHM), and the aerial temperatures T_c proportional to I_c, and T_L proportional to $(I_{\nu_o}-I_c)$ where ν_o is the frequency of the line centre. If $\theta(\nu)$ has the Gauss form, one obtains

$$\frac{I_L}{I_c} = \frac{1}{2}\left(\frac{\pi}{\ell n\ 2}\right)^{1/2}(\Delta\nu)\ \frac{T_L}{T_c} \tag{36}$$

which has often been used.

BS give an example of the difficulties which can arise. With $T_e = 10^4$ K and $T_D = 2 \times 10^4$ K one obtains

$$\frac{\Delta_L}{\Delta_G} = 50 \times \left(\frac{N_e}{10^4}\right) \tag{37}$$

for the H 220α line, with N_e in cm^{-3}. Optical and radio observations of Orion show that a substantial part of the total emission comes from regions with $N_e = 10^3$ to 10^4 cm^{-3}. It follows that the 220α line from Orion must have broad wings, and equation (36) cannot be correct.

Such broad wings are difficult to observe due to the difficulty of setting the continuum level, in consequence of the rather rapid variation with frequency of the response of high gain amplifiers (the problem of "baseline subtraction").

In comparing results from a model with observations, BS subtracted "baselines" from computed profiles in a fashion which they believed to be similar to that used by observers. Lockman and Brown (1975) suggest that it is better to use the intensity I_{ν_o} at the line centre, which is less sensitive to baseline subtraction procedures; this gives

$$\frac{I_{\nu_o} - I_c}{I_c} = \frac{T_L}{T_c} = \frac{I_L}{I_c} \times \theta(\nu_o) \quad . \tag{38}$$

In comparing this quantity with results of model calculations one uses the computed value of $\theta(\nu_o)$, the profile factor for the line centre.

6.2 Conspiracies

Observers sometimes ask whether nebulae are close to LTE, in the sense that:

(i) $b_n = 1$ for all n;

(ii) there is no maser action;

(iii) there is no pressure broadening.

Attempts are made to answer the question by considering observed line-to-continuum ratios, and observed ratios of intensities for lines which happen to come close in frequency, such as 109α and 137β. I shall show that the different factors can conspire to give ratios in apparent agreement with LTE.

Let us first consider the case of low nα lines, for which pressure broadening is negligible and continuum optical depths, τ, are small. Solution of the transfer equation for an isothermal homogeneous nebula gives

$$\frac{I_L}{I_c} = \frac{\kappa_L(\mathrm{LTE})}{\kappa_c} \times Y \tag{39}$$

where

$$\kappa_L(\mathrm{LTE}) = \int \kappa_\nu^L(\mathrm{LTE})d\nu \quad , \tag{40}$$

κ_c is the continuum opacity and the factor Y is a measure of departures from LTE. For $\tau \ll 1$ we obtain

$$Y = b \times \left(1 - \frac{1}{2}\beta\tau\right) \qquad (41)$$

which is a product of two factors: (i) the departure coefficient b; (ii) $[1-(1/2)\beta\tau]$ which is a measure of maser action. Let us now consider numerical results for a case similar to that considered by Pauls and Wilson (1977) for the central regions of Orion: $n = 66$, $T_e = 10^4$ K, $N_e = 5 \times 10^3$ cm^{-3} and $\tau = 3.7 \times 10^{-3}$. We obtain: (i) $b_{66} = 0.922$; (ii) $\beta_{66} = -42.0$ and hence $[1-(1/2)\beta_{66}\tau] = 1.078$. These give $Y = 0.994$. Two departure factors of 8% from (i) and (ii) have conspired to give a combined departure factor of 0.6%!

Let us now consider larger values of n, for which the b_n factors are close to unity. Figure 6 shows results adapted from Figure 1 of BLK. The quantity L is defined by

$$L = \frac{100}{\nu} \times \frac{T_L}{T_c} \qquad (42)$$

with ν in GHz; this differs from the quantity in (38) only by a convenient frequency-dependent normalizing factor. Results are given for two homogeneous models, with length 1 psc, $T_e = 10^4$ K and $N_e = 10^3$ and and 10^2 cm^{-3}. For each model, three curves are given:

> Curve (a) allows for departure coefficients (our effect (i)) but not for maser action (effect (ii)) or pressure broadening (effect (iii)). With the exception of the smallest values of n -- which we have already discussed -- curve (a) is very close to the curve for LTE.
>
> Curve (b) allows for effects (i) and (ii) but not (iii).
>
> Curve (c) allows for effects (i), (ii) and (iii).

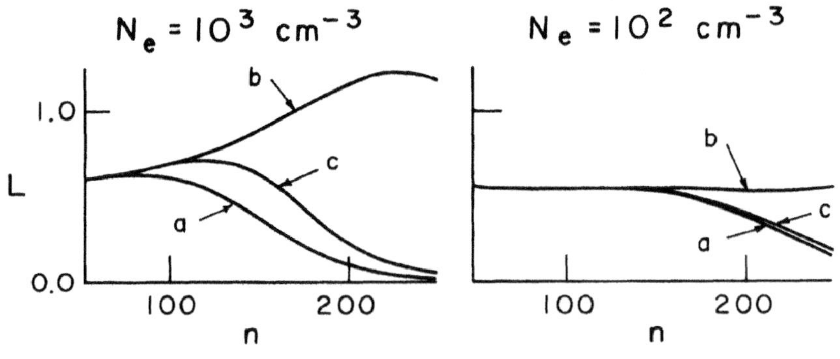

Figure 6. L, defined by (42), for homogeneous models of length 1 psc.

The conspiracy we see here is that maser action tends to increase the intensity at the line centre, giving large differences between curves (a) and (b), but that pressure broadening decreases the central intensity and brings curve (c) back much closer to curve (a).

I conclude that observations can closely mimic the predictions of LTE theory, even when the LTE assumptions (i), (ii) and (iii) are false.

6.3 Inhomogeneous Models

In order to have satisfactory interpretations of radio line observations it is necessary to consider models of variable density and possibly also variable temperature: but these problems are discussed by BS and BLK and in other papers in the present volume; hence they will not be discussed further here.

7. ELECTRON TEMPERATURES

7.1 The Importance of Accurate Temperature Determinations

Accurate electron temperatures are required for the determination of abundances in nebulae. Consider a line such as [O III] λ5007. The observed flux is proportional to

$$\int N_e N(O^{+2}) T_e^{-1/2} \exp\left[-\frac{E_{21}}{kT_e}\right] dV \tag{43}$$

where the integral is over the observed volume, and E_{21} is the excitation energy. For the λ5007 line, $E_{21}/(kT_e) = 28\ 700/T_e$ with T_e in K. Comparison with the flux in the H 2β line, λ4861, gives a measure of the abundance ratio $N(O^{+2})/N(H^+)$ but this is sensitive to the adopted value of T_e. One can use a second line of [O III], λ4363, with excitation energy E_{31} such that $E_{31}/(kT_e) = 61\ 700/T_e$. The observed ratio of the fluxes in λ4363, λ5007 gives a temperature $T_e(FL)$ defined by

$$\exp\left[-\frac{E_{32}}{kT_e(FL)}\right] = \frac{\int N_e N(O^{+2}) T_e^{-1/2} \exp\left[-\frac{E_{31}}{kT_e}\right] dV}{\int N_e N(O^{+2}) T_e^{-1/2} \exp\left[-\frac{E_{21}}{kT_e}\right] dV} \tag{44}$$

with $E_{32} = E_{31} - E_{21}$. If T_e varies throughout a nebula, (44) gives the greatest weight to the hottest regions. In order to obtain accurate abundancies, it is important to establish the possible importance of such variations.

Let us consider the determination of temperatures from hydrogen nα lines. Inserting expressions for $\kappa_L(TE)$ and κ in (39) we obtain

$$\frac{I_L}{I_c} = \frac{M}{T_e(LC)} \times Y \qquad (45)$$

where

$$M = \frac{N(H^+)}{N_e} \times \frac{2.925 \times 10^{-2}\ \nu^2}{\frac{3}{2}\log T_e(LC) - 1.31 - \log \nu}\ \mathrm{kHz} \quad , \qquad (46)$$

$T_e(LC)$ is the line-to-continuum temperature in K, and ν is in GHz. If temperature variations occur, we must consider the integrals over the line and continuum emissivities. Neglecting slowly varying functions of T_e we obtain

$$T_e(LC) = \frac{\int N_e^2\ T_e^{-1/2}\ dV}{\int N_e^2\ T_e^{-3/2}\ dV} \quad . \qquad (47)$$

This gives greater weight to the cooler regions and differences between $T_e(FL)$ and $T_e(LC)$ provide a measure of temperature variations. Earlier work gave $T_e(LC) < T_e(FL)$, with $T_e(LC)$ determined both from radio observations and optical observations of H 2β relative to the Balmer continuum. This led Peimbert (1967) to introduce a mean temperature T_o defined by

$$T_o = \int T\ du / \int du \qquad (48)$$

where $du = N_e^2\ dV$, and a temperature fluctuation parameter t^2 defined by

$$t^2 = \int (T-T_o)^2\ du / \{T_o^2 \int du\} \quad . \qquad (49)$$

The earlier work appeared to be confirmed by later work, quoted by Peimbert and Torres-Peimbert (1977, to be referred to as PTP), who adopted $t^2 = 0.035$.

7.2 Results for Orion

Recent work gives values of $T_e(LC)$ for Orion which are closer to $T_e(FL)$. If $T_e(LC) = T_e(FL)$ we have $t^2 = 0$ and the oxygen to hydrogen ratio adopted by PTP, assuming $t^2 = 0.035$, must be reduced by a factor of 0.6 (which would give the oxygen abundance in Orion to be less than the best estimate of the solar oxygen abundance). The measurement of $T_e(LC)$ therefore requires careful scrutiny.

I consider the region within about 2' of θ^1 Ori C, since in this region the observations are most accurate. The observations of PTP give $T_e(FL) = 8900$ K for [N II] and 8700 for [O III]: I adopt

$$T_e(FL) = 8800\ \mathrm{K} \quad .$$

Observations of H 66α, in the region considered, have been made by Chaisson and Dopita (1977) who obtain $T_e(LC) = 9000 \pm 500$ K and by

Pauls and Wilson (1977) who obtain $T_e(LC) = 8200 \pm 300$ K. In both papers, Y in equation (45) is taken to be unity. Pauls and Wilson estimate that $Y = 1.00 \pm 0.03$. Their argument has already been discussed in section 6.2 above. Their adopted density, $N_e = 5 \times 10^3$ cm^{-3}, is in good agreement with that obtained from optical observations by PTP. I have checked that a variable density model, similar to that of BS, also gives Y close to unity.

The results of Chaisson and Dopita and of Pauls and Wilson are just consistent with $T_e(LC) = 8500$ K, to within the combined error estimates. But even the lower value, $T_e(LC) = 8200$ K from Pauls and Wilson (which is probably the more accurate), is a good deal higher than $T_e(LC) = 7150$ K adopted by PTP. With this value they obtained $T_o = 7550$ K and $t^2 = 0.035$. With $T_e(LC) = 8200$ K we obtain $T_o = 8360$ K and $t^2 = 0.013$. This smaller value of t^2 gives an oxygen abundance which is smaller than that adopted by PTP, by a factor of 0.7.

8. CONCLUSIONS

Radio astronomers observe regions in which a low density gas fills a space which is very great and hence, in agreement with Bohr's prediction, observe a great many lines.

The frequencies are given, to a very good approximation, by simple classical formulae together with a quantum condition which determines the periods of revolution in the quantized orbits. The theory for the atomic processes, both radiative and collisional, involves the use of classical laws of dynamics and electromagnetism, followed by the imposition of quantum conditions. I believe the atomic data, used for recombination line studies, to be reliable. If the spectrum of a source has been computed, using the atomic data and a model for the structure of the source, and the results obtained do not agree with reliable observations, then I think it far more likely that the fault lies in the model rather than in the atomic data.

Further precision work is desirable on the measurement of electron temperatures using low $n\alpha$ lines. This is important both for the radio studies of other recombination lines, for which maser action and pressure broadening become more important, and for the determination of abundances using results of optical observations.

ACKNOWLEDGMENT

The present paper was written on a visit to the Joint Institute for Laboratory Astrophysics, supported in part by National Science Foundation Grant PHY76-04761 through the University of Colorado.

REFERENCES

Balmer, J.J.: 1885, Ann. der Physik und Chemie 25, p. 80.
Bohr, N.: 1913, Phil. Mag. 26, p. 1.
Brocklehurst, M.: 1970, Mon. Not. R. Astr. Soc. 148, p. 417.
Brocklehurst, M.: 1971, Mon. Not. R. Astr. Soc. 153, p. 471.
Brocklehurst, M. and Leeman, S.: 1971, Astrophys. Lett. 9, p. 35.
Brocklehurst, M. and Salem, M.: 1977, Comp. Phys. Comm. 13, p. 39.
Brocklehurst, M. and Seaton, M.J.: 1972, Mon. Not. R. Astr. Soc. 157, p. 179.
Brown, R.L., Lockman, F.J. and Knapp, G.R.: 1978, Ann. Rev. Astr. and Astrophys. 16, p. 445.
Burgess, A. and Percival, I.C.: 1968, Adv. Atom. Molec. Phys. 4, p. 109.
Burgess, A. and Summers, H.P.: 1969, Astrophys. J. 157, p. 1007.
Chaisson, E.J. and Dopita, M.A.: 1977, Astron. and Astrophys. 56, p. 385.
Einstein, A.: 1905, Ann. d. Physik 17, p. 132.
Giles, K.: 1977, Mon. Not. R. Astr. Soc. 180, p. 57P.
Goldberg, L.: 1966, Astrophys. J. 144, p. 1225.
Griem, H.: 1967, Astrophys. J. 148, p. 547.
Huggins, W.: 1880, Phil. Trans. R. Soc. 171, p. 669.
Landau, L.D. and Lifshitz, E.M.: 1962, "The Classical Theory of Fields" (Pergamon Press).
Lockman, F.J. and Brown, R.L.: 1975, Astrophys. J. 201, p. 134.
Menzel, D.H.: 1968, Nature 218, p. 756.
Menzel, D.H.: 1969, Astrophys. J. Suppl. Ser. 18, p. 221.
Mezger, P.G.: 1968, "Physics of One- and Two-Electron Atoms," ed. F. Bopp and H. Kleinpoppen (North-Holland), p. 801.
Miller, J.S.: 1971, Astrophys. J. 165, p. L101.
Nicholson, J.W.: 1912, Mon. Not. R. Astr. Soc. 72, pp. 49, 139, 677, 693 and 729.
Oster, L.: 1961, Rev. Mod. Phys. 33, p. 525.
Paschen, F.: 1908, Ann. d. Phys. 27, p. 565.
Pauls, T. and Wilson, T.L.: 1977, Astron. Astrophys. 60, p. L31.
Peach, G.: 1972, Astrophys. Lett. 10, p. 129.
Peimbert, M.: 1967, Astrophys. J. 150, p. 825.
Peimbert, M. and Torres-Peimbert, S.: 1977, Mon. Not. R. Astr. Soc. 179, p. 217.
Pengelly, R.M. and Seaton, M.J.: 1964, Mon. Not. R. Astr. Soc. 127, p. 165.
Percival, I.C.: 1968, "Physics of One- and Two-Electron Atoms," ed. F. Bopp and H. Kleinpoppen (North-Holland), p. 252.
Percival, I.C.: 1978, "Electronic and Atomic Collisions," ed. G. Watel (North-Holland), p. 569.
Percival, I.C. and Richards, D.: 1975, Adv. Atom. Molec. Phys. 11, p 1.
Pickering, E.C.: 1896, Astrophys. J. 4, p. 369.
Rutherford, E.: 1911, Phil. Mag. 21, p. 669.
Seaton, M.J.: 1978, Mon. Not. R. Astr. Soc. 185, p. 5P.
Seaton, M.J.: 1979, Mon. Not. R. Astr. Soc. 187, p. 785.
Thomson, J.J.: 1897, Phil. Mag. 44, p. 293.

LABORATORY STUDIES OF RYDBERG ATOMS

Daniel Kleppner
Department of Physics and Research Laboratory of Electronics
Massachusetts Institute of Technology,
Cambridge, Massachusetts 02139

Time and again the world of atomic physics has been illuminated and stimulated by astrophysical discoveries, and though it may be an exaggeration to claim that the field of Rydberg atom research sprang from the discovery of recombination radiation in 1965, there is no doubt that my own involvement started from that point. News of Höglund and Metzger's discovery was reported with great excitement at Harvard, where I was an assistant professor, by Ed Purcell. A few matchbook calculations were all it took to show that highly excited atoms represented an essentially different species from atoms in their low-lying states, the traditional stock of atomic physics. Cross sections, dipole moment, radiative lifetime, density of states - all the familiar properties are fantastically distorted. Unfortunately, a few more matchbook calculations were all it took to show that simply because radio astronomers had a galaxy of highly excited atoms more or less waving for attention, it did not necessarily follow that the atoms could be studied in isolation in the laboratory. In fact, by the techniques of the 1960's the task of preparing highly excited atoms seemed formidable. Radiative association is too slow to be practical in the laboratory, while optical excitation of ground state atoms into Rydberg states would require spectral sources far brighter than anything then available. So for about a decade the field was left essentially to the radioastronomers. During that decade, however, tunable lasers were developed, and about five years ago several atomic physicists independently set about using the new lasers to create and study Rydberg atoms in preselected quantum states. Interest spread rapidly, and now more than a dozen laboratories are involved with Rydberg atom studies. To provide some idea of the research I shall review some of our work at M.I.T., though I should emphasize that this represents the efforts of but one of the many research groups.

First, a brief description of the techniques. We usually work with alkali atoms rather than hydrogen, for hydrogen is difficult to prepare in Rydberg states. To avoid collisions, we employ atomic beams. Pulsed tunable dye lasers are used to excite the atoms. Such lasers, which operate throughout the optical region, typically provide a few kilowatts of power in a spectral line 0.1 cm^{-1} wide, with a pulse length of 5

P.A. Shaver (ed.), Radio Recombination Lines, 23–33.

nsec, and a repetition rate of 10-200 pps. The atoms are excited in a series of steps: one or two steps among the low-lying optical levels, and a final step into a Rydberg state. We detect the atoms by field ionization. A field of a few hundred to a few thousand volts per centimeter is applied one or two microseconds after the laser pulse, and the ions are detected by an electron multiplier or other charge sensitive device. Field ionization is close to 100% efficient, highly selective, and free from intrinsic noise. It works so well that experiments can be carried out with less than one atom per pulse, though 10^4 atoms per pulse is not unusual.

By way of an introduction, let us look at plots of the computed charge density for two of the n = 8 state of hydrogen. Contrary to many people's intuition, the charge density in a Rydberg state looks nothing like an electron in a Bohr orbit. The reason, of course, is that a localized particle must be described by a wave packet, and a wave packet has little in common with an eigenfunction. (So far no one has actually prepared an atom in a state with a localized electron, though such an experiment may be possible.) The eigenstates of hydrogen look like standing waves, which, of course, they are. Figure 1a shows a cross section of the charge density in the 8s state. To make things visible the density has been multiplied by r^2, where r is the radius. Figure 1b shows the charge density in the state n = 8, l = 7, m = 0.

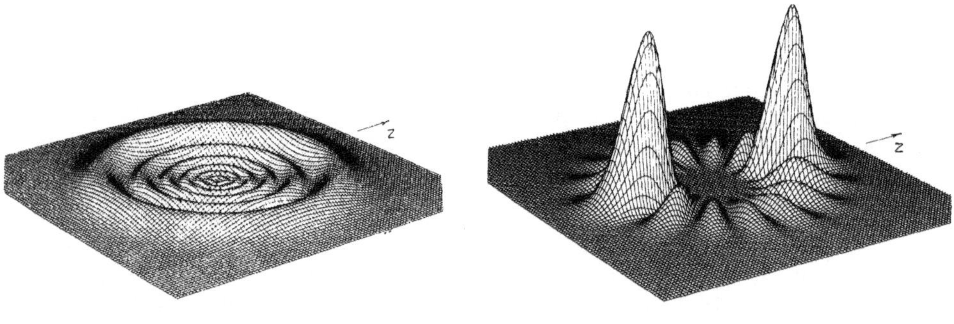

1a 1b

Fig. 1. Charge density of two n = 8, m = 0 Rydberg states, in a plane through the z axis.
(a) n = 8, l = 0; (b) n = 8, l = 7.

One of the most conspicuous features of the Rydberg states is their linear Stark effect. The Stark effect of the n = 2 state of hydrogen is perhaps the most frequent textbook example of degenerate perturbation theory (provided the Lamb shift and relativistic effects are neglected). High-lying states display much the same behavior except that the multiplicity increases with n. The "magnetic" quantum number m remain good

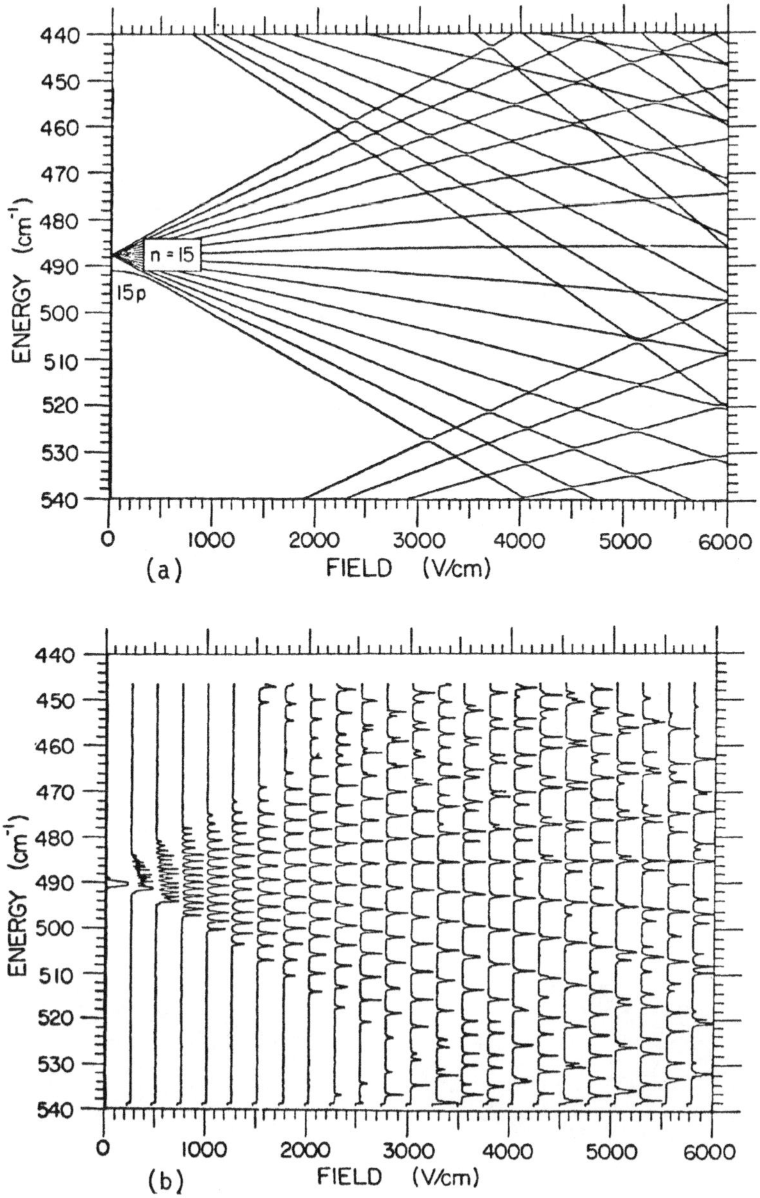

Fig. 2. (a) Calculated Stark structure of lithium m = 1 states in the vicinity of n = 15.
(b) Observed Stark structure. The horizontal peaks represent field ionization signals. (From ref.1)

in an electric field, and each term splits into a series of manifolds with n - m sublevels, where m = 0, 1,..n - 1. Because of the dynamical symmetry of the Coulomb field, Stark levels from different terms can cross, contrary to the usual "no crossing" rule.

Figure 2a shows the calculated Stark structure (1) in the vicinity of n = 15, m = 1. The calculation is actually for lithium. It differs from hydrogen in one minor respect: the p-state (l = 1) penetrates the core shell of 1s selections and as a result its energy is slightly lowered. The energy for the alkali metal atoms is given very accurately by Rydberg's formula: $E_n = (1/2)/n - \delta_l)^2$ (atomic units). δ_l, the quantum defect, is a constant which depends on l but is independent of n to good approximation. For lithium, $\delta_1 = 0.05$, and $\delta_l = 0$ for l > 1. As a result, the p state has a second order Stark shift at low fields, but it rapidly merges with the linear manifold as the field is increased. Incidentally, although the levels in Fig. 2 do not actually cross the anti-crossings are too small to see. (The apparent anti-crossings are artifacts of the plotting procedure.)

Figure 2b is an experimental Stark map of the m = 1 level lithium in the vicinity of the term n =15. The map was generated by sweeping the laser frequency at successively higher fields and plotting the field ionization signal as horizontal peaks. The Rydberg states were excited from the 3s level so that at zero field only the p states are visible. With even a small field, however, the p-character is mixed throughout the term, and all the levels are excited.

The linear Stark effect of a Rydberg atom means that the atom has a permanent electric dipole moment. Such, indeed, is the case as Figs. 3a and 3b show. The nodal lines in Fig. 3b can be seen to be parabolas, reflecting the fact that the problem is separable in parabolic coordinates.

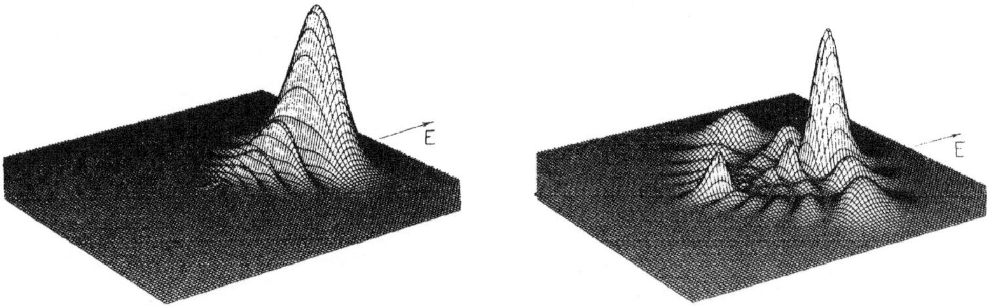

3a 3b

Fig. 3. Charge density of two n = 8, m = 0 Rydberg states in an electric field.

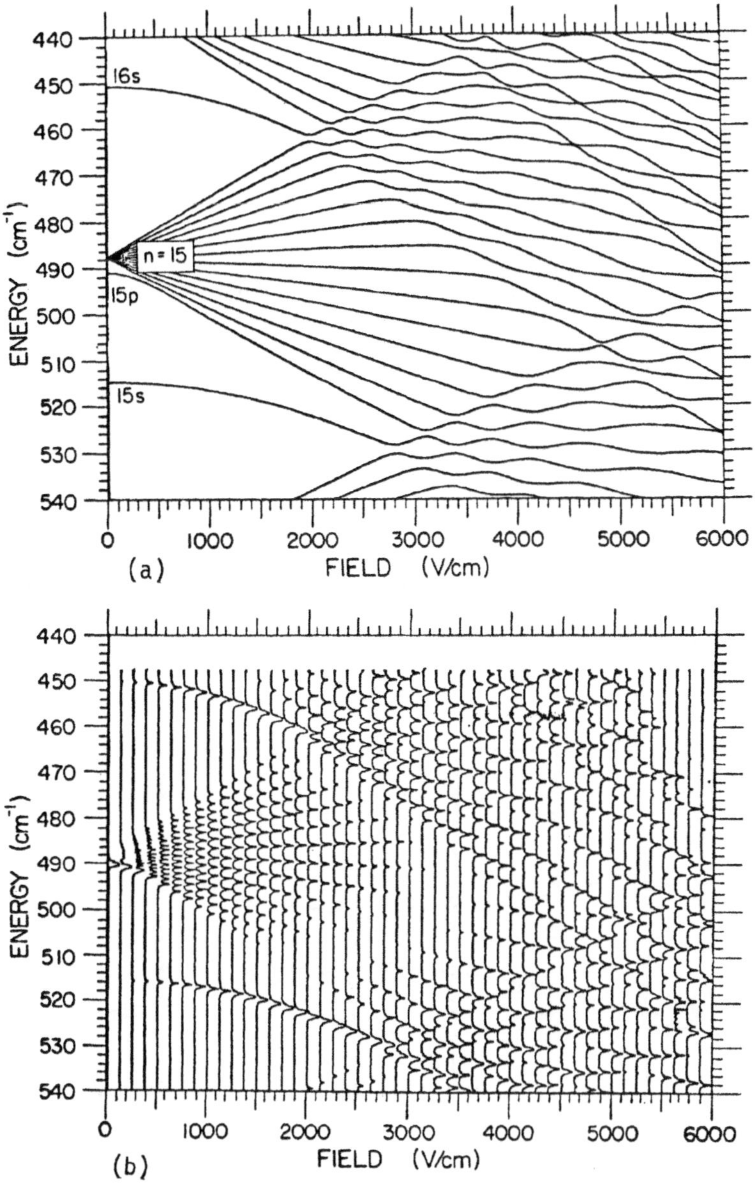

Fig. 4. (a) Calculated Stark structure of lithium m = 0 Stark states in the vicinity of n = 15.
(b) Observed Stark structure. The energy level structure and the variation in oscillator strength (not visible in 4a) agree with the calculated values. (From ref. 1)

The m = 0 Stark maps of lithium (Figs. 4a and 4b) are typical of the Stark structure of the alkali metal atom, and present a marked contrast to the hydrogenic appearance of the m = 1 maps. The s state has a large quantum defect, $\delta_0 = 0.35$, which completely distorts the Stark structure. The levels repel strongly at high field, and the oscillator strength varies markedly. The calculated and observed structure are nonetheless in good agreement.

We have carried out similar calculations for all of the alkali metal atoms, and have measured enough of the atoms to give us confidence that we understand the important feature of Stark structure.[1] Although the motivation for this work was primarily to understand a rather interesting system, there was also a practical objective: Stark states are often the most natural eigenstates for Rydberg atoms. Being able to identify and excite a given state is a prerequisite for many experiments.

We mentioned earlier that the levels in Fig. 2 do not actually cross. Fig. 5 shows a high resolution study of an anti-crossing. The separation is 0.07 cm^{-1}, too small to see in Fig. 2, but in excellent agreement with the calculated structure. Recall that at an anticrossing the eigenstates are the symmetric and antisymmetric combinations of the two interacting states. This is the reason that the oscillator strength of the upper state vanishes at the anticrossing, and all the intensity is transferred to the lower level.

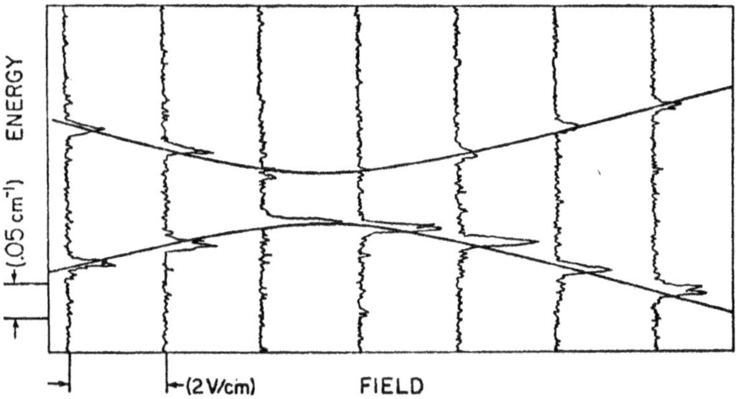

Fig. 5. An anticrossing between two Stark levels of lithium, m = 1 states. For a pure Coulombic potential, and neglecting relativistic effects, such levels would cross. The solid lines are the calculated energy levels. (From ref. 1)

Field ionization is an invaluable technique for detecting Rydberg atoms. It is also of considerable interest in its own right.[2] The basic mechanism, tunneling, is one of the most fundamental quantum mechanical processes. Because the potential is known exactly, the tunneling rates for Rydberg atoms can be calculated with far greater precision than in any other system. The increase of tunneling rate with the applied field is truly dramatic; the rate typically increases by a factor of 10^6 when the field is increased 20%. Fig. 6 shows experimental and calculated tunneling rates for one of the Stark level of sodium.[3]

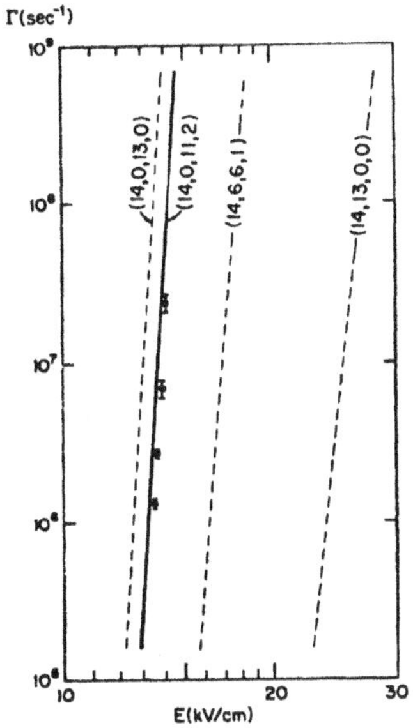

Fig. 6. Measured ionization rate vs. field for a hydrogenic state of sodium. The labels give the parabolic quantum numbers (n, n_1, n_2, $|m|$). The solid line is the calculated rate. The displacement of the experimental points is within the accuracy of the electric field calibration. (From ref. 3)

An alternative view of field ionization follows from elementary energy considerations. The potential $-1/r - Ez$ (atomic units) has a saddlepoint on the z axis. States with energy above the saddlepoint can escape over it, and this leads to a simple law for the onset of ionization: ionization can occur when $W > -2\sqrt{E}$. This locus is shown by the dashed line in Fig. 7. In taking the data the collection gate was delayed a few microseconds so that atoms which ionized within that time escape observation. The disappearance of the signal is in excellent agreement with the dashed line.

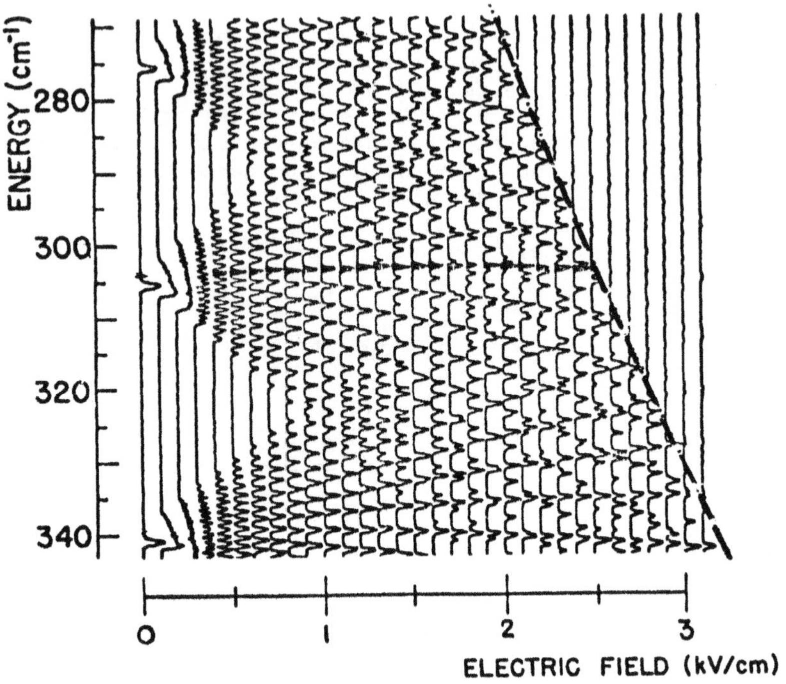

Fig. 7. Experimental demonstrations of the onset of field ionization at the "saddlepoint" field (shown dashed). Disappearance of the signal indicates that the atom has spontaneously ionized in less than 3 microsec. (From ref. 2)

It turns out that the two descriptions of ionization, tunneling and escape over the saddlepoint, are in serious conflict notwithstanding the excellent data confirming each. The resolution of the conflict is discussed in reference 2.

As a final example of our laboratory study of Rydberg atoms, we present some results on the structure of Rydberg atoms in a magnetic field. The atomic units of magnetic field is 5×10^5T (5×10^9 gauss). The realization that fields this large exist on white dwarf stars, and fields a thousand times larger exist on neutron stars, has motivated considerable theoretical work on the structure of atoms in very strong fields. Although laboratory studies of such fields are impossible, Rydberg atoms can provide the opportunity to study atoms under conditions where the magnetic interaction exceeds the Coulombic interaction. This can occur because the magnetic energy arises from the diamagntic interaction which varies as r^2B^2, whereas the Coulombic interaction varies as $1/r$. Since $r \sim n^2$, the ratio of magnetic energy to electric energy varies as n^6B^2. For a field of 10T, the ratio is unity at n = 40.

The charge distributions for two n = 8, m = 0 Rydberg states in low magnetic fields are shown in Fig. 8. Although m is always a good quantum number, l is never a good quantum number in a magnetic field (provided that relativistic effects are ignored) and the eigenstates are mixtures of all possible values of l having the same parity. "Low" here means that the diamagnetic interaction is not yet large compared to the term separation, so that n remains a good quantum number.

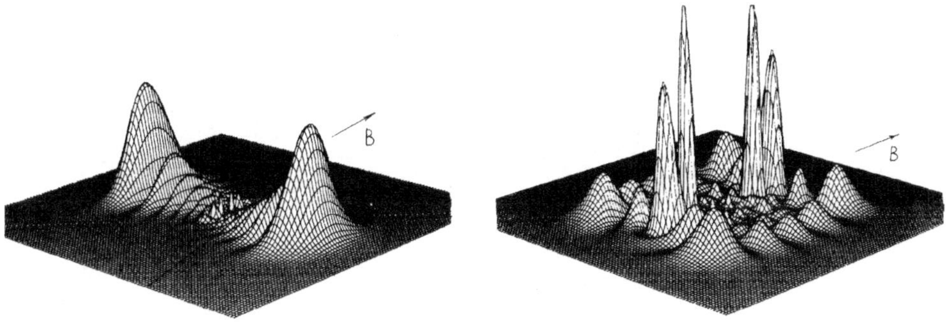

8a 8b

Fig. 8. Charge density for two even parity n = 8, m = 0 Rydberg states in a magnetic field. (a) The state with the largest "circulating currents", having the maximum diamagnetic energy shift. (b) A typical magnetic field state showing cylindrical geometry at large distances from the origin and spherical geometry near the origin.

Figure 9 shows calculated and experimental results for levels of sodium in the vicinity of n = 28 in a field up to 6T.(4) The energy is plotted as a function of B^2 in order to linearize the plots. The region is "low field" in the sense that the Coulombic energy still dominates, but it is high field in the sense that the diamagnetic energy vastly exceed the paramagnetic energy, and even exceeds the term separation. Theory and observation are in good agreement here, though this fortunate state of affairs seems likely to deteriorate when we extend the experiments to higher field and to higher value of n. At the present time there is no comprehensive theory for Rydberg atoms in magnetic fields of arbitrary strength. Quite possibly this is the last unsolved problem of elementary atomic physics.

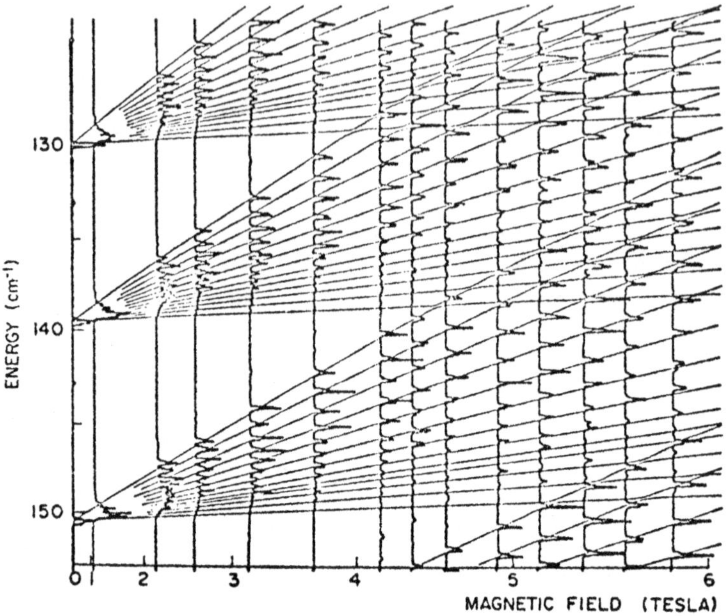

Fig. 9. Theoretical and experimental energy levels for even parity m = 1 levels of Na in the vicinity of n = 28. (From ref. 4)

The focus of the research described here is the structure of Rydberg atoms in strong fields, but the scope of contemporary Rydberg atom research is far broader than this. Other topics of current interest include collisions of Rydberg atoms with atoms, molecules and electrons; high resolution optical, infrared and millimeter wave spectroscopy; radiative phenomena including stimulated emission and superradiance; autoionization and dielectronic recombination and applications ranging from optical wavelength standards to new types of infrared detectors. Some review articles are listed in references 5-7.

I would like to acknowledge the contributions of Michael G. Littman and Myron L. Zimmerman who did much of the work displayed here. The charge distribution pictures were created by William P. Spencer. This work was sponsored by the National Science Foundation, the Joint Services Electronics Program and the Department of Energy, Contract No. EG-77-S-02-4370.

References

1) M. L. Zimmerman, M. G. Littman, M. M. Kash, and D. Kleppner, Phys. Rev. A., to be published.

2) M. G. Littman, M. M.Kash, and D. Kleppner, Phys. Rev. Lett. 41, 103 (1978).

3) M. G. Littman, M. M. Kash, and D. Kleppner, Phys. Rev. Lett. 37, 486 (1976).

4) M. L. Zimmerman, J.C. Castro, and D. Kleppner, Phys. Rev. Lett. 40, 1083 (178).

5) R. F. Stebbings, Science 193, 537 (1976)

6) T. F. Gallagher, Advances in Atomic and Molecular Physics, Vol. 14, Ed. B. Bederson, Academic Press (1978), p. 365.

7) D. Kleppner, Progress in Atomic Spectroscopy, Ed. Hanle and Kleinpoppen, Plenum Press (1979), p. 713.

PART II

RADIO RECOMBINATION LINES FROM HII REGIONS

"Observers should not be discouraged
by theoretical predictions"

- L. Goldberg

INTERPRETATION OF HII REGION RADIO RECOMBINATION LINES

C.M. Walmsley
Max-Planck-Institut für Radioastronomie
Auf dem Hügel 69, 5300 Bonn 1, Federal Republic of Germany

I. INTRODUCTION

There have been several recent reviews of the subject of radio recombination line from HII regions (Brown et al., 1978; Viner et al. 1979) and I see no need for another review in the traditional sense. There does however seem to be a certain amount of confusion, at least among observers, as to whether or not theory and observations are consistent with one another. One problem is that simple models of the recombination line transfer do not appear to explain the observed run of line-to-continuum-ratio (T_L/T_C) with frequency. In fact, the observed values of T_L/T_C seem to vary only slightly from those which one would expect in the ideal situation of an optically thin L.T.E. gas without impact broadening. This has led some observers (see e.g. Chaisson and Dopita (1977)) to claim that, for practical purposes, L.T.E. may be assumed and electron temperatures derived on this basis. The claim is however hotly contested by Lockman and Brown (1978), who point out that there are difficulties fitting low frequency data ($n > 85$) on the basis of an "L.T.E." model. More fundamentally, it is clear that the lines are simply not in L.T.E. in the sense that, unless we completely misunderstand the physics of highly excited atoms in HII regions, the b_n factors are not equal to unity and the line absorption coefficients are negative. Hence stimulated emission should occur and the question remains as to why its effects are not more apparent.

A partial answer is that, as pointed out in a "classical" paper by Brocklehurst and Seaton (1972), impact broadening has the effect of causing the low density extended portions of HII regions to make the main contribution to the line emission at low frequencies. This effect is heightened by the fact that telescope angular resolutions are worse at low frequencies. Hence, one is in effect observing different gas at 1 GHz, than at, say, 20 GHz and the frequency dependence of the line intensity is accordingly dependent on the difference between, for example, the electron temperature in high density and low density gas. This is probably not the whole story however. The variable density models of Brown et al. (1978) show consistently a maximum in T_L/T_C at intermediate values of n

P.A. Shaver (ed.), Radio Recombination Lines, 37–51.

(n is principal quantum number of a transition $n+1 \rightarrow n$), which is caused by stimulated emission. The observations show little sign of this behaviour and this has led to some confusion.

In this paper, I will attempt both to clarify the observational situation and to critically analyse what can be said on the basis of theoretical models. I will lean heavily on a recent article by Shaver (1980) in which a model fit is made to line and continuum observations of the Orion nebula. Shaver makes the point that part of the confusion concerning recombination line modelling has been due to the fact that model builders have not felt themselves constrained by the properties of real galactic HII regions. The simplest such property is that regions observed with sufficient angular resolution have excitation parameters $u \lesssim 100$ pc cm^{-2}. This presumably corresponds to an upper limit to O star masses. High excitation parameter HII regions should show more markedly the effects of stimulated emission and hence models with unrealistically high excitation parameters are misleading. Additionally, it is known from optical observations that clumping is common within HII regions. The local electron density is thus higher than the R.M.S. value obtained from continuum observations. This has the effect both of increasing the relativ importance of impact broadening and of decreasing departures from L.T.E. The stimulated emission bump is thus reduced.

The present paper takes Shaver's (1980) model of Orion and illustrates the consequences of variations of the basic model. The insights thus gained are used to make some statements concerning the derivation of physical parameters of HII regions from recombination line observations. In particular, I conclude that there is no reason not to use high frequency line observations to obtain electron temperatures. Observations at frequencies where the continuum optical depth is less than or of order 0.01 can hardly be importantly affected either by stimulated emission or by impact broadening. The level population can then be reasonably estimated and the derived electron temperature should be reliable. Such estimates can be checked by comparison with other independent methods of determining electron temperature and the results appear to be consistent with one another. Another application of recombination line measurements might be to determine clumping or volume filling factors. This is much less straightforward and does require the use of models. However, some limits can be set and these are discussed.

In section II of this paper, I give a qualitative discussion of the factors determining the relative importance of collisional broadening and stimulated emission. Section III attempts to make this more precise by giving the results of some model calculations which use Shaver's (1980) paper as a point of departure. Finally in section IV, I survey some of the observational data and discuss the applications of recombination line measurements to the determination of the physical properties of HII regions.

II THE RELATIVE IMPORTANCE OF COLLISION BROADENING AND STIMULATED EMISSION

Before discussing model calculations it is useful to estimate the critical quantum number or frequency beyond which collisional broadening, on the one hand, and stimulated emission on the other become important. At high frequencies, or low quantum numbers, both effects may be neglected. At lower frequencies, the behaviour of the line-to-continuum ratio is dependent to a considerable degree upon which of the two processes dominates. To compare the two, I will assume an isothermal nebula ($T = 10^4$ K) of given excitation parameter u. Assuming spherical symmetry, one can then determine the local electron density N_e from the equation:

$$N_e = f^{-0.5} \, (E/u)^{0.75} \tag{1}$$

where f is the volume filling factor and E is the radial emission measure (from the centre) of the HII region. I assume here that the gas is clumped on a small scale and that only the fraction f of the volume which is filled contributes to the emitted continuum and line radiation. By applying equation (1) and interpolating in tabulations of departure coefficients (Brocklehurst (1970)), one can determine the optical depth through the nebula of a given recombination line. One can also determine the non-LTE factor Y which gives, for a constant density nebula, the ratio of the line-to-continuum ratio to that which would obtain under conditions of L.T.E. (see Brocklehurst and Seaton (1972)). In the limit of low continuum optical depth τ, Y may be written as:

$$Y = b \left\{ 1 + \frac{1}{2} \, (1 - \beta) \, \tau \right\} \tag{2}$$

In this equation, b and β are the usual departure coefficients. The condition that stimulated emission becomes important is essentially that the term $(1-\beta)\,\tau$ becomes large. With the help of equation (1), one can solve for the critical quantum number n_{CR} at which $-\beta\,\tau = 0.2$ (i.e. stimulated emission is a ten percent effect). The results of such a solution for given f, u and E are shown in figures 1a and 1b. One can in similar fashion determine for what value of n_{CR} is the impact broadening line width $\Delta\nu_I$ comparable to the Doppler width $\Delta\nu_D$. For a typical case (Doppler temperature = 2 10^4, kinetic temperature 10^4), the ratio of these quantities is given by (Brocklehurst and Seaton (1972))

$$\Delta\nu_I/\Delta\nu_D = 0.14 \, (n/100)^{7.4} \, (N_e/10^4) \tag{3}$$

Also shown in figures 1a and 1b is the critical value of n for which $\Delta\nu_I = 0.1 \, \Delta\nu_D$ obtained by combining equations (1) and (3).

Figure 1a shows that, with increasing emission measure, stimulated emission becomes relatively more important than collision broadening for the case with unity filling factor and with the excitation parameter $u = 50$ pc cm^{-2}. However, more striking is the fact that, with a fairly modest filling factor of 0.1, collision broadening appears always to dominate. If collision broadening becomes important at lower n than

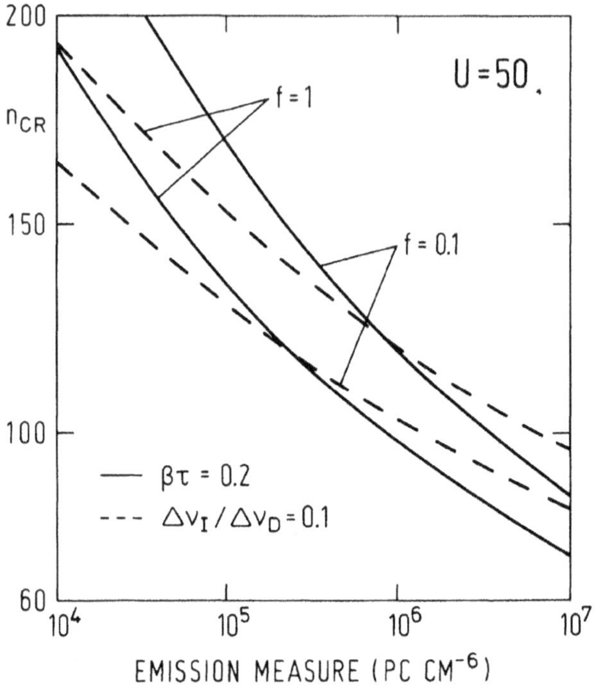

Fig. 1a. The critical quantum number n_{CR} at which stimulated emission (full lines) and collision broadening (dashed lines) become "ten percent effects" is plotted as a function of the HII region emission measure. The excitation parameter u of the central star has been assumed to be 50 pc cm^{-2}. The defining equations for the curves are given in the lower left corner of the diagram. Results have been computed for two filling factors f equal to 1 and 0.1.

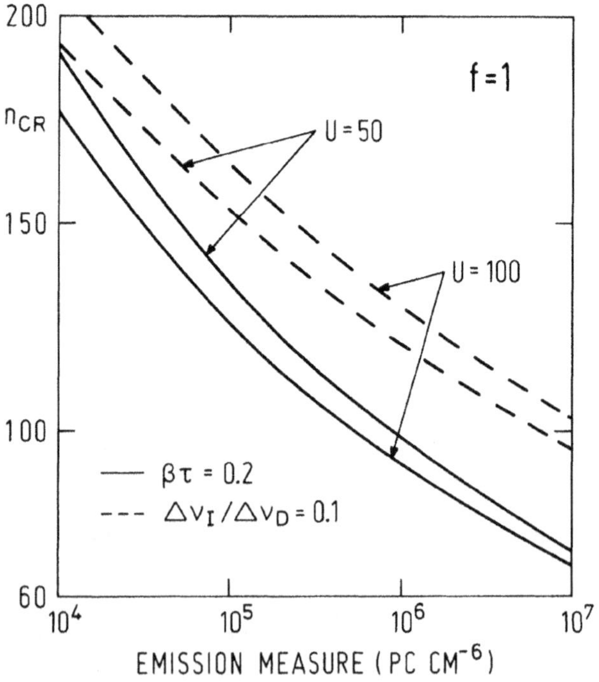

Fig. 1b. This is similar to figure 1a except that the effect of varying the excitation parameter u between 50 and 100 is investigated. The filling factor f is equal to unity.

stimulated emission, the lines with n greater than n_{CR} (coll) will deposit all their energy in weak Lorentzian wings which are, for practical purposes, undetectable. Hence, one can infer from figure 1a that small scale clumping can have a drastic effect on the question of whether stimulated emission effects in radio recombination line emission from HII regions are observable or not. Figure 1b is intended to illustrate by contrast the effect of increasing the excitation parameter u. High excitation parameter sources are better candidates for stimulated emission. In the case with f=1 and u=100, there is a large range of n where stimulated emission is important and collision broadening is not.

This approach of course ignores a variety of complications due to geometry, density gradients, and the like which may be of some importance. Lockman and Brown (1978) describe some of the problems which may occur. However, the qualitative picture that one derives from figures 1a and 1b should be correct. Collisional broadening and stimulated emission both have a rather strong inverse frequency dependence and typically become important at frequencies which are not greatly different from one another. Small scale clumping favours a situation in which collisional broadening dominates. A large excitation parameter on the other hand will tend to cause the stimulated term to be important. The competition between these effects is decisive for the low frequency behaviour of the line-to-continuum ratio.

III MODEL CALCULATIONS

Details of the transport equation and of the physics of highly excited atoms are given by various authors (e.g. Brown et al. (1978)) and do not need to be rediscussed here. The basic model used in this paper has been discussed by Shaver (1980) but, for the sake of clarity, I will describe its main features here. The model is isothermal with a temperature of 8200 K assumed in all regions. This appears to be justified in the case of the Orion nebula. Bonn measurements both on the peak of Orion and at offset positions show that no striking temperature gradients exist within the HII region (Pauls and Wilson (1977), Pankonin et al. (1979, 1980)). The model further supposes that a core of R.M.S. electron density 7000 cm^{-3} and diameter 0.13 pc is surrounded by four lower density outer regions. For simplicity, the geometry is supposed to consist of face-on cylindrical slabs. A cross-sectional view of the model is given in the lower section of figure 2. The length of each cylinder is, for the simplest case (filling factor = 1), supposed equal to its diameter. However, as discussed earlier, I wish to simulate the effect of small-scale clumping upon the line transfer. For the case where the gas is clumped and the filling factor f is less than unity, I suppose that the situation is adequately described by multiplying the R.M.S. electron densities of figure 2 by $f^{-1/2}$ and multiplying the line of sight depths of the slabs by f. Thus, all emission measures remain unchanged. It is not evident that this "equivalent pancake" model is a good representation of a real clumped nebula. In particular, the method ignores the presence of inter-clump gas which presumably is present. However, inter-clump gas in one slab may be approximately accounted for by lower density gas in an

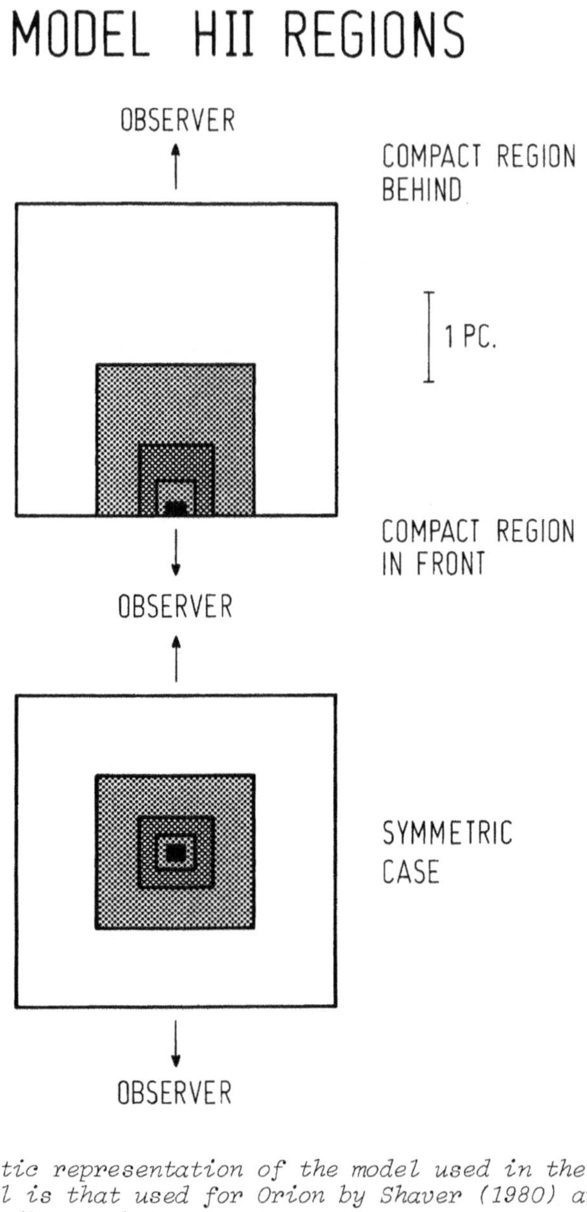

Fig. 2. *Schematic representation of the model used in the computations. The basic model is that used for Orion by Shaver (1980) and is labelled symmetric case (bottom half). It has a central region of R.M.S. electron density 7000 cm $^{-3}$ and diameter 0.13 pc. Successive outer layers, indicated by varying shading, have progressively lower density and larger extent. The upper section of the figure shows a variation of this model in which the high density regions are on one side of the nebula. This does not affect the radio continuum properties but does have an effect on the line transport which is dependent upon which side the nebula is viewed from. The R.M.S. electron densities of the slabs, starting from the center, are: 7000, 3000, 800, 200, 60 cm^{-3}.*

exterior slab. Shaver (1980) has used filling factors of 0.2 in the two central slabs and 0.1 in the three outer slabs in his Orion model.

One approach to estimating the importance of stimulated emission in a model such as that of figure 2 is to change the order of the slabs. Thus one can consider the configuration shown in the upper half of figure 2 in which the high density regions are at one end of the cylinder. This does not affect the radio continuum properties of the region since the total free-free optical depth along all lines of sight remains unchanged. It does however affect the stimulated emission from the lines. For an observer at the top of figure 2, the continuum emission from the bright core now has a longer path length through lower density foreground gas. One can also invert the situation and consider a model with a high density core in the foreground and the lower density slabs to the rear. In this situation, stimulated emission should be less important although it will not be entirely absent. I shall suppose that the difference between the two extreme cases (compact region in front and behind) gives a qualitative idea of the importance of stimulated emission in the symmetric model. The latter is evidently intermediate between the extremes.

Figure 3 shows the difference between the two extremes in a plot of T_e^* against n where T_e^* is the "L.T.E. temperature" that an observer would derive from the line-to-continuum ratio. Brown et al. (1978) have discussed at some length the definition of T_e^*. T_e^*, as used in this article, has been obtained from the computed line-to-continuum ratio and half-power width assuming a gaussian profile. This thus ignores the energy lost in the line wings. The justification for this is that it is a reasonable approximation to observational practise. It is also worth noting that the calculations assume an observing telescope of 100-m diameter and hence the beam gets progressively larger at larger n. One sees that all curves show a pronounced minimum at values of n $\sim$110 - 130 which is due to stimulated emission. Comparison of the case with the compact region in front and the compact region behind certainly shows that stimulated emission is not negligible. For larger n, impact broadening becomes important and T_e^* exceeds the actual kinetic temperature T_e. For small n or high frequency however, we note that the "non-LTE correction" is not large.

Brown et al. (1978) have pointed out that stimulated emission also makes itself apparent in the variation of T_e^* for a given transition as a function of offset position from the center of a nebula. To examine this, one can take the model of Orion and make 110α and 138β computer observations at offset positions. This procedure has the advantage that the model has been constructed to reproduce the continuum and line properties of Orion. Figure 4 shows the results of such computer offset observations. There is a fall of $\sim$1500 K at the centre of the nebula in H110α (model S) but the H138β behaviour is much flatter and impact broadening is clearly important for the β transition. Unfortunately, observations of Orion at 6 cm show no evidence for this predicted dip in T_e^* (110α) (see Pankonin et al. (1980) - Fig. 4) and hence we conclude that the model is still overestimating the amount of stimulated emission. An

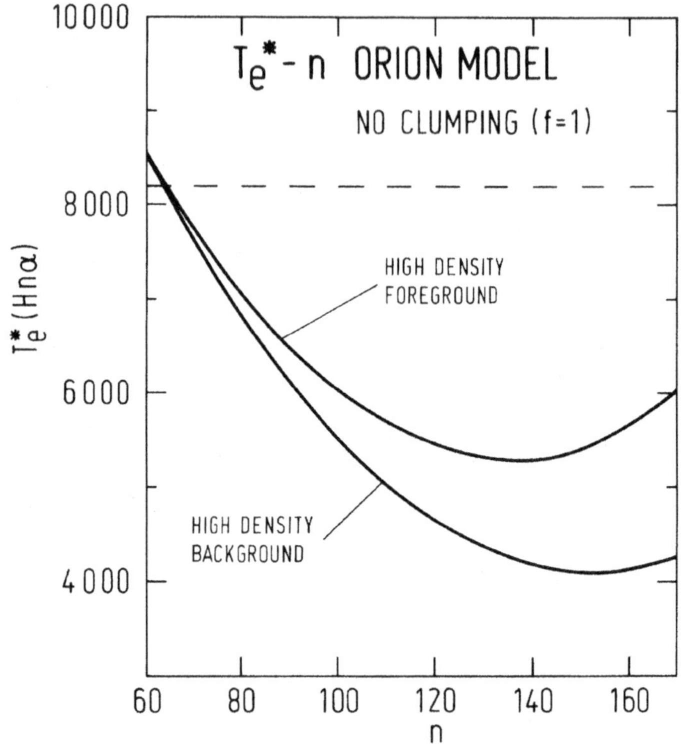

Fig. 3. *The values of T_e^* which an observer at a distance of 500 pc from the model of figure 2 (top) wou obtain as a function of quantum number. The observing beams are determined supposing a telescope of 100 m diameter and the two curves correspond to the two cases schematically depicted in the top half of figure 2. The dashed horizontal line shows the assumed kinetic temperature (8200 K).*

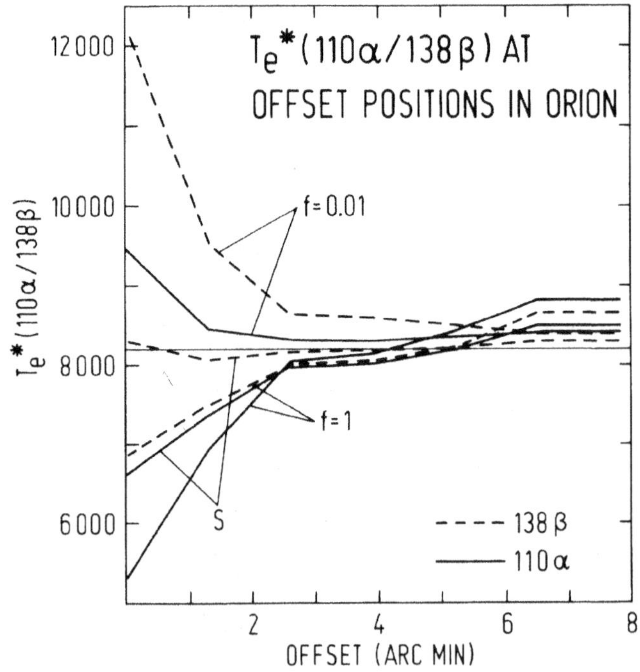

Fig. 4. *Computed values of $T_e^*(110\alpha)$ and $T_e^*(138\beta)$ for the case of a telescope pointed at an offset position from the centre of the model. A distance of 500 pc and an angular resolution of 2.6 arc min have been assumed. Results are shown for different values of the filling factor f (same for all layers). The cuves marked S assume the values of f (0.1 - 0.2) proposed by Shaver (1980).*

alternative might be that the temperature decreases outwards in Orion. However, the measured temperature of the core of the nebula (Pauls and Wilson (1977)) is 8200 K and a median value for the outer regions is 7700 K (Pankonin et al. (1980)). This gradient cannot explain the observations. What is more likely is that the model has underestimated the clumping in the nebula. Models, which attempt to fit the optical data (Simpson, 1973) suggest filling factors as low as 0.003. To investigate the effect of this for offset measurements, we have run models of offset 110α/138β measurements with f taken as 0.01 and as 1 for all layers. The f = 1 models merely show that, without clumping, stimulated emission effects are very large. f = 0.01 on the other hand is clearly too clumped and causes $T_e^*(110\alpha)$ to rise towards the centre of the nebula due to impact broadening. Thus it seems likely that the filling factor in Orion is of order 0.1 thus quenching stimulated emission in the core of the nebula.

One can attempt to verify this by computing T_e^*-n curves for various filling factors. Figure 5a shows the results of such calculations for f = 0.01, 0.04 , 1 as well as for the Shaver (1980) model. Observed values in Orion A are shown as filled circles. f = 1 and f = 0.01 produce too much stimulated emission and too much impact broadening respectively and need not be discussed further. The model with f = 0.04 provides a good fit to observations for n <120 but the lines are too weak at higher values of n. The Shaver (1980) model has, as noted previously, too much stimulated emission at intermediate n (∿100) but provides a reasonable

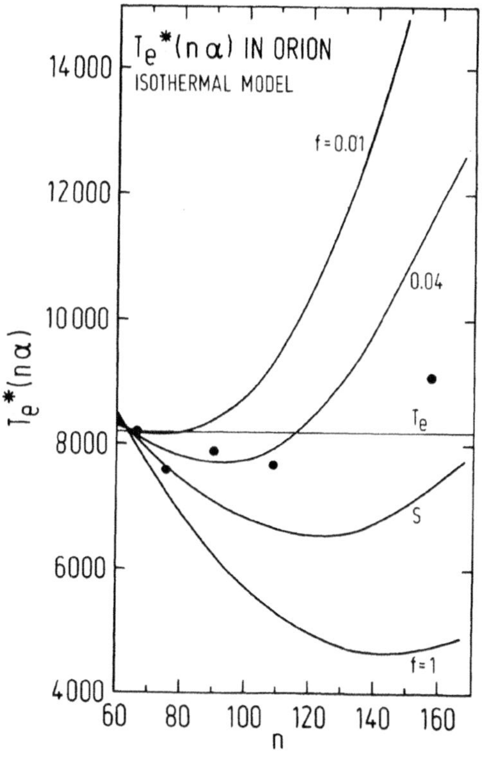

Fig. 5a. Similar to figure 3 except that the symmetric model (bottom half of figure 2) has been used. The curves have been computed assuming different values of the filling factor f for all layers (S is the model of Shaver (1980)). The filled circles correspond to recent observations of the Orion nebula with the Bonn telescope (66α; Wilson, T.L. et al. (1979a), 76α; Pankonin et al. (1979), 109α; Churchwell et al. (1978), 90α and 157α; Pankonin and Walmsley (unpublished)). The horizontal line marked T_e shows the kinetic temperature assumed (8200 K).

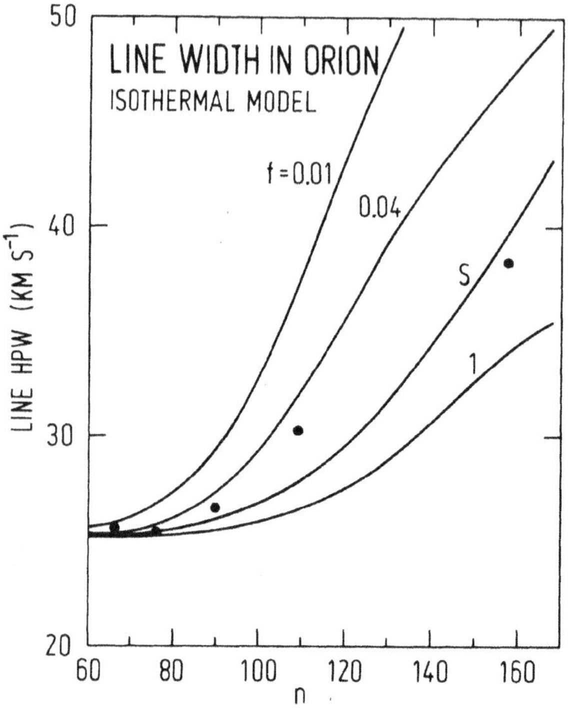

Fig. 5b. Line widths obtained from the set of calculations discussed in the caption of figure 5a as a function of n. Observations (see caption to figure 5a) are represented by filled circles.

low frequency fit. Presumably, an amalgam of these two models should be an improvement on either. Figure 5b shows the results of the same computations for the line width. Some observations are again plotted as filled circles. One should note however that, for line widths, agreement or disagreement between theory and observation is not very significant due to the large spread in angular resolutions employed. The observed line profile is influenced by systematic motions within the HII region and hence, even in the absence of collision broadening, one would expect to observe broader lines when one samples larger regions at lower frequencies. The computations, incidentally, assume a constant "turbulent velocity" in all regions.

One method of solving the problem of differing angular resolutions at differing frequencies is to compare α ($\Delta n = 1$) and β ($\Delta n = 2$) transitions which fall close to each other in frequency and hence can be studied with essentially identical beams. I show in figure 6 a calculation of T_e^* and line width for both α and β transitions using Shaver's (1980) model for Orion. One sees that, if clumping is invoked to explain the fairly flat behaviour of $T_e^*(\alpha)$ with frequency, the consequence is a fairly rapid rise in $T_e^*(\beta)$ and in the β line widths towards lower frequencies. One notes also that $T_e^*(\beta)$ is usually larger than the real electron temperature and hence there is not much ground for the assumption, which is occasionally made, that $T_e^*(\beta)$ is closer to T_{kin} than $T_e^*(\alpha)$. This result is a consequence of collision broadening and the fact that I have ignored the (essentially unobservable) line wings.

IV OBSERVATIONS AND CONCLUSIONS

In the previous section, I made a crude comparison of recent observational data for Orion with model predictions. The objective was to show

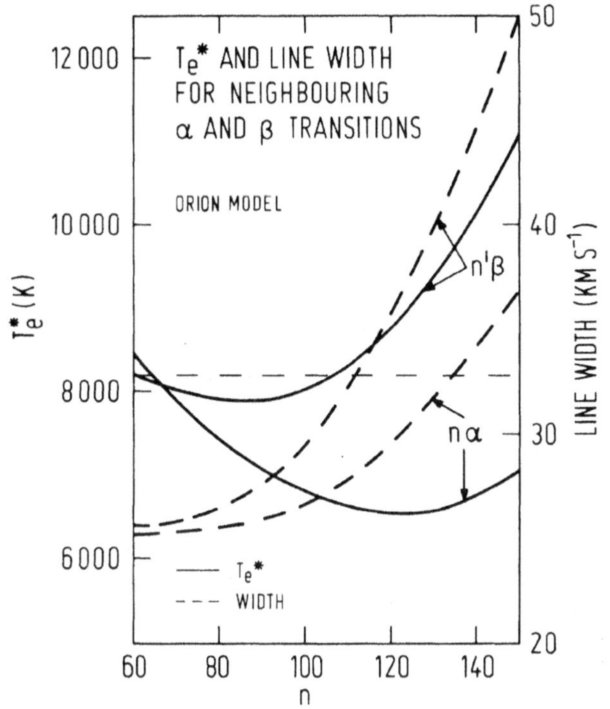

Fig. 6. *T_e^* (full curves) and line width (dashed) as a function of α quantum number for nα transitions and n'β transitions which lie at a neighbouring frequency. The Orion model of Shaver (1980) has been assumed. The horizontal dashed line denotes the assumed kinetic temperature (8200 K).*

the tendency implied by the data rather than to attempt a detailed comparison. One might reasonably ask if the picture changes when one considers the total quantity of data available for Orion. There is also the question of whether Orion is a typical case. There have been a series of model fits of Orion recombination line data (Brocklehurst and Seaton (1972), Lockman and Brown (1975), Shaver (1980)). I do not believe that another such attempt is presently useful. The main conclusion from the data is that there is no evidence for the predicted minimum in T_e^* as a function of n (This corresponds to the maximum in the curves of Brown et al. (1978) for $L = (100/\nu)\ (T_L/T_C)$). The effect appears not to be confined to Orion. Figure 7 shows a comparison of $T_e^*(66\alpha)$ and $T_e^*(109\alpha)$ derived from recent Bonn surveys (Churchwell et al. (1978), Wilson et al. (1979a), Wilson et al. (1979b)). There is no evidence of a general tendency for $T_e^*(109)$ to be smaller than $T_e^*(66)$. Hence, small scale clumping with consequent collisional broadening is probably a general feature of HII regions. It should be stressed however that the angular resolutions employed at the two frequencies are utterly different (2.6'(109α) as compared with 0.7'(66α)). Hence, one is often observing completely different volumes of gas at the two frequencies. A fairly direct demonstration of this is given by Van Gorkom et al. (1979) in their recent 6 cm aperture synthesis study of W49 and W51 (see also this volume). Interestingly enough, these observers do observe a high line-to-continuum ratio in one of the compact components in W51. By comparison with single dish results at 66α, they conclude that this is a case where collisional broadening is relatively less important and stimulated emission is the dominant effect. Such compact components make a very small contribution to the line emission seen in 109α single dish observations, which presumably is formed in relatively extended regions. Hence the information con-

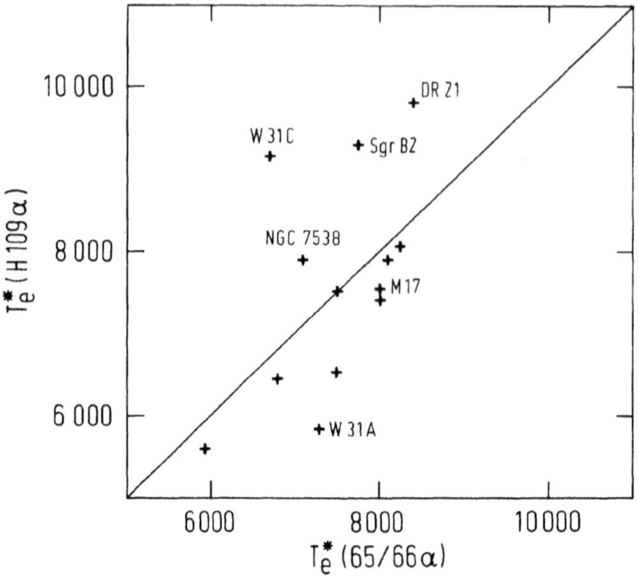

Fig. 7. A comparison of observed values of T_e^*(H109α) and T_e^*(65/66α) for several HII regions based upon the data of Churchwell et al. (1978) and of Wilson et al. (1979a, 1979b). The diagonal line shows the locus of T_e^*(109) equal to T_e^*(66).

tained in figure 7 should be treated with some scepticism.

As mentioned earlier, a more fruitful approach is probably the measurement of α and β transitions which occur at neighbouring frequencies. Churchwell et al. (1978) give evidence that the integrated intensity ratio H137β/H109α is correlated with emission measure in the sense that it becomes smaller than the L.T.E. value for emission measures larger than 10^6 pc cm^{-6}. Churchwell et al. point out that the observed dependence is compatible with simple estimates of the deviation from the L.T.E. ratio due to stimulated emission. However, they also find a tendency for H137β line widths to be larger than H109α widths. This is displayed in figure 8 which shows a plot of the observed line width differences as a function of emission measure. There is some evidence for correlation of line width difference with emission measure as one might expect, since impact broadening will be unimportant for low electron densities and hence low emission measures. Hence collision broadening may often be the overriding factor in causing low 137β/109α ratios (Shaver (1979)). However, one should note the case of M17 with a large emission measure, low 137β/109α ratio, and essentially equal line widths for 137β and 109α. This suggests that stimulated emission can be relatively important. In reality, both effects matter, and sorting out which one is dominant in individual cases is probably not very interesting. Model calculations become necessary and the only practical application may be to determine the amount of small scale clumping in radio HII regions.

Recombination line measurements are nevertheless useful for electron temperature determination. So long as measurements are made at frequencies higher than the critical values discussed in section II, they should be as reliable as most astronomical measurements. In essence, high emission measure objects require high frequency observations and low emission mea-

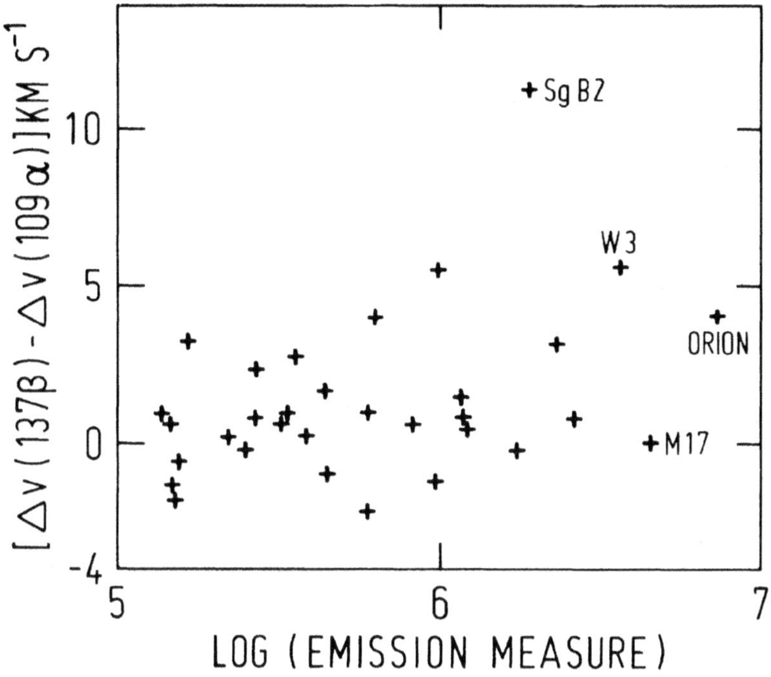

Fig. 8. The difference in the observed line widths of 137β and 109α plotted as a function of emission measure for the HII regions studied by Churchwell et al. (1978).

sure objects low frequency observations. For example, Pedlar (1980) has used H166α measurements to derive electron temperatures in some low brightness HII regions with emission measures less than 10^4 pc cm^{-6}. It is not however recommended to use low frequency measurements for high emission measure objects even if, as evidenced by figure 7, T_e^* in many sources shows very weak frequency dependence. Stimulated emission and collision broadening may often cancel each other but cannot be guaranteed to do so. On the other hand, at high frequencies, non L.T.E. effects are still present but are easily corrected for when making a temperature determination.

One might finally ask how well recombination line results compare with other independent determinations of the electron temperature. One such is to compare helium and hydrogen line widths measured at high frequency. Due to the greater mass of the helium atom, one can in principle separate the thermal and turbulent contributions to line broadening. Figure 9 shows the results of such an attempt based upon recent high frequency Bonn observations by Thum et al. (1980). There is some tendency for the line-width temperatures T_{LW} to be larger than the line-to-contin-

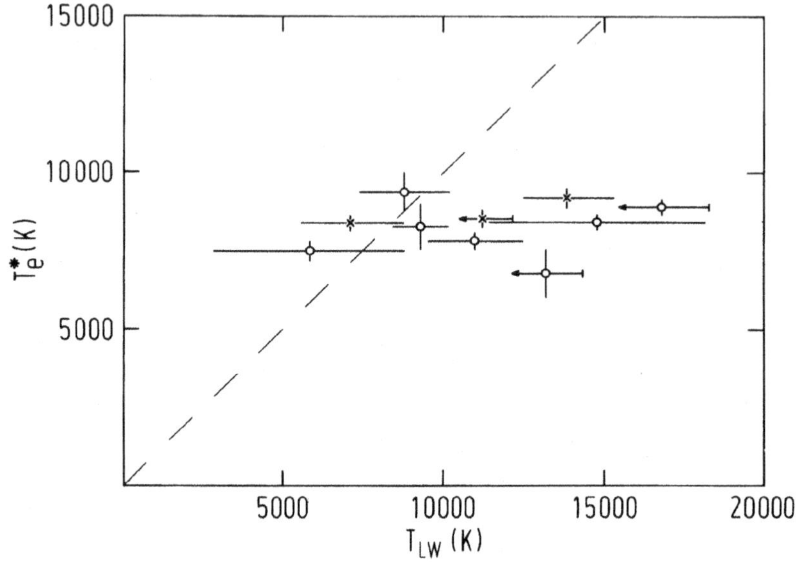

Fig. 9. A comparison of T_e^* derived from H66α and H76α measurements with estimates of the temperature derived from measurements of the widths of hydrogen and helium lines at the same frequencies. The data is taken from the work of Thum et al. (1980) and the diagram was kindly made available to me by Dr. C. Thum. Circles denote H66α measurements and crosses H76α measurements.

uum temperatures T_e^*. This might be due to the fact that H^+ regions are often more extended than He^+ regions and hence there are extra contributions to the broadening of the observed H lines. It is difficult to assess how important this effect is. One can apply a similar technique to optical line profiles and, for example, Chaisson and Dopita (1977) have made temperature measurements at several positions in Orion using Hα and He I 6678 Å. These authors find that their results agree well with T_e^*(H66α) measurements made at the same positions. Forbidden line temperatures in Orion seem also to agree reasonably well with recombination line results (McCall (1979), Pankonin et al. (1980)). Hence, while it must be realised that high frequency recombination line temperature measurements, like other measurements, involve an average over a large and often inhomogeneous volume of gas, there seems no reason to discard such results because of "non L.T.E. effects".

In conclusion, it seems worth pointing out that the future of recombination line research almost certainly lies in interferometry. High angular resolution will hopefully allow studies of the dynamics of HII regions. It should be possible to determine the internal velocity structure of young HII regions thus providing a check on dynamical calculations. Single dish measurements of H76α in Orion (Pankonin et al. (1979)) already suggest that the stellar wind bubble around the Trapezium is having an appreciable influence upon the flow of the ionized gas. Presumably, interferometry will allow many more detailed studies of this kind. In addition, it will allow the study of line-to-continuum ratios over a large range of frequency with similar angular resolutions. The results of van Gorkom et al. (1979) are a promising first step in this direction.

References

Brocklehurst, M. (1970) Mon. Not. Roy. Astron. Soc. 148, 417
Brocklehurst, M., Seaton, M.J. (1972) Mon. Not. Roy. Astron.Soc. 157, 179
Brown, R.L., Lockman, F.J., Knapp, G.R. (1978) Ann. Revs. Astron. Astrophys. 16, 445
Chaisson, E.J., Dopita, M.A. (1977) Astron. Astrophys. 56, 385
Churchwell, E., Smith, L.F., Mathis, J., Mezger, P.G., Huchtmeier, W. (1978) Astron. Astrophys. 70, 719
Lockman, F.J., Brown, R.L. (1975) Astrophys. J. 201, 134
Lockman, F.J., Brown, R.L. (1978) Astrophys. J. 222, 153
McCall, M.L. (1979) Astrophys. J. 229, 962
Pankonin, V., Walmsley, C.M., Harwit, M. (1979) Astron. Astrophys. 75, 34
Pankonin, V., Walmsley, C.M., Thum, C. (1980) Astron. Astrophys.(submitted)
Pauls, T., Wilson, T.L. (1977) Astron. Astrophys. 60, L31
Pedlar, A. (1980) Submitted to Mon. Not. Roy. Astron. Soc.
Shaver, P.A. (1979) Astron. Astrophys. 78, 116
Shaver, P.A. (1980) Astron. Astrophys. (in press)
Simpson, J.P. (1973) Pub. Astron. Soc. Pacific 85, 479
Thum, C., Mezger, P.G., Pankonin, V. (1980) Submitted to Astron. Astrophys.
Van Gorkom, J.H., Goss, W.M., Shaver, P.A., Schwarz, U.J. (1979) Astron. Astrophys. (submitted)
Viner, M.R., Vallée, J.P., Hughes, V.A. (1979) Astrophys. J. Supplement 39, 405
Wilson, T.L., Bieging, J., Wilson, W.E. (1979a) Astron. Astrophys. 71, 205
Wilson, T.L., Pauls, T.A., Ziurys, L.M. (1979b) Astron. Astrophys. 77, L3

THE IMPORTANCE OF NON-LTE EFFECTS TO THE INTERPRETATION OF RADIO RECOMBINATION LINES

Robert L. Brown
National Radio Astronomy Observatory
Green Bank, West Virginia 24944

Abstract. We review the ramifications of non-LTE effects on the determination of nebular properties made by interpreting radio recombination line observations. In particular we focus on the effect that departures from $b = 1$ and $\tau_c << 1$ have on the determination of nebular temperatures, and we discuss the effect of impact broadening on the interpretation of radio recombination lines (hydrogen and helium) from inhomogeneous nebulae.

I. INTRODUCTION

Over the course of the last decade, radio recombination lines have proven to be a readily available and indeed very useful probe of the physical conditions in Galactic HII regions. For nearby, optically visible HII regions, the results from radio recombination line observations have provided a global measure of nebular physical conditions (viz. temperature and velocity) that is highly complementary to the detailed optical observations. For more distant, optically obscured HII regions, the radio studies provide essentially the only information available on nebular parameters. Moreover, because radio recombination line emission depends sensitively on many of the fundamental properties of a nebula--temperature, temperature gradient across the nebula, mean density, density distribution and even nebular geometry--it is possible, given a sufficient body of observational data and an adequate theoretical analysis, to obtain considerable insight into the physics of Galactic HII regions. The task in this case, while straightforward, is not without its attendant pitfalls.

In this review I will address myself specifically to the peculiar effects encountered when one attempts to use radio recombination line observations to determine the two most frequently sought nebular parameters: nebular temperature and the abundance of helium relative to that of hydrogen. In doing so I will lump all such effects under the convenient heading of "non-LTE effects", although many have nothing precisely to do with departures from local thermodynamic equilibrium

P. A. Shaver (ed.), Radio Recombination Lines, 53–61.

(e.g. impact broadening, geometric effects and the like). Throughout the discussion I will regard the basic physics of ionized gas as understood--reviews of these matters are given by Dupree and Goldberg (1970), Brocklehurst (1970), Brocklehurst and Seaton (1972), Brown et al. (1978), and Shaver (1979).

II. RADIO RECOMBINATION LINE DETERMINATIONS OF THE TEMPERATURE OF HII REGIONS

A determination of the temperature of an HII region from radio recombination line observations is formally quite straightforward and is done as follows. The observable, the antenna temperature of a radio recombination line when the telescope is pointing toward a position (x,y) on a nebula, is a convolution of the antenna response function with the nebular brightness

$$T_L(x,y) = \eta_\nu \iint f_\nu(x'-x,y'-y) \int_0^{\ell(x',y')} j_L(x',y') e^{-t_L(x',y')} \, ds \, dy' dx' \quad (1)$$

where η_ν includes information specific to the antenna, f_ν is the normalized response function, j_L is the line emissivity, $t_L = \int \kappa \, ds'$, κ_L is the line absorption coefficient and $\ell(x,y)$ is the extent of the nebula along a line of sight through (x,y). All these quantities except $\ell(x,y)$ are, of course, frequency dependent. To eliminate the telescope-specific terms η_ν and f_ν, an often employed procedure is to simultaneously determine the continuum brightness $T_C(x,y)$--a quantity precisely like equation (1) except that the emissivity and absorption coefficients are those appropriate to radio continuum emission (cf. e.g., Brown et al. 1978)--and then use as a basic observable the ratio T_L/T_C.

T_L/T_C depends on the nebular parameters, electron density, electron temperature, and the distribution of these quantities across the beam and along the line of sight. To obtain the temperature, therefore, the proper procedure is to model the geometry and density of the nebula (as determined from radio interferometric observations for example) and then determine the temperature T_e which brings the expression for T_L/T_C into agreement with the observed T_L/T_C. In those few cases in which this procedure has been followed (Brocklehurst and Seaton 1972; Lockman and Brown 1975; Shaver 1980), the results (in this case those obtained most recently by Shaver) should be quite reliable.

Unfortunately, however, most frequently this procedure is not followed. Instead approximations are used, especially the "LTE approximation", which ease considerably the computational burden but exacerbate the interpretation of the derived temperatures in terms of the real nebular kinetic temperature.

Instead of properly modeling and evaluating the expression (1), it has become commonplace to invoke the "LTE approximation", which is really 4 distinct approximations.

1. The lines and continuum are formed in a plane parallel, isothermal, homogeneous medium.

2. The line and continuum optical depths are small, $\tau_C << 1$ and $|\tau_C + \tau_L| << 1$.

3. Recombination lines are formed and transferred under conditions of LTE, b=1 and β=1.

4. Impact broadening is negligible.

With these 4 approximations, equation (1) (and its counterpart for T_C) provide an expression for T_L/T_C which can be solved for the "LTE" electron temperature T_e^*,

$$T_e^* = \left[6.35 \times 10^3 \, \nu^{1.1} \frac{T_C}{T_L \, \Delta v} \right]^{0.87} \tag{2}$$

where Δv is the FWHM line width in km s^{-1} (cf. Lockman and Brown 1978). If the 4 approximations listed above are valid, T_e^* will be the true electron temperature T_e. On the other hand, the extent to which they are invalid--note especially that in expression (2) the nebular path length and the nebular density are entirely missing although they are certainly present in equation (1)--will be reflected in discrepancies in T_e^* from the true T_e. A few simple examples illustrate this point.

Consider first a homogeneous N_e = 100 cm^{-3}, isothermal T = 10^4 K, spherical nebula and let us "observe" (with an appropriate computer model, Lockman 1979) this nebula, obtain T_L/T_C from the observations, compute T_e^* and compare it to T_e. The results are illustrated in Figure 1.

Here, even in this simple example, it can be seen that T_e^* is not a good measure of T_e. In fact we expect this because in such a case T_e^* can be expressed in terms of T_e by

$$T_e^* \simeq T_e \, [b(1 - \beta\tau_C/2)]^{-0.87} \tag{3}$$

(Lockman and Brown 1978). At high frequencies b < 1 and we expect, and see, $T_e^* > T_e$, while at lower frequencies $b \rightarrow 1$ and $T_e^* \rightarrow T_e$. But as the path length through the nebula increases (i.e. R is larger) τ_C increases and $T_e^* < T_e$ (β < 0 remember). Therefore T_e^* is in error because the expression (2) for T_e^* includes neither nebular density N_e (low n, b $\neq$ 1) nor nebular path length (high n, τ_C is significant).

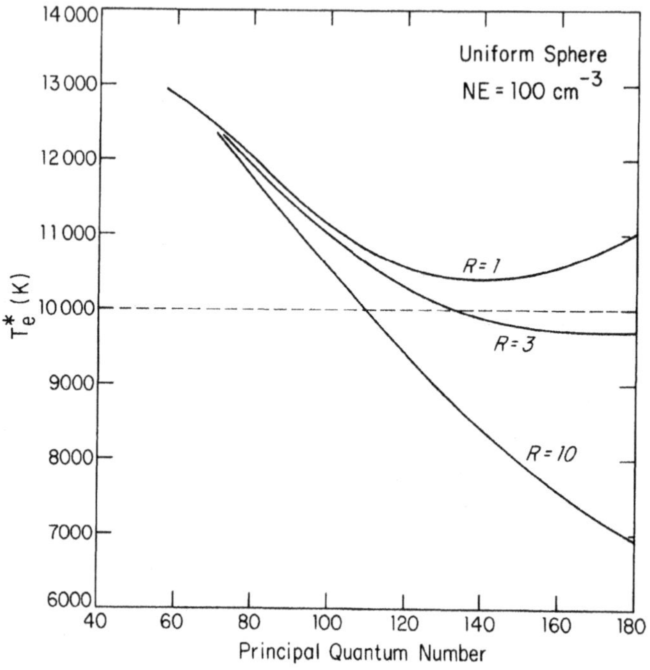

Fig. 1. The "observed" LTE electron temperature T_e^* as a function of principal quantum number for an isothermal 10^4K (dashed line), homogeneous, spherical nebula. The radius of this nebula is chosen to be R = 1, 3, 10 pc (solid lines).

The same discrepancies can be seen if we fix the path length R and vary only the density N_e; this is illustrated in Figure 2. Here we choose a nebular radius, R = 1 pc, and vary the density between 10 cm^{-3} and 1000 cm^{-3}.

Again, in this example, at low principal quantum number (high frequency) where $b < 1$ we see $T_e^* > T_e$, while at larger n when $b \to 1$ we see $T_e^* \to T_e$ for the $N_e = 10$ and 100 cm^{-3} cases but $T_e^* < T_e$ when $N_e = 1000$ cm^{-3} because τ_C is becoming significant. Also for these same models we have plotted the "observed" α/β ratio for α and β lines observed at essentially the same frequencies; here the dashed line represents the LTE value of α/β.

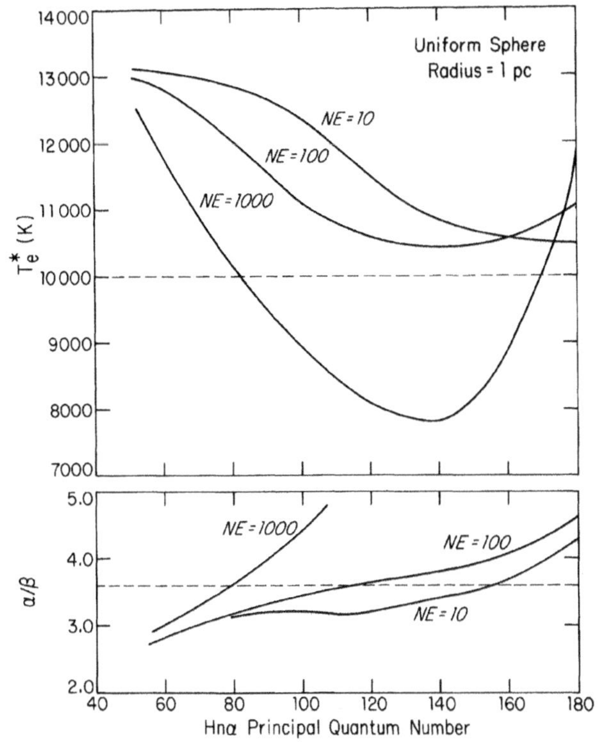

Fig. 2. The upper panel shows the "observed" LTE electron temperatures as a function of principal quantum number for a spherical R = 1 pc, isothermal $T = 10^4$K, homogeneous nebula. The solid lines are solutions for N_e = 10, 100, 1000 cm^{-3}. The lower panel shows the "observed" α/β ratios for the same models.

The point of these very simple, isothermal, homogeneous models is to demonstrate how sensitive T_e^* is to those nebular properties, particularly continuum optical depth and departures from b = 1, which are not part of its definition (eqn. 2). This obtains even in the simplest of cases: for more realistic situations Te^* can be equally deceptive (cf. Lockman and Brown 1978 for a complete discussion). It is important to note, however, that we need not necessarily be burdened with the deceptions of T_e^*--we know how to properly obtain reliable measures of nebular electron temperatures from radio recombination line observations; what we lack is the industry to gather the needed data and carry out the necessary computations.

III. RADIO RECOMBINATION LINE EMISSION FROM INHOMOGENEOUS NEBULAE

In a classic paper on radio recombination line radiation, Brocklehurst and Seaton (1972) emphasized the dominant role that impact broadening plays in inhomogeneous nebulae. Briefly, since impact broadening has such an extreme dependence on principal quantum number

$$\Delta\nu_I \propto n^{7.4} N_e, \qquad (4)$$

it overwhelms all other broadening mechanisms at low frequencies, and it effectively removes the power in the line core and distributes it in the far line wings. The net result is a line unobservably weak. Moreover, as can be seen in eqn. (4), the higher the electron density of a nebula the higher the frequency at which this effect becomes important. Hence if we observe radio recombination lines from an inhomogeneous nebula at low frequencies, we can expect to sample only the low density nebular gas; line emission from the high density gas, if any, will be "erased" by impact broadening.

The observational consequences of the importance of impact broadening can be simply stated as follows:

1. High frequency recombination line observations sample the dense gas in a nebula but provide little or no information on the low density nebular gas.

2. Low frequency recombination line observations sample the tenuous nebular gas but provide no information on the high density gas.

A simple example illustrates this point. Consider an isothermal, $T = 10^4$K, inhomogeneous nebula that has the density structure shown in Figure 3. On this figure is plotted both the run of electron density with nebular radius (left ordinate) and the cumulative run of fractional mass with nebular radius (right ordinate).

We may now compute the fraction of the line that comes from each part of this inhomogeneous nebula assuming that such a nebula is at the distance of Orion, for example, as seen by a 100 m telescope. We do this at several frequencies: 0.81 GHz (n = 200), 1.9 GHz (n = 150), 4.8 GHz (n = 110), and 12.6 GHz (n = 80). The results are shown in Figure 4.

Here it can be seen that at the highest frequencies, n = 80, more than 90% of the recombination line intensity comes from the inner 15% of the mass of the nebula--that is, from gas with densities greater than 3000 cm^{-3}. On the other hand, at the lower frequencies, n = 200, less than 20% of the line emission comes from this inner 15% of the nebular mass and instead more than half of the n = 200 line comes from material with a density of less than 1000 cm^{-3}.

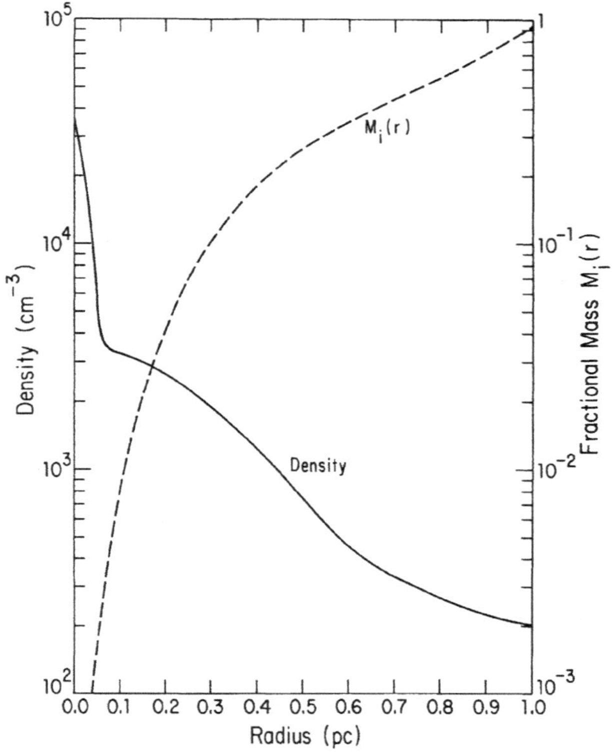

Fig. 3. A model inhomogeneous spherical nebula. Shown by a solid line is the variation of electron density with nebular radius (left ordinate) and by a dashed line is the cumulative variation of fractional nebular mass (right ordinate).

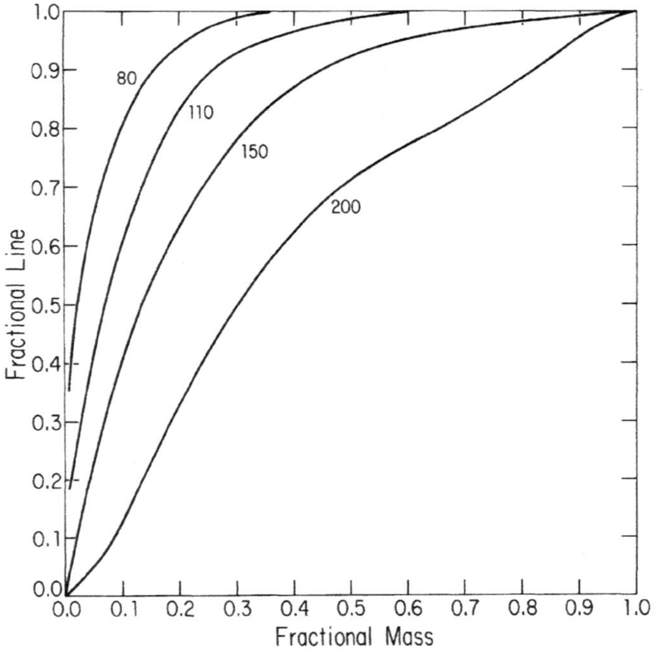

Fig. 4. The fractional power in recombination lines of n = 80, 110, 150, 200 as a function of the fractional mass of the nebula. The results here refer to the model nebula shown in Figure 3.

One cannot emphasize this conclusion too strongly. In an inhomogeneous nebula, high frequency radio recombination line observations provide information (temperature, abundances or whatever) on the dense nebular gas but they provide little or no information on the tenuous nebular gas; low frequency observations preferentially weight the low density gas and provide little information on the more dense nebular material.

IV. THE ABUNDANCE OF HELIUM IN HII REGIONS

Since a detailed discussion of the observations of helium radio recombination lines is given elsewhere in this volume, I will not review the extent of this subject. Instead, and in light of the preceding discussion, let me simply mention that the salient new observational result is that in many, if not all, HII regions the observed He^+/H^+ ratio is higher when measured at high frequencies ($\nu \gtrsim 15$ GHz) than it is when it is measured at lower frequencies ($\nu \lesssim 8$ GHz) (Thum et al. 1980; Brown and Lockman 1980). Since, again, the high frequency observations selectively weight the higher density nebular gas and ignore the low density material whereas the lower frequency observations weigh these two regions in the converse sense, 3 conclusions are immediate:

1. Most, if not all, Galactic HII regions are inhomogeneous.
2. Non-LTE effects (in this case impact broadening) are demonstrably influencing the observational results.
3. The fractional ionization of helium He^+/He (where He = $He^o + He^+ + He^{+2}$) is lower in the lower density nebular material than it is in the high density nebular gas.

The elucidation of these effects from both an observational and physical standpoint remain to be explored.

V. CONCLUDING REMARKS

Radio recombination lines are quite useful and powerful probes of the global properties of Galactic HII regions. Moreover, in this field of research we are in the fortunate position of understanding the basic physics of the formation and transfer of radio recombination line emission. What remains as undone is the realistic application of this understanding to the observations. Strangely, in this particular disciple of astronomy the theory is well ahead of either the observations or of the proper application of the theory to the observations. Motivated by the increasingly large body of available data, and the prospect for exciting new observations such as aperture synthesis maps of radio recombination lines (Van Gorkom et al. 1980), this is a situation which should soon be rectified.

References

Brocklehurst, M. (1970) Mon. Not. Roy. Astron. Soc. 148, 417.
Brocklehurst, M., and Seaton, J. J. (1972) Mon. Not. Roy. Astron. Soc. 157, 179.
Brown, R. L., Lockman, F. J., and Knapp, G. R. (1978) Ann. Rev. Astron. Astrophys. 16, 445.
Brown, R. L., and Lockman, F. J. (1980) Astrophys. J. (submitted).
Dupree, A. K., and Goldberg, L. (1970) Ann. Rev. Astron. Astrophys. 8, 231.
Lockman, F. J. (1979) Ph.D. Thesis, Univ. Mass. (unpublished).
Lockman, F. J., and Brown, R. L. (1975) Astrophys. J. 201, 134.
Lockman, F. J., and Brown, R. L. (1978) Astrophys. J. 222, 153.
Shaver, P. A. (1979) Astron. Astrophys. 78, 116.
Shaver, P. A. (1980) Astron. Astrophys. (in press).
Thum, C., Mezger, P. G., and Pankonin, V. (1980) Astron. Astrophys. (submitted).

DETERMINATION OF ELECTRON TEMPERATURES USING RADIO RECOMBINATION LINES

P.A. Shaver
European Southern Observatory c/o CERN CH-1211 Geneva 23
Switzerland

ABSTRACT Methods of determining electron temperatures of HII regions using radio recombination lines are reviewed. It is stressed that accurate and reliable temperatures can easily be derived from individual radio recombination line observations, if those observations are made at appropriate frequencies.

Radio recombination lines provide a very important addition to our arsenal for determining electron temperatures of HII regions. Indeed, they make it possible to obtain electron temperatures which cannot be measured optically at all, for example in HII regions which are obscured (distant), faint, or cool (the critical temperature-sensitive optical lines are very weak, especially at low temperatures). In addition, radio recombination line observations have the advantage of high intrinsic velocity resolution, and they are not affected by reddening.

However, electron temperatures determined from radio recombination lines have been subject to much criticism (cf. Brown et al.,1978, and references therein), far too pessimistic in my view. I would like to briefly review the methods of determining electron temperatures using radio recombination lines and the problems associated with them.

(a) Line Intensities. By far the most common method involves the line-to-continuum ratio, which is simply related to the electron temperature under LTE conditions (T_e^*) as follows,

$$\frac{\int \Delta T_L d\nu}{T_C} \propto T_e^{*\,-1.15}$$

The difficulties lie in accurately determining the continuum temperature T_C consistent with the line temperature ΔT_L, and in allowing for non-LTE effects which cause ΔT_L to differ from its LTE value. The first problem can be easily overcome with the use of proper observing techniques. And

P.A. Shaver (ed.), Radio Recombination Lines, 63–66.

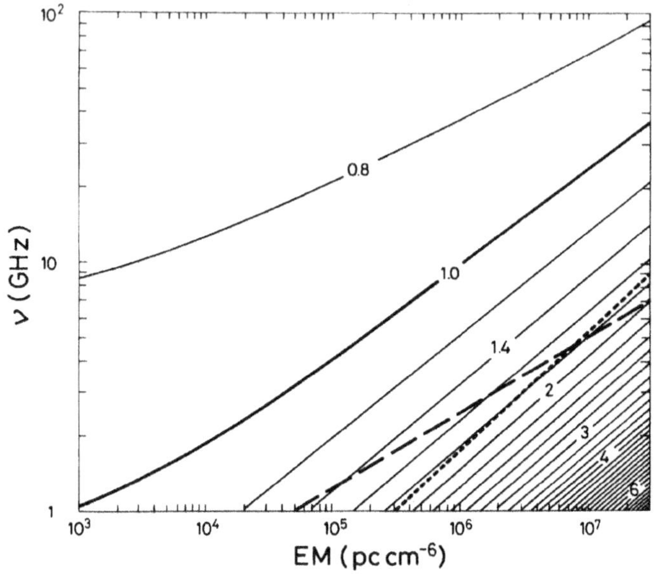

Figure 1. Observing frequency against emission measure, for given values of T_e/T_e^* as indicated. A homogeneous, optically-thin HII region has been assumed, with $T_e = 10^4$ K and U = 100 pc cm^{-2}. The galactic and cosmic background radiation has been neglected. b_n-factors are from Brocklehurst (1970). The dotted and dashed lines represent $\tau_C = 0.1$ and $\gamma = 0.1$ respectively, where γ is the ratio of pressure to Doppler broadening.

the second problem can be largely eliminated simply by the appropriate choice of observing frequency.

Figure 1 illustrates this latter point. The relevant atomic theory is well known; we can therefore reliably predict the deviation of T_e^* from T_e for given conditions. Figure 1 shows contours of the ratio T_e/T_e^* on a graph of observing frequency versus emission measure, for $T_e = 10^4$ K and U = 100 pc cm^{-2} (corresponding to an O5 star). Below the line $T_e/T_e^* = 1.0$, stimulated emission is the dominant non-LTE effect, and can be very important for high emission measure and low frequency; pressure broadening and optical depth effects are also important in this part of the diagram. The interpretation can in principle be very model-dependent, and it is to this corner of the diagram that the criticism of radio-determined electron temperatures has been directed.

However this criticism simply does not apply in the broad band around $T_e/T_e^* = 1.0$ (which is far broader and flatter still when density inhomogeneities are taken into account). In this region the derived temperature is relatively model-independent, and any non-LTE corrections are small. Thus, by using a recombination line at the appropriate frequency for a given emission measure of the source, one immediately obtains a value for T_e^* which is close to the true value, and accuracies of 5-10% and better are readily obtainable in the final estimate of T_e (including all observational errors and non-LTE uncertainties). These advantages of using radio recombination lines at high frequencies to determine electron temperatures were clearly recognized a decade ago (e.g. Sorochenko et al., 1969).

Table 1. Electron Temperatures Determined for the Orion Nebula

Method	Old Value	Ref.	New Value	Ref.
Radio Continuum	3000	1	8550	4
Radio Recomb.Line (T_e^*)	5170	2	8200	5
[OIII] Lines	9300	3	8800	6
[NII] Lines	12200	3	8340	6

References for Table 1: (1) Terzian et al., (1968); (2) Höglund and Mezger (1965); (3) Peimbert (1967); (4) Shaver and Goss (1970); (5) Pauls and Wilson (1977); (6) Peimbert and Torres-Peimbert (1977).

Table 1 shows electron temperatures obtained by various methods for the Orion Nebula. They have converged markedly over a decade. The Pauls and Wilson (1977) value is from H66α observations; that line is at 22 GHz, corresponding to $T_e/T_e^* \sim 1.0$ with the emission measure of $\sim 5.10^6$ pc cm^{-6}, from Fig. 1. It is striking that the value of T_e^* for Orion from recent H109α observations at 5 GHz, 8100 K (Churchwell et al., 1978), is also close to the true value, whereas according to Fig. 1 it should be much lower; the reason is that in reality density inhomogeneities greatly reduce the observable effects of stimulated emission, as discussed by Shaver (1979). Thus, even T_e^* values determined for high emission measure sources from low-frequency single-dish observations do not differ greatly from T_e; however there is no guarantee of this in individual cases, and the most reliable temperatures are obtained by observing at frequencies where $T_e/T_e^* \sim 1$ in Fig. 1.

(b) Line Widths. The widths of radio recombination lines are determined principally by the electron temperature and turbulence. Other factors, such as systematic mass motions and pressure effects, increase the line widths still further. Thus the line width gives a reliable upper limit to the electron temperature. This is the simplest approach to T_e using radio recombination lines. It takes advantage of the inherently high velocity resolution, and only requires moderately high sensitivity. Usually turbulence is dominant, and such limits are not significant. However there are some HII regions for which unusually low electron temperatures (< 5000 K) are implied by the line widths (Shaver et al., 1979), and these objects are of special interest.

The widths of the H and He radio recombination lines can also be used together in order to determine T_e:

$$T_e \, \alpha (\Delta V_H^2 - \Delta V_{He}^2).$$

This method, however, is fraught with difficulties and uncertainties. It requires very high sensitivity (about 40 times higher than the Hnα line-to-continuum method, for comparable accuracy in T_e); it assumes the H and He lines originate in the same volume of gas; there is often contamination of the He line by a narrow carbon line; and at low frequencies there are further complications due to both stimulated emission and pressure broadening. This method will always be restricted to the brighter HII regions, and is of limited interest only.

In conclusion, it is easy to avoid the problems associated with stimulated emission and pressure broadening simply by working at high enough frequencies. Accurate electron temperatures can certainly be obtained in a straightforward manner from line-to-continuum ratios, when the appropriate lines are observed with due care. The importance of radio recombination lines in determining electron temperatures cannot be exaggerated, since in many cases there is no option.

REFERENCES

Brocklehurst, M.: 1970, Monthly Notices Roy.astr.Soc. 148,417
Brown, R.L., Lockman, F.J., and Knapp, G.R.: 1978, Ann.Rev.Astron. Astrophys. 16,445
Churchwell, E., Smith, L.F., Mathis, J., Mezger, P.G. and Huchtmeier, W.: 1978, Astron.Astrophys. 70,719
Höglund, B., and Mezger, P.G.: 1965, Science 150,339
Pauls, T., and Wilson, T.L.: 1977, Astron.Astrophys. 60,L31
Peimbert, M.: 1967, Astrophys.J. 150,825
Peimbert, M., and Torres-Peimbert, S.: 1977, Monthly Notices Roy.astr. Soc. 179,217
Shaver, P.A.: 1979, Astron.Astrophys. (submitted)
Shaver, P.A., and Goss, W.M.: 1970, Aust.J.Phys.Astrophys.Suppl.No.14,133
Shaver, P.A., McGee, R.X., and Pottasch, S.R.: 1979, Nature 280,476
Sorochenko, R.L., Puzanov, V.A., Salomonovich, A.E., and Shteinshleger, V.B.: 1969, Astrophys.Letts. 3,7
Terzian, Y., Mezger, P.G., and Schraml, J.: 1968, Astrophys. Letts. 1,153

APERTURE SYNTHESIS OBSERVATIONS OF RECOMBINATION LINES FROM COMPACT HII REGIONS

J.H. van Gorkom
Kapteyn Astronomical Institute
Groningen, The Netherlands

INTRODUCTION

I shall present recombination line observations obtained with the Westerbork Synthesis Radio Telescope (WSRT). My collaborators on this work are W.M. Goss, P.A. Shaver, R.H. Harten and U.J. Schwarz.

The advantage of doing recombination line observations with a synthesis telescope is twofold. The high angular resolution enables you to study compact components separately and with a synthesis instrument there is reduced sensitivity to line emission from extended components. Thus you can study the lines from the compact regions without the contaminating influence of the surrounding medium. The talk consists of two parts. First I shall discuss results obtained with the 80 channel filter spectrometer, an analogue correlator, which has been in use with the WSRT for eight years. After that I shall show a few maps, that were recently obtained with the digital line backend receiver, which is operational since a year.

H109α OBSERVATIONS OF W49A AND W51A

H109α line observations were made of the compact components of W49A and W51A with an angular resolution of 6" arc in RA. The 80 channel filter spectrometer obtains simultaneously maps at eight frequencies, which are spaced 11 kms^{-1} apart at 5GHz. Eight maps were obtained for W49A, sixteen for W51A, giving a total velocity coverage of 80 and 108 kms^{-1} respectively. A typical example of the maps is shown in Fig. 1. Left the continuum emission of W49A , right the line emission at a velocity of V = 8 kms^{-1}. The line map is obtained after subtracting a continuum formed by the mean of the line free channels. The map is at the velocity where the maximum line emission occurs. In the line map the observed peak line to continuum ratio and halfwidth of the profile is indicated for each component. In Fig. 2 the same is shown for W51A. The line map is at a velocity of V = 63 kms^{-1}. The most striking result of these observations is the great variation of line to continuum ratio from one component to the other. For example the very bright and compact source W49.4 has a ratio of less than 1%. On the other hand the rather weak source W51.b has a ratio of 9%. So very unlike single dish results where the ratios are usually measured to be 4% or 5%,

P.A. Shaver (ed.), Radio Recombination Lines, 67–73.

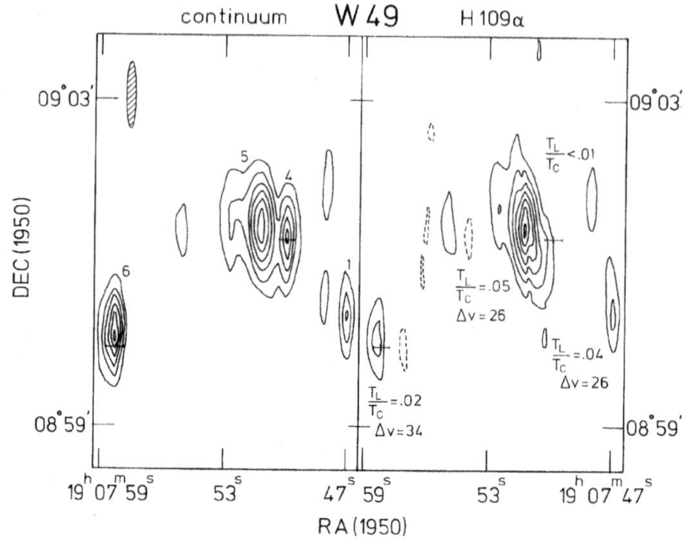

Figure 1. Maps of W49A. Left the mean continuum. The synthesized half power beam is indicated on the upper left. Right a line map at V = 8 kms^{-1} The contour interval is (left) 350, (right) 15 mJy per synthesized beam. Negative contours are dashed.

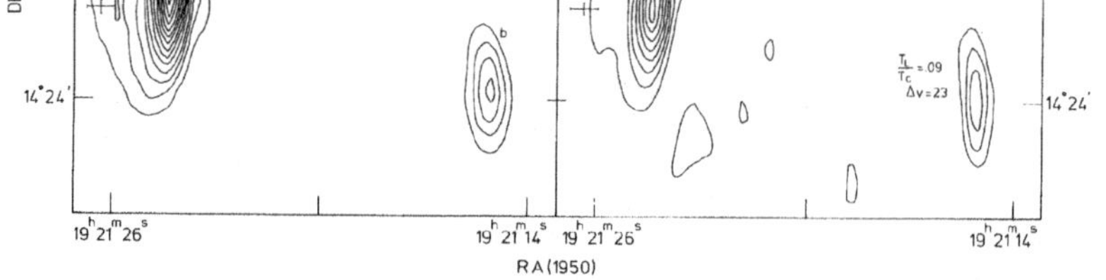

Figure 2. Maps of W51A. Left the mean continuum. The synthesized half power beam is on the upper right. Right a line map at V = 63 kms^{-1}. The contour interval is (left) 375, (right) 37.5 mJy per synthesized beam.

we find a range from less than 1% to 9%. It appears that the more compact sources have the lowest line to continuum ratios and the largest linewidths.

DISCUSSION

Do the Lines Arise from the Compact Components?

Since it is well known that these compact components of W49A and W51A are embedded in an extended low density medium the first question that arises is: do the lines originate from the compact sources or are they (partly) formed in the low density gas along the line of sight. The answer can be found by comparing single dish observations with the WSRT results. The integrated continuum flux density in the WSRT maps is ∿35% of the single dish value, while the power in the line is ∿25% of that seen by single dish (Churchwell et al. 1978). Due to the absence of the shortest spacings the WSRT is insensitive to extended structures. This explains the lower flux densities measured with the WSRT. It also shows that most of the line emission seen by the single dish must arise in extended regions. The only line emission from the halo that would be detectable by the WSRT is stimulated emission in the direction of the compact components. We can estimate the strength of these lines by making a reasonable model of the halos based on the single dish continuum and line observations. For n_e = 70 cm^{-3} and Te = 8000° K the optical depth in the line towards the centre of the halo is only $\tau_L \approx -.007$, thus the peak line to continuum ratio of these lines would be ∿0.7%. This is below the sensitivity of our observations. Therefore the line emission measured with the WSRT must arise from the compact components itself.

What H109αLine Intensities do we Expect from the Compact Components?

To answer this question we made some model calculations for homogeneous spherical regions using the full solution of the equation of transfer and taking pressure broadening into account. The results are shown in Fig. 3. The solid curves illustrate how for a region with fixed excitation parameter the peak line to continuum ratio varies when the continuum optical depth τ_c increases. Increasing τ_c implies increasing the electron density n_e since $\tau_c \propto u\, n_e^{4/3}$. Initially as τ_c increases stimulated emission becomes more important. As a radiative transfer effect it builds up with optical depth. Thus the ratio goes up. Above a critical density pressure broadening becomes important and the peak line intensity drops sharply. Compact HII regions have excitation parameters between 50 and 100 and at 5 GHz a continuum optical depth close to unity, clearly they provide the conditions for both stimulated emission and pressure broadening in the H109α line.

Interpretation of the Observed Line Intensities

To interpret the observed line intensities we calculated the expected peak line to continuum ratios assuming homogeneous sources with rms electron densities and emission measures as derived from the continuum radiation. Electrontemperatures and Dopplerwidths of the lines were taken

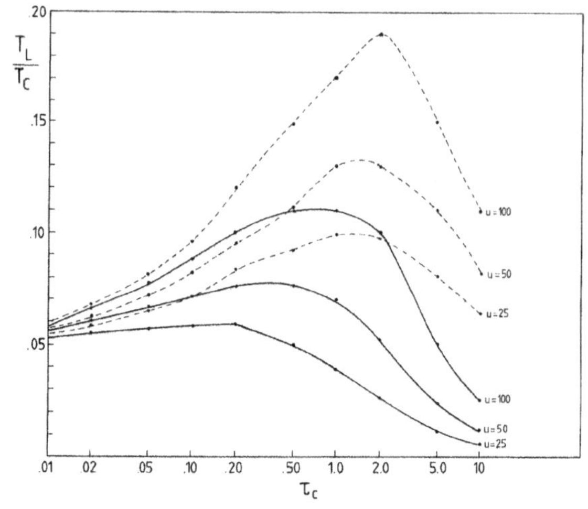

Figure 3. Calculated line intensities for model HII regions. Each curve gives for a region with fixed excitation parameter, u(cm^{-2}pc), the peak line to continuum ratio as a function of continuum optical depth of the region. The dashed curves represent the same calculations with no pressure broadening.

Table 1. H109α line parameters

source	observations T_L/T_c	model calculations T_L/T_c ($n_e=n_{e_{rms}}$)	T_L/T_c ($n_e=3n_{e_{rms}}$)
W49.1	.04 ± .01	.09	.05
4	.01 ± .005	.09	.04
5	.050 ± .001	.13	.06
6	.020 ± .005	.08	.05
W51.b	.090 ± .005	.09	
d	.026 ± .004	.12	.04
e	.062 ± .002	.11	.05

from the H66α observations made by Wilson et al. (1979). The observed and predicted ratios are given in Table 1. Only one source W51.b is represented well by the homogeneous model. The very high ratio of 9 % indicates that inside this sources stimulated emission is the dominant process, and as such it is the first HII region where stimulated emission inside the source itself can be observed directly. All the other sources have observed line intensities that are considerably less than the predicted values. Since the rms electron densities of these sources are close to the critical density where pressure broadening becomes dominant, already a moderate clumping can have a significant effect on the peak line intensities. Assuming a filling factor of ten percent would give the ratios given in the last column of Table 1. There are still three sources W49.4, W49.6 and W51.d for which the observed line intensities are considerably lower than the model values. Further decreasing the filling factor would not help since then the calculated halfwidths would be much larger than the observed ones, due to pressure broadening. A possible explanation is that the density distribution is inhomogeneous. The observed lines arise then from the less dense parts. The lines produced in the highest density parts are so broad that they remain undetected. This would happen if the densities were as high as 10^5 cm^{-3}.

CONCLUSION

Comparison of single dish observations with aperture synthesis results shows that at 5 GHz most of the line emission in W49A and W51A seen by single telescopes is due to spontaneous emission from the extended low density regions. The lines seen by the WSRT arise from the compact components.

The observed line intensities indicate that both stimulated emission and pressure broadening occur inside the compact regions. Which process dominates depends on the electrondensity. In one out of seven cases stimulated emission can be observed directly, while in three cases the line intensities are extremely low, implying electron densities in excess of 10^5 cm^{-3}.

H166α AND H110α OBSERVATIONS OF W3

Finally I would like to show you some preliminary results that were recently obtained on W3. These observations were made with the 5120 channel digital line backend receiver. The great advantage of this receiver for recombination line work is the much better channel to channel stability. Furthermore the total velocity coverage and velocity resolution can be chosen per observation. A range in total bandwith is available from 80 kHz to 10 MHz with 16 to 256 frequency points.

W3 was observed in the H166α and the H110α line. In Fig. 4 the 1.4 GHz continuum map and a H166α line map at V = -42 kms^{-1} are shown. This source has previously been observed in the H166α line by Sullivan and Downes (1973) with the WSRT. That observation was actually the first detection of a recombination line with an aperture synthesis telescope. In one channel a 4σ line signal was detected from component 8. With the present observations the sensitivity is a factor 3 better and we detect the line in all components. The intensities show the same trend as for the H109α results discussed above: the more compact components have lower line to continuum ratios. Intriguingly enough the profiles indicate again that there must be density structure on a smaller scale. The most dense parts remain undetected. The compact core of the region, consisting of components 8, 5 and 3 was also mapped in the H110α line. These results are shown in Fig. 5. The top map is the continuum emission at 5 GHz.

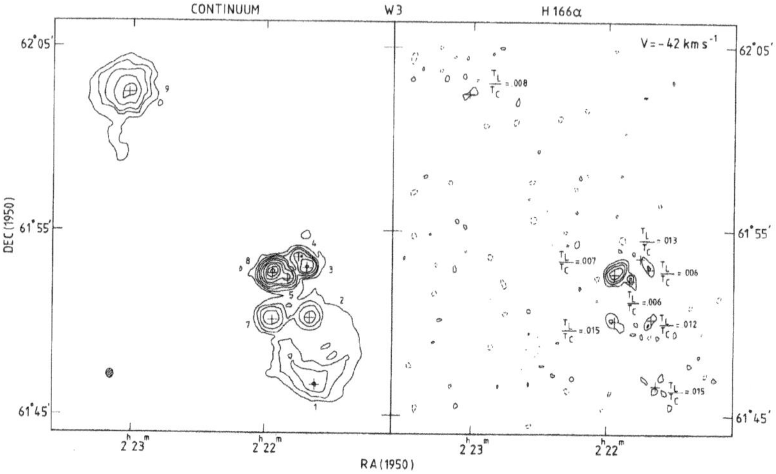

Figure 4. Maps of W3. Contours continuum map: .04, .08, .15, .3, .5, 1.5, 2.5, 3.75, 5 Jy per synthesized beam. Contours line map 3.5, 6.5, 10, 20, 30 mJy per synthesized beam. Negative contours are dashed.

Component 8 (or W3A) shows the well known shell structure. Below that are two line maps at velocities -44 and -39 kms^{-1}. It shows that component 3 has a peak velocity of -44 kms^{-1}, while the other components peak at -39 kms^{-1}. This is in exact agreement with a velocity gradient found in molecular lines from that region (Dickel et al. 1979), indicating that the HII regions must be physically associated with the molecular gas. Most interesting is the bottom map, which shows the line emission at -39 kms^{-1} divided by the continuum emission. The shell structure of component 8 can clearly be seen in the relative intensities. The peak line to continuum ratio from the shell is almost twice as high as from the low brightness central region. This could possibly indicate that the temperature in the cavity is considerably higher than in the shell.

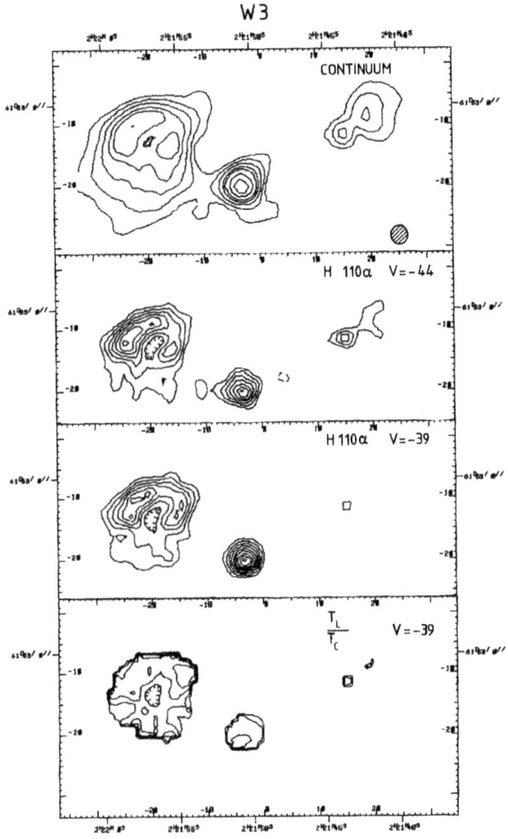

Figure 5. Maps of W3. Top-continuum map; contours .075, .225, .45, .75, 1, 1.5, 2 Jy per synthesized beam. Below that-two line maps; contours: 15, 25, ..., 95 mJy per synthesized beam. Bottom-line map divided by continuum map. Contours: 2, 3, 4, 5%.

REFERENCES

Churchwell, E., Smith, L.F., Mathis, J., Mezger, P.G. and Huchtmeier, W.: 1978, Astron. Astrophys. 70, 719.
Dickel, H.R., Dickel, J.R., Wilson, W.J. and Werner, M.W.: 1979, preprint.
Sullivan, W.T., III and Downes, D.: 1973, Astron. Astrophys. 29, 369.
Wilson, T.L., Bieging, J. and Wilson, W.E.: 1979, Astron. Astrophys. 71, 205.

RADIO RECOMBINATION LINES FROM PLANETARY NEBULAE

Yervant Terzian
Cornell University, NAIC
Ithaca, New York

I. INTRODUCTION

In general it has been very difficult to study the radio recombination lines from planetary nebulae primarily because of their very weak radio emission (Terzian 1973, 1978). Only in recent years it has been possible to detect these lines with adequate signal-to-noise ratios. The largest radio telescopes have been used for such observations with long integration times. Successful results have been obtained for Hnα lines with n $\lesssim$ 110, and only upper limits exist for lines originating from higher principal quantum levels.

In addition to Hnα lines, the H95β lines have been detected from NGC 7027 and IC 418; and the H113β line from NGC 7027. Until the present time no Cnα lines have been seen from planetary nebulae. However, the He76α, He^+105α, and He^+121α lines have been detected from NGC 7027.

The electron densities of planetary nebulae are on the average higher than the densities of normal interstellar diffuse HII regions. Many young and compact planetary nebulae have densities $\gtrsim 10^4$ cm^{-3}. Under such conditions the radio recombination lines should show the effects of electron collisional broadening. Such line broadening was detected for the first time by Churchwell, Terzian and Walmsley (1974) from NGC 7027; and most recently by Walmsley, Churchwell and Terzian (1979) for M1-78. Since the theoretical predictions of the electron collisional broadening made by Griem (1967), and Brocklehurst and Leeman (1971) seem essentially correct, we should try and use the comparison of the theory and observations to derive accurate electron densities of interstellar ionized objects. Hence, many more Hnα line observations are essential with n $\gtrsim$ 100 for such studies.

P.A. Shaver (ed.), Radio Recombination Lines, 75–80.

II. THE AVAILABLE OBSERVATIONS

Seven radio recombination lines have been observed from NGC 7027, in the principal quantum number range H66α to H110α (22.4 to 4.9 GHz). These results together with results for Hnβ, Henα, and $He^{+}n\alpha$ are given in Table 1.

TABLE 1

NGC 7027

Radio Recombination Lines

Line	Freq. (GHz)	T_1/T_c	ΔV (km/s)	V(LSR) (km/s)	Ref.
H66α	22.4	0.087	37±1	26±1	1
H76α	14.7	0.051	48±2	23±2	2
		0.039	48±3	24±1	3
		0.030	41±7	24±3	4
H85α	10.5	0.023	43±1	23±3	5
H90α	8.9	0.017	49±5	25±2	6
H94α	7.8	0.010	44±12	17±4	7
H109α	5.0	0.0035	65±12	24±2	6
H110α	4.9	0.0097	71±3	20±1	3
H95β	14.9	0.007	29±5	24±1	2
H113β	8.9	0.0035	77±14	29±6	6
He76α	14.7	0.0083	27±4	17±2	2
$He^{+}105\alpha$	22.4	0.0160	36±4	25±1	1
$He^{+}121\alpha$	14.7	0.0098	31±4	17±2	2
		0.0061	48±12	20±4	3

1. Pankonin, Thum, and Mezger (1979); 2. Walmsley, Churchwell and Terzian (1979); 3. Chaisson and Malkan (1976); 4. Bignell (1974; 5. Terzian and Balick (1972); 6. Churchwell, Terzian, and Walmsley (1976); 7. Goad and Chaisson (1973).

This table indicates the line to continuum temperature ratios, the half widths of the observed lines, and the LSR velocities of the line centers. In Table 2 a summary of the observations of the radio recombination lines from other planetary nebulae is presented with the same entries as in Table 1. The H140α observations were made with the 305-m radio telescope at Arecibo, Puerto Rico, however only upper limits of this line were derived from NGC 6210, NGC 6572, and NGC 6853. Figure 1 shows the H76α observations from NGC 7027, IC 418, and NGC 6572, performed with the Bonn 100-m radio telescope (Walmsley, Churchwell and Terzian 1979). The spectra also show the He76α, and He^{+}121α lines from NGC 7027, and possibly the He76α line from IC 418.

TABLE 2

Planetary Nebulae

Radio Recombination Lines

Planetary Nebula	Line	T_1/T_c	ΔV (km/s)	V(LSR) (km/s)	Ref.
IC 418	H76α	0.150	26±4	42±1	2
		0.075	26±8	41±4	4
	H85α	0.096	39±3	46±3	5
	H95β	0.041	26±4	42±1	2
	He76α	(0.019)	—	—	2
NGC 6210	H140α	≲0.0104	—	—	2
NGC 6543	H76α	0.122	41±12	-45±5	4
	H85α	0.057	57±12	-55±3	8
	H94α	0.078	34±10	-50±4	7
	H109α	0.036	46±9	-57±2	6
NGC 6572	H76α	0.069	33±3	8±2	2
	H140α	≲0.0030	—	—	2
NGC 6853	H140α	≲0.0044	—	—	2
BD+30°3639	H76α	0.124	50±5	-14±2	2
M1-78	H76α	0.048	37±5	-72±2	2
	H85α	0.030	54±12	-77±3	8
	H109α	0.018	65±12	-76±4	6
NGC 7662	H76α	(0.05)	—	(5)	2
	H85α	0.039	49±24	-4±4	8

8. Terzian, Higgs, MacLeod, and Doherty (1974); see Table 1 for other references.

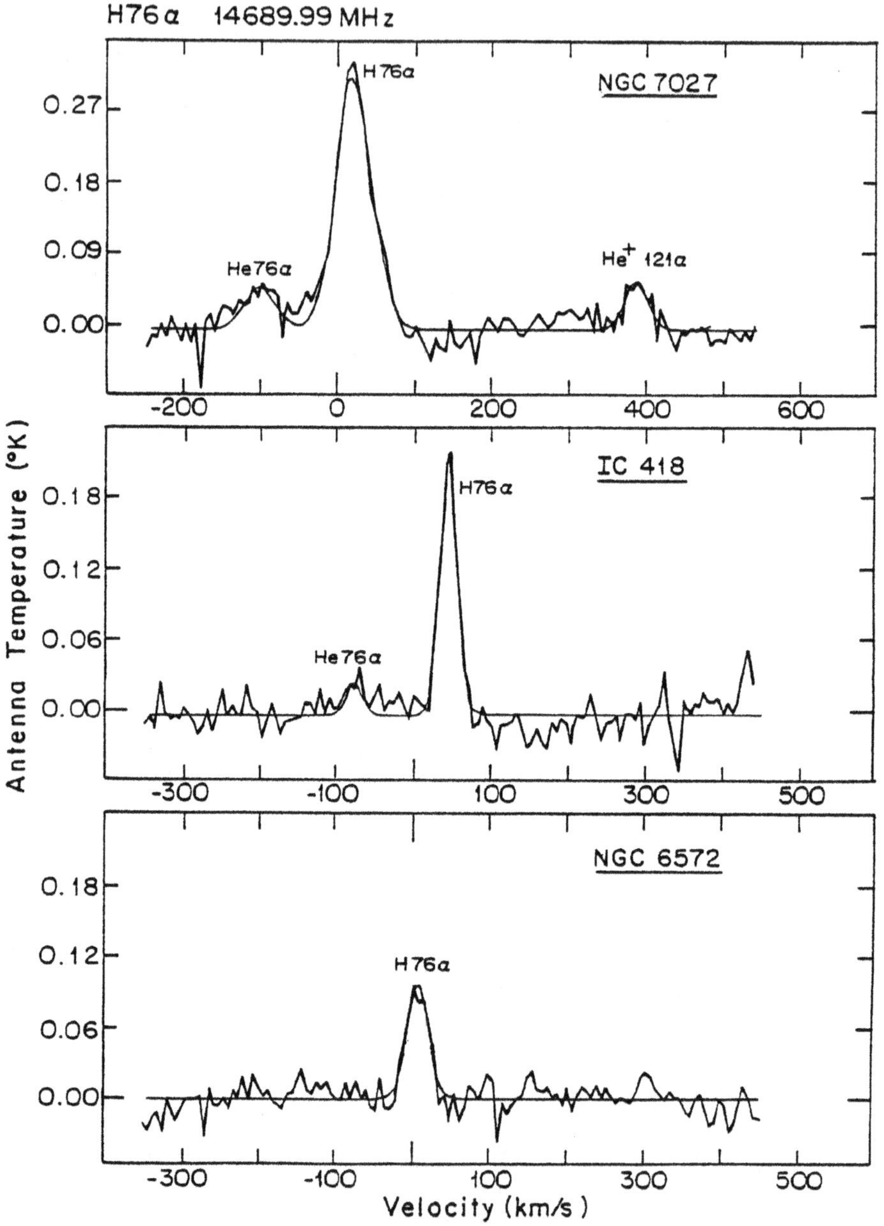

Fig. 1. The H76α spectra for NGC 7027, IC 418, and NGC 6572 observed with the Bonn 100-m telescope by Walmsley, Churchwell and Terzian (1979). These spectra also show the He76α, and He^{+}121α lines for NGC 7027, and possibly the He76α line for IC 418.

III. LINE BROADENING

The data in Tables 1 and 2 clearly indicate that the radio recombination lines become broader for NGC 7027 and M1-78, as the principal quantum level n increases. We attribute this to electron collisional broadening, and use the theory to derive the mean electron densities of these nebulae from the observations.

The ratio of the line width due to electron collisional broadening $\Delta\nu_e$, to the doppler broadening $\Delta\nu_D$, depends strongly on the principal quantum number n, and is proportional to the first power of the electron density N_e of the medium. For this ratio we can write (Griem 1967)

$$\frac{\Delta\nu_e}{\Delta\nu_D} \simeq 9.5 \times 10^{-16} \frac{N_e\, n^7}{(T_e\, T_D)^{1/2}} (\ln T_e\, n - 11.47).$$

Clearly variations of the electron temperature T_e, and doppler temperature T_D, do not play a strong role, and line width measurements at relatively high n values can be used to derive the mean electron densities of the ionized hydrogen clouds. The observed line width, $\Delta\nu_{obs}$, in the presence of electron collisional broadening, will be a convolution of the Gaussian unbroadened line with the Lorentzian function, and the observed line width can be expressed as

$$\Delta\nu_{obs} \simeq (\Delta\nu_D^2 + \Delta\nu_e^2)^{1/2}.$$

Applying the above procedure to the data for NGC 7027 and M1-78 from Tables 1 and 2 we deduce the following electron densities:

NGC 7027 $\quad N_e \simeq 5 \times 10^4\ \mathrm{cm}^{-3}$

M1-78 $\quad N_e \simeq 7 \times 10^4\ \mathrm{cm}^{-3}$.

Miller and Mathews (1972) have derived a density value of $(5.9 \pm 0.7) \times 10^4\ \mathrm{cm}^{-3}$ for NGC 7027 from the optical [OIII] and [NII] observations, in excellent agreement with the results presented above.

ACKNOWLEDGEMENTS

The author would like to thank M. Seaton, E. Churchwell, and M. Walmsley for fruitful discussions. This work was supported by the National Astronomy and Ionosphere Center, which is operated by Cornell University for the National Science Foundation.

REFERENCES

Bignell, R. C., 1974, Ap. J., 193, 687.
Brocklehurst, M. and Leeman, S., 1971, Astrophys. Letters, 9, 35.
Chaisson, E. and Malkan, M., 1976, Ap. J., 210, 108.
Churchwell, E., Terzian, Y., and Walmsley, M., 1976, Astron. and Astrophys., 48, 331.
Goad, L. and Chaisson, E., 1973, Mem. Soc. Royale Des Sciences de Liège, V, 115.
Griem, H. R., 1967, Ap. J., 148, 547.
Miller, J. S. and Mathews, W. G., 1972, Ap. J., 172, 593.
Pankonin, V., Thum, C., and Mezger, P., 1979 (private communication).
Terzian, Y. and Balick, B., 1972, Astrophys. Letters, 10, 41.
Terzian, Y., 1973, Mem. Soc. Royale Des Sciences de Liège, V, 109.
Terzian, Y., Higgs, L., MacLeod, J., and Doherty, L., 1974, Astron. J., 79, 1018.
Terzian, Y., 1978, Planetary Nebulae: Observations and Theory, Y. Terzian (ed.), (D. Reidel Publ. Co., Holland), 111.
Walmsley, M., Churchwell, E., and Terzian, Y., 1979, (preprint).

HELIUM RECOMBINATION LINES

P.G. Mezger
Max-Planck-Institut für Radioastronomie, Bonn, FRG

I. ASTROPHYSICAL GOALS

Due to the difference in the reduced mass, the radio recombination line emitted by a ^{4}He atom is shifted towards higher frequencies relative to the corresponding hydrogen line by a factor $1+3m_e/4m_p = 1.0004085$; with m_e and m_p the mass of electron and proton, respectively. This shift allows one in most galactic HII regions to separate the two lines. If He is singly ionized throughout the H^+ region*) the ratio of integrated line profiles $\varepsilon(He^+)/\varepsilon(H^+) = \int T_L(He)d\nu/\int T_L(H)d\nu$ immediately yields the number $N(^4He)/N(H) \equiv y = y^+$. In general, however, the determination of the He-abundance requires the knowledge of three terms

$$y = y^o + y^+ + y^{++} \tag{1a}$$

with $$\varepsilon(He^i)/\varepsilon(H^+) = y^i = \int_{H^+} n_e\, n(He^i)dV / \int_{H^+} n_e\, n(H^+)dV \tag{1b}$$

Here $\int_{H^+} \ldots dV$ means the volume integral over the H^+ region while n_e, $n(He^i)$ and $n(H^+)$ are the volume densities of electrons, He ions and protons, respectively. While the last two terms can be determined from radio recombination lines the fraction, y^o, of neutral He inside the H^+ region can only be estimated.

One defines the quantity

$$R = \int_{He^+} n^2(H^+)dV / \int_{H^+} n^2(H^+)dV \tag{2}$$

as the ratio of He^+ and H^+ Strömgren spheres, weighted with the proton density. In galactic HII regions ionized by MS stars $R \lesssim 1$ and $y^{++} << y$, so that

*) I refer to an "HII region" in the historical meaning as an ionized region of interstellar matter. I use the notation H^+ and He^+ region specifically for the volumes (= Strömgren spheres) in which H and He are ionized.

P.A. Shaver (ed.), Radio Recombination Lines, 81–97.

$$y \approx y^o + y^+ \approx R^{-1}y^+ \tag{3}$$

If $R < 1$, non-LTE effects, continuum opacity and Stark broadening can affect the observable ratio $\varepsilon(He^+)/\varepsilon(H^+)$. However, as Churchwell, Mezger and Huchtmeier (1974; hereafter referred to as Paper I) have shown, the observed ratio can never exceed the true He abundance, i.e.

$$\varepsilon(He^+)/\varepsilon(H^+) \leq y;$$

and, if highly unlikely density distributions are excluded, this inequality reduces to

$$\varepsilon(He^+)/\varepsilon(H^+) \leq y^+$$

From observations of He recombination lines one can either deduce the He abundance of the interstellar gas (by estimating y^o or R, respectively); or one can deduce the quantity R and hence the ionization structure of the HII region (by adopting a value for the He abundance y). The present galactic He abundance y is thought to be the sum of primordial He, y_p, and He enrichment as a result of the chemical evolution of the Galaxy, Δy. The possibility to determine He abundances y of the interstellar matter in various parts of the Galaxy and the hope to find systematic variations in the evolution term Δy motivated my colleagues and me to carry out a number of surveys of He recombination lines with the NRAO 140-ft telescope, the CSIRO 210-ft telescope and more recently with the MPIfR 100-m telescope.

II. RESULTS OF PREVIOUS SURVEYS

Paper I summarized the observations of He recombination lines up to 1973, which yielded the following results:

1) $y^+ \lesssim 0.10$ throughout the Galaxy.
2) y^+ decreases with decreasing galactocentric distance D_G.
3) There exists a correlation between y^+ and the IR excess (IRE) = $L_{IR}/N'h\nu_\alpha$.

The interpretation of these results led us to the conclusion that the present He abundance is $y \approx 0.10$ throughout the Galaxy, of which $y_p \approx 0.08$ is the primordial contribution. Values of $y^+ < 0.10$ were explained by the presence of neutral He in the HII region or, in terms of eq. (3), by $R < 1$. Since model atmospheres of O stars by Auer and Mihalas (1972) predicted that O stars earlier than O9 should ionize He throughout the H^+ region, and since O6.5 stars are the prime contributors to the ionization of compact radio HII regions, it was concluded that selective absorption of Lyman continuum (Lyc) photons by dust is responsible for non-coincidence of He^+ and H^+ Strömgren spheres, i.e. for $R < 1$ (Mezger, Smith and Churchwell, 1974). Selective dust absorption could also explain the correlation between decreasing y^+ and increasing (IRE).

Once the telescope beam (HPBW Θ_A) resolves the H^+ region (HPW Θ_s), i.e. $\Theta_A < \Theta_s$, eq. (1b) assumes the form (Paper I)

$$\frac{\varepsilon(\mathrm{He}^+)}{\varepsilon(\mathrm{H}^+)} = \frac{\int_{\Omega m} f(\Omega) \int_{s_1'}^{s_2'} n_e\, n(\mathrm{He}^+)ds\, d\Omega}{\int_{\Omega m} f(\Omega) \int_{s_1''}^{s_2''} n_e\, n(\mathrm{H}^+)ds\, d\Omega} \qquad (4)$$

Here $f(\Omega)$ is the antenna pattern, Ωm is the main beam solid angle and s_1, s_2 are the intersection points of the line of sight, s, with the Strömgren sphere of He^+ (') and H^+ ("), respectively. The important point is that the telescope beam supresses part of the line radiation from the outer H^+ region where He is neutral, so that the observed He^+ abundance increases with increasing telescope resolution. This is referred to as the "geometrical effect". With $\xi = \Theta_A/\Theta_S$, eq (3) becomes

$$y^+(\xi) = R(\xi)y \qquad (5)$$

For $\xi \leq 1$, $R(\xi)$ increases monotonically with decreasing ξ and, for a homogeneous spherical HII region, attains the limiting value $R(\xi \to 0) = R^{1/3}(\xi \gg 1)$. For more realistic HII regions with a centrally peaked density distribution the increase of R is even more pronounced.

In 1977 high quality He recombination line measurements for n = 109 and 91 became available, made with the Effelsberg 100-m telescope, which warranted a new analysis similar to that in paper I. The results were published by Churchwell, Smith, Mathis, Mezger and Huchtmeier (1978, hereafter referred to as paper II). The decrease of y^+ with decreasing D_G found in paper I was confirmed (Fig. 1a, b) and was, as in paper I, interpreted by an increasing fraction of neutral He in the H^+ region or, quantitatively, by decreasing values of R. The newly discovered increaseof y^+ from ~ 0.07 for $D_G \gtrsim 12$ kpc to ~ 0.09 at $D_G = 9$ kpc was interpreted as reflecting a true increase of y, i.e. the conclusion of paper I of a constant He abundance $y \simeq 0.10$ throughout the Galaxy was abandoned.

In the past two years 76α (14.7 GHz) and 66α (22.4 GHz) lines of H and He were observed with the 100-m telescope. At these high frequencies opacity and non-LTE effects as well as Stark broadening are negligible except for the most compact HII regions. The interpretation of these lines thus is much more straightforward. With telescope HPBW's of $\Theta_A = 0.9$ and 0.7 arc min, respectively, many H^+ regions are resolved and the derived He^+ abundance, as a result of the geometrical effect, yields a much better lower limit to the true He abundance y. Results of 76α and 66α He line surveys are in press (Thum, Mezger and Pankonin, 1979; hereafter referred to as paper III) and will be dealt with in the following sections.

III. WHAT DETERMINES THE IONIZATION STRUCTURE OF HII REGIONS?

It was shown by Mathis (1971) that R depends only on the ratio $\gamma = N_{He}/N_H$ of He ionizing to H ionizing photons and that for y = 0.10,

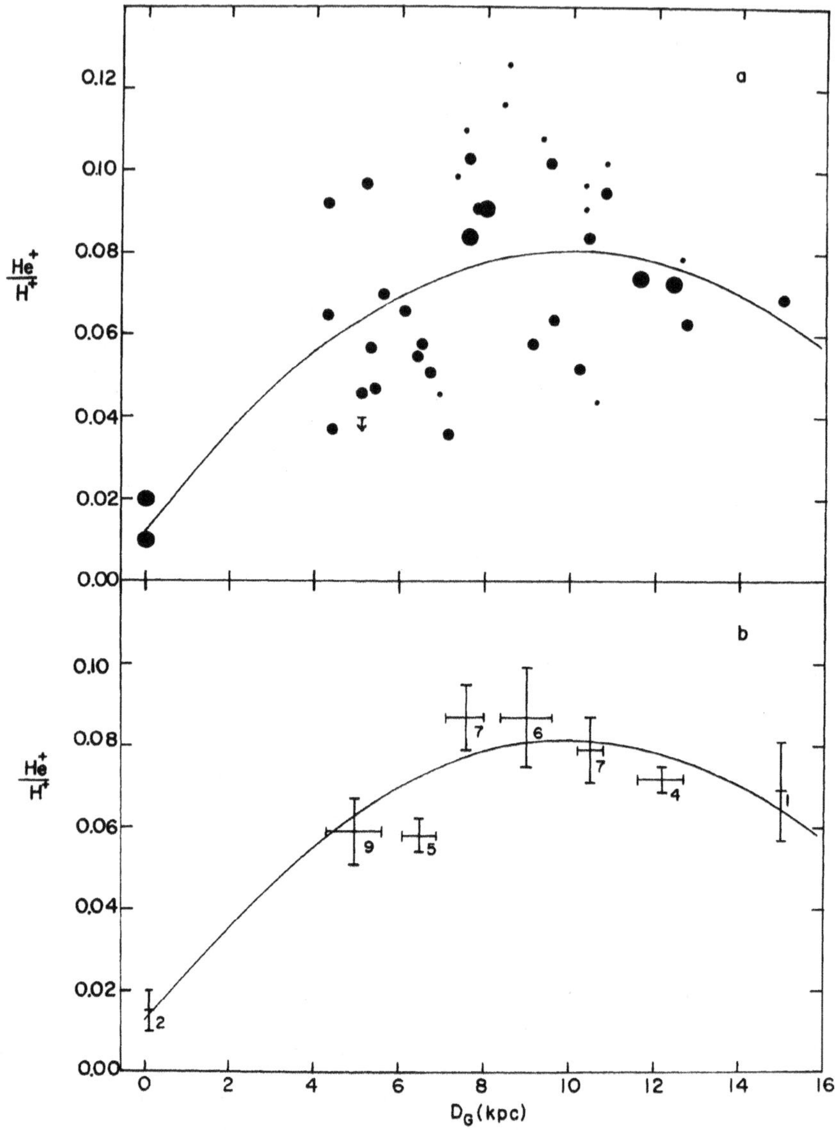

Fig. 1. Variations of the ionized He abundance (Paper II). (a): Results obtained for individual HII regions; sizes of dots refer to the accuracy of the measurements. (b): Average values; numbers refer to the number of individual measurements used for computing the average value.

$R \simeq 1$ for $\gamma \gtrsim 0.20$. In Fig. 2, γ is plotted as a function of the effective star temperature T_{eff} for (a) a black body, (b) the non-LTE atmospheres by Auer and Mihalas (1972) and (c) recent model atmospheres by Kurucz (1979) which include line blanketing. The relation between T_{eff} and spectral type (Sp.T) is from Conti (1975). Arrows on the right side of the diagram indicate, for different values of the He-abundance y, at which γ the quantity R attains the value 0.95 (rather than unity, which is only reached asymptotically).

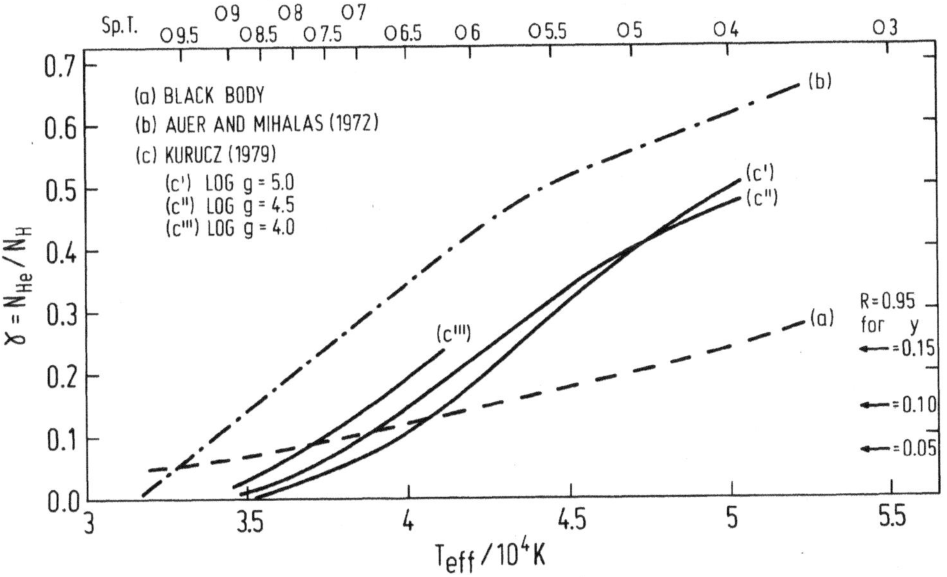

Fig. 2. Ratio of He ionizing photons to H ionizing photons emitted by O stars. Curves a, b, c refer to a blackbody and to two different model atmospheres, respectively.

In the solar neighborhood and for a Salpeter Initial Mass Function (IMF) O6.5 stars with an effective temperature of 40 000 K contribute most to the galactic Lyc radiation. On the basis of the Auer and Mihalas atmospheres (curve b in Fig. 2) these stars have a $\gamma \simeq 0.40$. A softening of the Lyc radiation of an O6.5 star to the extent that the effective (for the ionization of the gas) $\gamma' < 0.20$, can only be done either by a drastic change of the IMF (as compared to the solar neighborhood) or by selective dust absorption. This latter effect has been investigated by Mezger et al. (1974) who showed, that besides strong selective absorption, values of the optical depth $\tau_{Lyc} \gtrsim 1$ are required to reduce γ substantially. This means that R and hence y^+ should decrease with increasing τ_{Lyc}. The infrared excess (IRE) = $L_{IR}/N'_c h\nu_\alpha$ is a quantity which can be determined for an HII region from FIR and radio observations. For an ionization bounded HII region surrounded by a dense neutral cloud $L_{IR} \approx L_*$; furthermore, for an O6.5 star, $L_*/N_c h\nu_\alpha \approx 5$. Hence

$$(\text{IRE}) \approx e^{\tau_{Lyc}} \frac{L_*}{N_c h\nu_\alpha} \approx 5\ e^{\tau_{Lyc}}$$

i.e. (IRE) should increase with increasing τ_{Lyc}. The corresponding observed correlation of decreasing y^+ with increasing (IRE) first noted by Mezger et al. (1974) was the strongest support for the hypothesis of selective dust absorption in the Lyc.

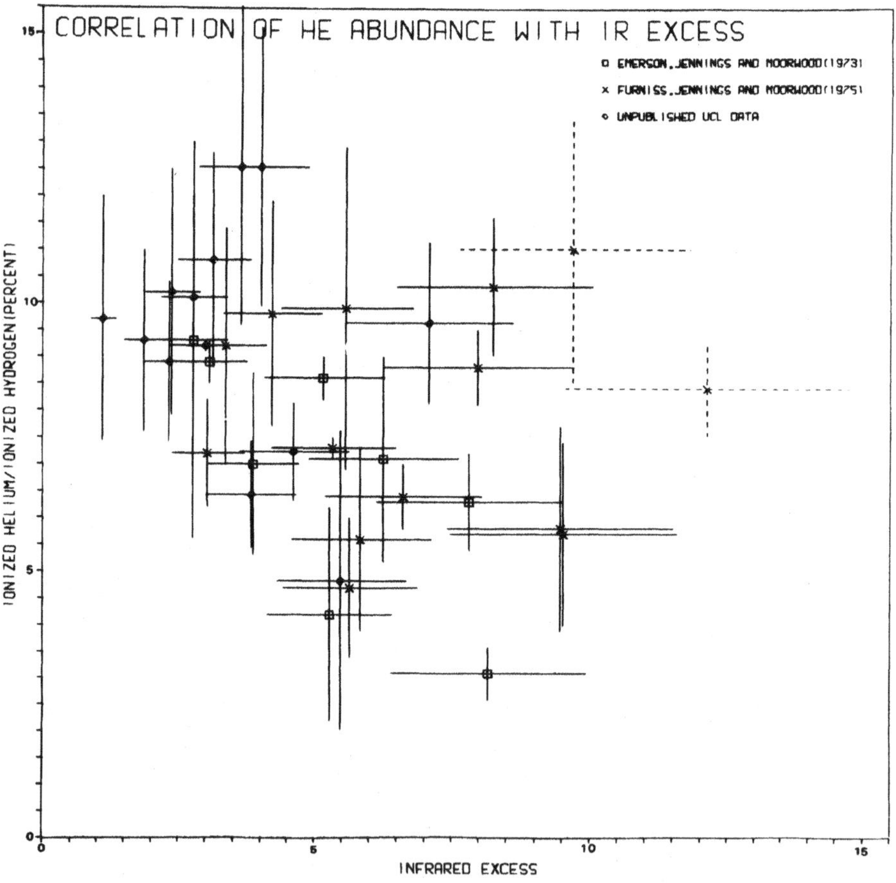

Fig. 3. Observed relation between He$^+$ abundance and Infrared Excess (Emerson and Jennings, 1978).

Fig. 3 shows the correlation of y^+ and (IRE) as observed by Emerson and Jennings (1978) which clearly shows the predicted effect. However, Panagia and Smith (1978), in their interpretation of the observations by Emerson and Jennings in terms of selective dust absorption cautioned already, that the Auer and Mihalas O-star atmospheres may not be correct, since they neglect line blanketing in the Far UV. In fact, in the model atmospheres by Kurucz (1979), which include line blanketing, the energy emitted beyond the He ionization limit is strongly reduced, and so are the corresponding γ values (curves c in Fig. 2). Under these conditions, small changes in the effective temperature of O stars could drastically alter the H/He ionization structure of HII regions. This led Panagia (1979 and this volume) to suggest that variations in the mean effective temperature of O stars, weighted with the IMF, $\langle T_{eff} \rangle$, rather than selective dust absorption is responsible for the variation of y^+ and R. With $\langle T_{eff} \rangle$ increasing from 31 000 K at the galactic center to 39 000 K at D_G = 10 Kpc he can approximately reproduce the observed variation $y^+(D_G)$ shown in Fig. 3. The underlying physical cause for a decrease of $\langle T_{eff} \rangle$ with decreasing D_G is thought to be a corresponding increase of the metal abundance Z, which affects $\langle T_{eff} \rangle$ both through changes of the stellar opacity and changes in the high-mass end of the IMF. By assuming an increase of the dust-to-gas ratio according to $m_g/m_d \propto Z$ and a wavelength independent dust absorption for λ <912 Å Panagia can also explain the inverse correlation between y^+ and IRE shown in Fig. 3.

Both selective dust absorption and constant $\langle T_{eff} \rangle$ on the one hand and wavelength independent dust absorption but a stellar $\langle T_{eff} \rangle$ which decreases with Z on the other hand can explain the observed correlations of y^+ with (IRE) and D_G. It may well be that both effects play a role, but at present it appears to me that the variation of the effective temperature is the dominant effect. This conclusion also agrees with laboratory measurements of potential dust material, such as graphite or silicates, which preclude selective absorption in the Lyman continuum. A firmer conclusion, however, can only be drawn once we know more about the nature of interstellar dust grains.

IV. NON-LTE EFFECTS VS. GEOMETRICAL EFFECT

Optical spectroscopists who tried to determine He abundances were always aware that part of the He inside the H^+ region is neutral. They considered this fact when estimating the true He abundance from observed He^+ abundances. Nevertheless the explanation given in Paper I that y^+ = Ry <0.10 is due to the presence of neutral He, i.e. R <1, was at first not generally accepted. Brown and Gomez-Gonzales (1975) preferred an explanation, where the observed ratio of integrated line profiles, $\varepsilon(He^+)/\varepsilon(H^+)$, is primarily determined by non-LTE radiative transfer effects in a cool and only partially ionized layer surrounding the H^+ region ("external maser effect"). If this were correct, He abundance determinations from radio recombination lines would be rather meaningless. It therefore was important fo find out, whether the variation of y^+, observed in galactic HII regions, is due to the presence of neutral He or due to the external maser effect. In the ensuing debate observations of He lines from G0.7-0.0, a giant HII region close to the galactic center, played an essential role.

The early 109α observations of giant HII regions close to the galactic center, made with the NRAO 140-ft and with the CSIRO 210-ft telescope, showed no He lines down to a level $y^+ \lesssim 0.02$. The first He lines in the galactic center were found in G0.7-0.0. Brown and Lockman (1975) considered their detection of the He76α line with the 140-ft telescope as a confirmation of the external maser model. Mezger and Smith (1976) considered their detection of the He109α line with the MPIfR 100-m telescope as a result of the geometrical effect.

Since Brown and Lockman's model predicted variations of y^+ from 0.01 at low frequencies to 1 at high frequencies, as well as large variations of the quantity $(T_L/T_C)\Delta\nu \cdot \nu_L^{-2}$ which, for TE emission, is nearly constant, one can discriminate between non-LTE and geometrical effects on the basis of recombination line observations at different frequencies. Smith and Mezger (1976), supplementing the available observations by further high frequency data obtained with the 100-m telescope, showed that the available data were compatible with the predictions of the geometrical effect but not with the Brown and Gomez-Gonzales model. This was later confirmed by Chaisson et al. (1978) and Thum et al. (1978). Only the 76α value of $y^+ = 0.085 \pm 0.015$ by Brown and Lockman was difficult to include in reasonable fits of $y^+ = y^+(\xi)$. At high frequencies two compact components, G0.7S and N, dominate the emission. The 100-m telescope, which was used by Thum et al. (1978), resolves the two components at 15 GHz. Their result for the 76α lines, $y^+ = 0.050 \pm 0.015$, relates to the stronger component G0.7S. The 140-ft telescope used by Brown and Lockman (1975) included both G0.7S and N in its wider beam and hence it could not be excluded that their higher y^+ value is the result of a high ionized He abundance ($y^+ > 0.10$) of G0.7N. However, recently Thum, Mezger and Pankonin (1979) measured the He 66α line separately for G0.7S and N and derived $y^+ \simeq 0.06$ for both components. This shows that not only is the hypothesis of an external maser effect inapplicable to this HII region, but also that the He 76α observations, which apparently supported this hypothesis, cannot be correct.

Fig. 4, from paper III, shows y^+ observed for nine sources in four different transitions (n = 109, 90/91, 76 and 66) with the 100-m telescope. All sources close to or inside the solar circle show the geometrical effect, i.e. a monotonic increase of y^+ with decreasing telescope HPBW Θ_A. Outside the solar circle the variation of y^+, if present at all, is less than 0.01.

The case of the Orion nebula is of special interest since for this HII region both accurate optical and radio measurements of y^+ at various positions are available. Churchwell, Mezger and Smith (unpublished) mapped y^+ in the Orion nebula at 5 GHz. Their data were interpreted by Pankonin et al. (1979) who find good agreement with optical determinations of y^+ (Peimbert and Torres-Peimbert, 1977; McCall, 1979). This in itself is an interesting result since it shows that the radio data cannot be strongly

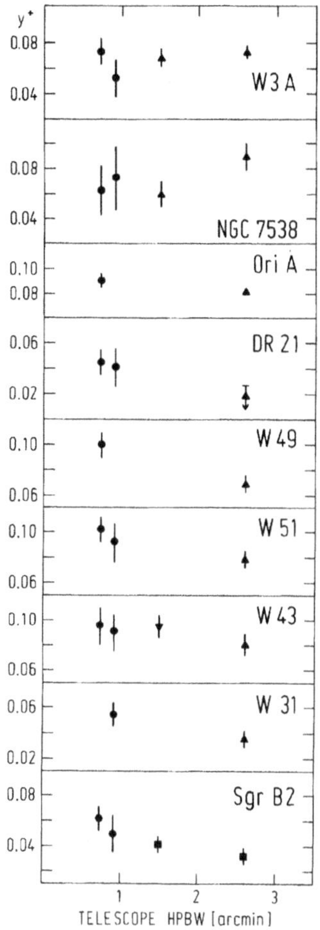

Fig. 4. Observed variation of the He abundance y^+ with telescope HPBW. Sources are arranged (from top to bottom) in order of decreasing galactocentric distance. Decreasing telescope HPBW corresponds to decreasing line frequency and principal quantum number n.

affected by non-LTE effects or Stark broadening. Fig. 5 shows y^+ as a function of distance r from the principal ionizing star of the Trapezium cluster, Θ^1C Orionis. Filled and open circles refer to two different reduction methods of the radio observations. It is evident that y^+ decreases with r, but it decreases much more rapidly to the east-northeast (fit to pos. 6, 7, 10 and 17 and to the center position 14) than to the west of the star (fit to pos. 1, 3, 5, 9, 11, 12, 13, 14 and 22).

V. CAN WE DEDUCE ASTROPHYSICALLY USEFUL DATA?

The astrophysial goals of He line surveys were outlined in Sect. I of this review. It has been argued that non-LTE effects would dominate the emission of radio recombination lines to an extent that practically no useful astrophysical information could be deduced. I therefore should comment on this problem before discussing, in the following section, He abundances in the Galaxy as derived from radio recombination lines.

Both radio and optical observations agree that there is neutral He in many HII regions so that the term y^o in eq. (1a) cannot be neglected in deducing the He abundance from measurements of the He^+ abundance. High

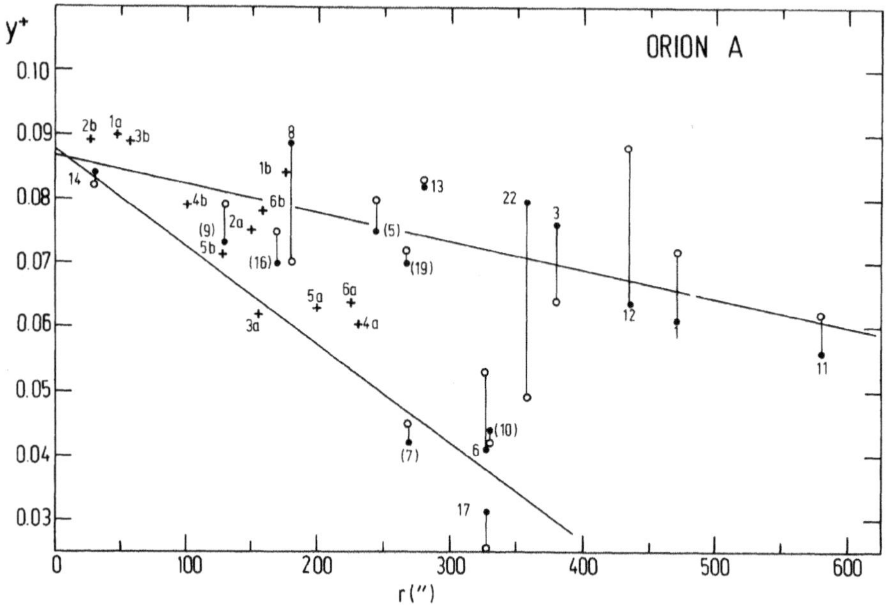

Fig. 5. Observed variation of the He^+ abundance y^+ in the Orion Nebula as a function of the distance r from the principal ionizing star.

resolution observations, as a result of the geometrical effect, tend to decrease the term y^0 and the corresponding values $y^+(\xi < 1)$ therefore yield better lower limits to the true He abundance y.

There is no observational evidence for the "external maser effect" suggested by Brown and Gomez-Gonzales (1975). More recently Brown, Lockman and Knapp (1978) published new model computations which refer to an "internal maser effect". The neutral He is thought to occur in the outer, low-density envelops of HII regions with compact cores. Brown et al. suggest that most of the variation of y^+ with telescope HPBW (such as shown in Fig. 4) is due to maser amplification of the hydrogen lines in those parts of the H^+ region where He is neutral. Some objections to these model computations have been formulated by Shaver (1979). In this context I am only considering to which extent the internal maser effect, if it should apply, could alter the observed ratio $\varepsilon(He^+)/\varepsilon(H^+)$ from its LTE value, i.e. the He^+ abundance y^+ as defined by eqs (1b) and (4), respectively. At lower frequencies the amplified hydrogen line emission in the outer envelop of the HII region would tend to decrease this ratio below its LTE value; with increasing frequency it would approach its LTE value y^+. Since usually higher frequency also means smaller telescope HPBW the net effect of the internal maser effect, in most cases, would result in an increase of y^+ with frequency, that would be somewhat steeper than could be accounted for by the pure geometrical effect.

What then are the consequences of the new model computations by Brown et al (1978) for the interpretation of He recombination lines? They are rather irrelevant since observers, due to the presence of neutral He in most H^+ regions, were always aware that they must interpret $\varepsilon(He^+)/\varepsilon(H^+)$ as a lower limit of the true He abundance. The only relevant result has already been stated in Paper I, and has been reiterated in Sect. I of this review. The ratio $\varepsilon(He^+)/\varepsilon(H^+)$, regardless of non-LTE effects or Stark broadening, can never exceed the true He abundance y; and for most real HII regions it will never exceed the LTE value y^+.

VI. GALACTIC HE ABUNDANCES

This section summarizes the results of paper III related to the ionized He abundance. Fig. 6, which is taken from this work, is a revised version of Fig. 1 supplemented by new high frequency, high resolution data. The boxes display the 109α data from paper II. The galactic radius interval of each box contains five sources. The outer boundary describes the total range of y^+ values observed at 109α in each box. In an attempt to reduce the effect of possible unnoticed experimental errors on the variation of y^+, the extreme values in each box have been omitted. This leaves the shaded range of y^+ in Fig. 6. The median value of each box is indicated by a dashed line. y^+ values of sources in the present high frequency survey are superimposed on the boxes together with the corresponding 91α and 109α data as dots connected by vertical lines. In the domain where the geometrical effect works (i.e. $D_G < 11$ kpc; see also Fig. 4) most 66α y^+ points fall at or above the upper boundary of the 109α values.

The interpretation of this diagram regarding its relation to the He abundance in the Galaxy is as follows. The upper limit of the shaded part of the figure qualitatively describes the variation of y with galactic radius for $D_G \gtrsim 7$ kpc. Despite the absence of a strong geometrical effect in HII regions outside the solar circle there is probably still a fraction $y^o \lesssim 0.01$ of He in the observed HII regions that is neutral so that y^+(radio HII) can only be considered as a good lower limit to the He abundance of the interstellar matter, y(ISM). This is also in accordance with estimates of y^o from measured abundances $[O^+/O]$ and $[S^+/S]$ by Peimbert et al. (1978). However, inside the solar circle, especially for $D_G \lesssim 7$ kpc, this fraction y^o increases strongly. Thus the true He abundance at $D_G \sim 4$-6 kpc may well be as high as 0.13 - 0.15. At galactic radii $D_G < 7$ kpc the He^+ Strömgren sphere shrinks so rapidly that this effect dominates over a possible relatively slow increase of the He abundance, i.e. the observable quantity $y^+ = Ry$ is dominated by the change of R rather than y. As mentioned in Sect. III this effect on the ionization structure can be explained by a decrease of the mean weighted effective temperature of O stars (Panagia 1979) which might be a consequence of an increase in the metal abundance Z from the edge to the center of the Galaxy, as derived from the measured variation of the electron temperature

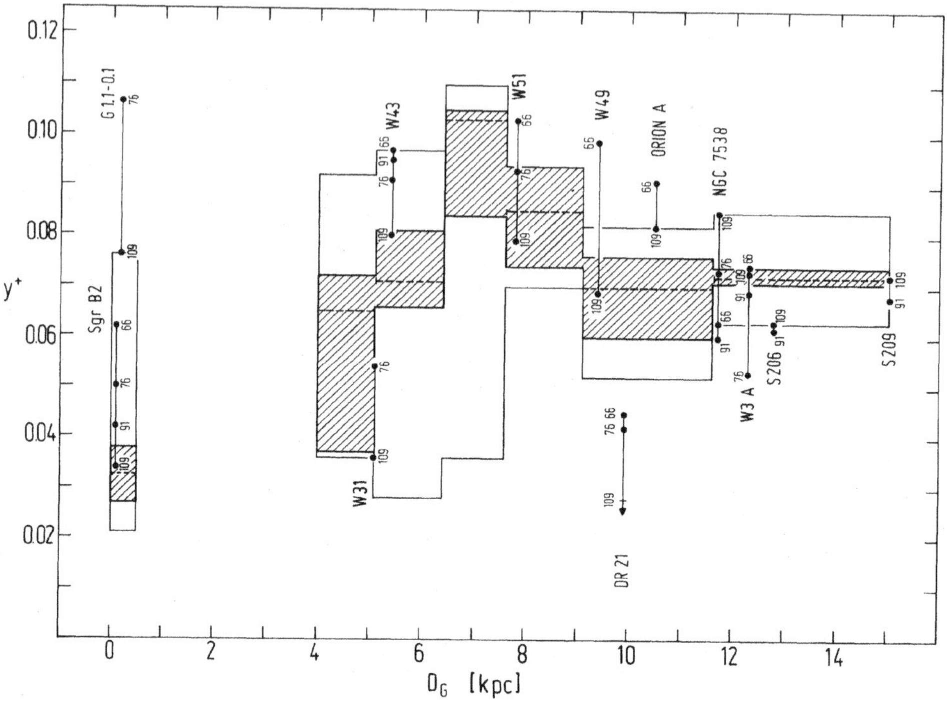

Fig. 6. Observed variation of the He$^+$ abundance y^+ with galactocentric distance D_G.

of galactic HII regions (Mezger et al., 1979).

The 76α value of ∿0.10 obtained by Wink and Mezger for G1.1−0.1 is crucial for this explanation since it shows that the very low values of y^+ observed for HII regions close to the galactic center are caused by small values of R rather than by a change of the total He abundance y.

For D_G >7 kpc the fraction of neutral He, y^o, appears to be small enough that the upper envelope of the ionized He abundance obtained with high angular resolution in radio HII regions and displayed in Fig. 6 is a meaningful lower limit for the He abundance of the interstellar matter. We suggest that He abundances of planetary nebulae (PN) provide a similar meaningful upper limit for y(ISM). In most PNe there is no neutral He and the terms y^+ and y^{++} can be determined with good accuracy directly from observations. The original chemical composition of the star from which the PN has evolved is that of the ISM at the time the star was born . However, the chemical composition of the PN has changed during the various nuclear burning phases of the star which now forms the nucleus of the PN. The implication of our suggestion, y(PN) $\gtrsim$y(ISM), is that the enrichment of the PN with He is equal to or larger than the average He enrichment of the ISM. This assumption is certainly correct in the solar vicinity where the He abundances of all accurately observed PNe are higher than that of the Orion Nebula, whose He abundance has been determined with high accuracy from both optical and radio observations.

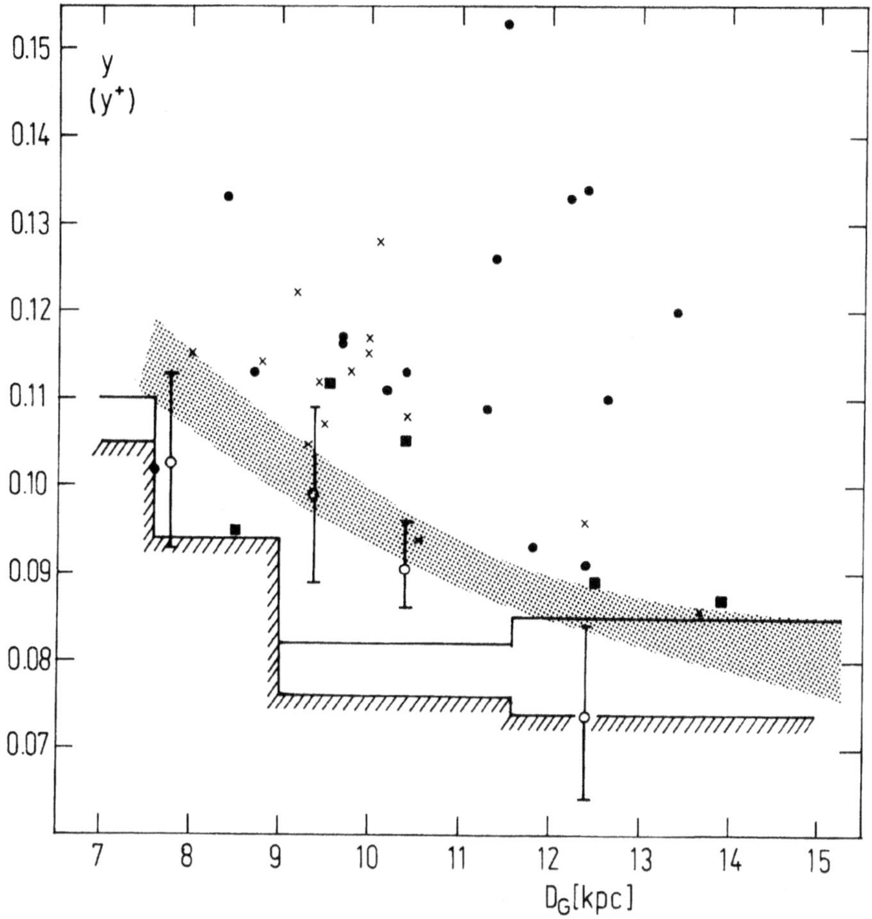

Fig. 7. The true He abundance of the interstellar matter, y(ISM) (shaded area), as a function of the galactocentric distance D_G. Shown in the diagram are: He abundances of PNe (filled dots; crosses for best observed PNe); of HII regions (squares refer to optical observations; upper boundaries refer to 109α data from Fig. 6; open circles with error bars refer to 66α data). From Paper III.

Following this line of argument we have combined in Fig. 7 values of y(PN) from Torres-Peimbert and Peimbert (1977) with values of y^+(radio HII) displayed in Fig. 6. The shaded region in Fig. 7 is defined by the upper envelop of y^+(radio HII) and by the lower envelop of y(PN). The He abundance of the ISM should lie between these upper and lower envelops within the shaded region; this leaves for y(ISM) a range of uncertainty of the order δy(ISM) = 0.005 to 0.01. Despite this uncertainty the variation of y(ISM) with galactocentric distance D_G between 8 and 15 kpc is clearly recognizeable. It compares well with He gradients derived by Peimbert et al. (1978) and Torres-Peimbert and Peimbert (1977).

At $D_G \simeq 15$ kpc y(ISM) assumes values as low as 0.08. This led us to the conclusion (Paper II) that at these large galactocentric distances the enrichment term, Δy, is negligible so that one actually observes the primordial He abundance y_p. However, if there exists a simple relation between He enrichment and metal abundance, as generally assumed, the relatively high oxygen abundance inferred from the electron temperature of HII regions at $D_G \simeq 15$ kpc (Mezger et al. 1979) is hard to reconcile with $\Delta y \simeq 0$.

VII. OBSERVATIONS OF HE^+ RECOMBINATION LINES

It was mentioned in Sect. I that $y^{++} < 0.01$ in HII regions ionized by MS stars. It is also known that in Planetary Nebulae, whose exciting stars can have much higher effective temperatures, a substantial fraction of He is doubly ionized. Radio observations of radio recombination lines which are formed after recombination of an electron with a He^{++} ion can be used for an investigation of the doubly ionized He abundance y^{++}. To minimize observational problems one uses a $He^+n'\alpha$ line which is close (in frequency) to the H/He$n\alpha$ lines used for the determination of y^+. One can show that the two lines coincide approximately in frequency if $n' \simeq 2^{2/3}n$. For example the $He^+ 173\alpha$ line is used in conjunction with H/He 109α lines, and the He^+ 105α line in conjunction with the H/He 66α lines. As shown e.g. in Paper I in this case

$$y^{++} = \frac{1}{4}\,[\varepsilon(He^{++})/\varepsilon(H^+)]$$

where the same notation is used as in Sect. I of this review. Radio measurements of y^{++} for galactic HII regions and especially for the giant HII regions with low y^+ abundances close to the galactic center are summarized in Paper I, and values y^{++} considerably less than 0.01 have been found.

In high excitation PNe one would expect to find about as much He^{++} as He^+ and therefore the He^+ radio recombination lines should be stronger than the adjacent He lines. However, both the flux densities and the high electron temperatures of PNe makes the detection even of H radio recombination lines difficult. A notable exception to this rule is NGC 7027, which is one of the strongest thermal sources in the cm- and mm-range. The first detection of a He^+ recombination line, viz. the He^+ 121α line, was reported for this PN by Chaisson and Malkan (1976). They derived a value of $y^{++} \simeq 0.04$ but no value for y^{+}.

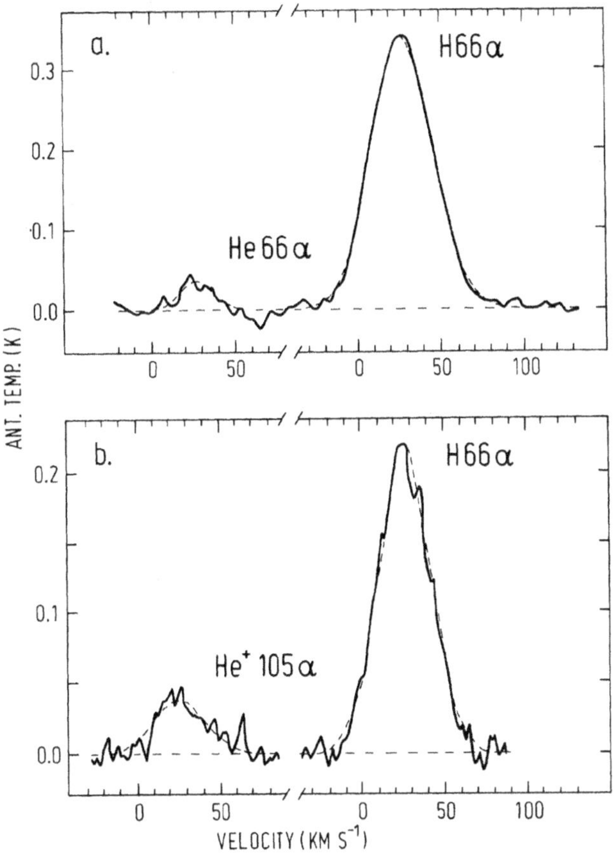

Fig. 8. He and He⁺ recombination lines from the Planetary Nebula NGC 7027, observed with the MPIfR 100-m telescope (Pankonin, Mezger and Thum, in prep.).

Recently Pankonin, Mezger and Thum (in prep.) observed both the H/He 66α and the adjacent He^{+} 105α line in NGC 7027 (see Fig. 8a, b) and derived abundances $y^{+} = 0.059 \pm 0.02$ and $y^{++} = 0.045 \pm 0.02$. The corresponding optical observations by Miller and Mathews (1972) yield $y^{+} = 0.085$ and $y^{++} = 0.042$, Torres-Peimbert and Peimbert (1977) give $y^{+} = 0.076$ and $y^{++} = 0.039$. With $y^{o} << 0.01$ the corresponding total He abundance of NGC 7027 is found to be $y = 0.104 \pm 0.028$ using radio measurements, and y = 0.127 and 0.115 using optical measurements. Within the accuracy of observations there is full agreement between optical and radio determination of y. I interpret this result as an indication, that the He abundance of NGC 7027 is higher by $\Delta y = 0.01-0.02$ than that of the interstellar matter located at the same galactocentric distance, y(ISM) = 0.095-0.10. This agrees with our suggestion that the He enrichment in PNe is somewhat higher than the general He enrichment of the ISM.

ACKNOWLEDGMENTS

I want to thank J. Mathis for the computation of curves (c) in Fig. 2; and N. Panagia for an interesting discussion on Planetary Nebulae. Last but not least I want to thank my colleagues V. Pankonin and C. Thum for letting me use in this review some of our results which are not yet published.

REFERENCES

Auer, L.H., Mihalas, D. 1972, Astrophys. J. Suppl. 24, 193
Brown, R.L., Lockman, F.J. 1975, Astrophys. J. 200, L155
Brown, R.L., Gomez-Gonzales, J. 1975, Astrophys. J. 200, 598
Brown, R.L., Lockman, F.J., Knapp, G.R. 1978, Ann. Rev. Astron. Astrophys. 16, 445
Chaisson, E.J., Malkan, A. 1976, Astrophys. J. 210, 108
Chaisson, E.J., Lichten, S.M., Rodriguez, L.F. 1978, Astrophys. J. 221, 810
Churchwell, E., Mezger, P.G. and Huchtmeier, W. 1974, Astron. Astrophys. 32, 283 (Paper I)
Churchwell, E., Smith, L.F., Mathis, J., Mezger, P.G., Huchtmeier, W. 1978, Astron. Astrophys. 70, 719 (Paper II)
Conti, P.S. 1975 in "HII regions and related topics" (T.L. Wilson and D. Downes, eds.) Springer Verlag, p. 207
Emerson, J.P., Jennings, R.E. 1978, Astron. Astrophys. 69, 129
Kurucz, R.L. 1979, Astrophys. J. Suppl. Ser. 40, 1
Mathis, J.S. 1971, Astrophys. J. 167, 261
McCall, M.L. 1979, Astrophys. J. 229, 962
Mezger, P.G., Smith, L.F., Churchwell, E. 1974, Astron. Astrophys. 32, 269
Mezger, P.G., Smith, L.F. 1976, Astron. Astrophys. 47, 134
Mezger, P.G., Pankonin, V., Schmid-Burgk, J., Thum, C., Wink, J. 1979, Astron. Astrophys. (in press)
Miller, J.S., Mathews, W.G. 1972, Astrophys. J. 172, 593
Panagia, N., Smith, L.F. 1978, Astron. Astrophys. 62, 277
Panagia, N. 1979, Mem. Soc. Astron. Ital. 50, 79
Pankonin, V., Walmsley, C.M., Thum, C. 1979 (in press)
Peimbert, M., Torres-Peimbert, S., Rayo, J.F. 1978, Astrophys. J. 220, 516
Shaver, P.A. 1979, Astron. Astrophys. (in press)
Smith, L.F., Mezger, P.G. 1976, Astron. Astrophys. 53, 165

Thum, C., Mezger, P.G., Pankonin, V., Schraml, J. 1978, Astron. Astrophys 64, L17
Thum, C., Mezger, P.G., Pankonin, V. 1979, Astron. Astrophys.,(in press) (Paper III)
Thum, C., Mezger, P.G., Pankonin, V. 1979 (in prep.)
Torres-Peimbert, S., Peimbert, M. 1977, Rev. Mex. Astron. Astrofis. 2, 187

SYSTEMATIC VARIATION OF THE HELIUM IONIZATION IN GALACTIC HII REGIONS

Nino Panagia
Istituto di Radioastronomia
Bologna, Italy

The existence of a gradient of the ionized helium abundance as a function of the galactic radius was first noted by Churchwell et al. (1974). This result has recently been confirmed by the much more complete survey of giant HII regions presented by Churchwell et al. (1978). In detail, the ratio He^+/H^+ is found to increase from $\sim$0.02 near the galactic center passing through $\sim$0.06 at 5 Kpc to reach a maximum of $\sim$0.09 near 9 Kpc and then to decrease for larger radii. This behaviour can be interpreted as reflecting a general decrease of the He abundance with the galactic radius (similar to that also found for planetary nebulae; Torres-Peimbert and Peimbert, 1977) plus a reduction of the ionization in the inner parts of the Galaxy.

The crucial aspect is that the observations refer to giant HII regions. This implies a total flux of Lyman continuum radiation well in excess of 10^{50} photons s^{-1} and thus requires the presence of a group of O type stars, possibly an OB cluster, in order to account for the ionization of the gas. Then, if the luminosity function as well as the properties of early type stars remain unchanged all throughout the Galaxy, the spectrum of the ultraviolet radiation field must be essentially the same for all regions. For the solar neighborhood one can estimate that more than 20% of the Lyman continuum radiated by the stars of an OB cluster is emitted at energies higher than 24.6 eV, i.e. the ionization potential of He. Therefore, the ionization degree of He should be the same as for H, which contradicts the observations. Since, on the other hand, it is clear that substantial amounts of dust grains are present within HII regions, Mezger et al. (1974) originally proposed selective absorption by internal dust as a possible explanation of the reduced ionization of He. However, in this hypothesis an absorption cross section of dust which increases with frequency faster than linearly is implied (Panagia and Smith, 1978). This requirement, although not contradicting any physical principle nor any astrophysical fact, is difficult to satisfy because no candidate material for interstellar grains is known whose absorptivity may behave in such a way.

A simple alternate solution is indicated by the direct observa-

P.A. Shaver (ed.), Radio Recombination Lines, 99–105.

tional result that the ionization of He decreases systematically when going toward the center of the Galaxy. This fact suggests that the correct explanation may be provided by a systematic variation of the spectrum of the ionizing radiation with the galactic radius. Clearly, if the effective temperature of OB stars is taken to decrease for small values of the galactic radius then the fraction of radiation emitted in the He-ionizing continuum ($\lambda<504$Å) will decrease and He will become less and less ionized when going toward the galactic center.

Such a systematic behaviour can be a direct consequence of the large scale gradient of heavy element abundance which is found for both the diffuse medium (Churchwell, this workshop; see also Panagia, 1979) and the stellar component of the Galaxy (see e.g. Janes, 1979; Panagia and Tosi, 1979). In fact, a systematic variation of the ultra-violet radiation temperature of O-type stars can be produced by the concurrency of three main effects, which all are a consequence of the metal abundance gradient, namely:

1) Blanketing by many atomic lines which are most numerous in the wavelength range 200-500 Å and can considerably depress the He-ionizing continuum.

2) Increase of the stellar radius and consequent decrease of the effective temperature with increasing metal abundance due to increased opacity in the sub-atmospheric layers of the stars. The effects of this process, which are already considerable when evaluated from model calculations based on Cox-Stewart opacity tables (Stothers, 1972), would become so extreme for models computed with Carson's opacities (Stothers, 1976) that even hydrogen would hardly be ionized around massive stars with metal abundance higher than twice the solar value. Since this appears a bit too extreme, only the set of calculations based on Cox-Stewart opacities is worth being considered.

3) Variation of the upper cut-off of the mass function of O-type stars. In fact, the mass of the most massive star which can form is determined by the heavy element abundance which controls both the cooling processes in the interstellar medium and the opacity of a collapsing cloud (Kahn, 1974). Since the higher the metal abundance the lower the mass of the earliest type star, the main sequence will be truncated at lower effective temperatures when metals become more abundant and the He-ionizing continuum will decrease accordingly.

The dependence of these effects upon the metal abundance can usefully be expressed in terms of the fraction ξ of Lyman continuum photons which are emitted at energies higher than the ionization potential of He, i.e.

$$\xi = \frac{\int_{\nu(He)}^{\infty} N(\nu)d\nu}{\int_{\nu(H)}^{\infty} N(\nu)d\nu}$$

where $N(\nu)$ is the photon flux of the stellar radiation field. $\nu(H)$ and

Table 1

Variation of the fraction of He-ionizing radiation with metal abundance

	Physical Process	$\frac{\Delta \log\xi}{\Delta \log Z}$
1.	Blanketing by UV Lines	-0.2
2.	Increase of Stellar Radius	-0.5
3.	Upper Cut-Off of Mass Function	-0.9
	Total Effect	-1.6

ν(He) denote the frequency of the ionization edge of H and He, respectively. A preliminary estimate of the above listed effects is summarized in Table 1. We see that the most important process is the reduction of the upper limit of the mass of O-type stars although none of the two others is negligible. The total effect amounts to a logarithmic gradient of $\Delta\log\xi/\Delta\log Z = -1.6$. This is to say that, for example, the fraction of He-ionizing photons will be reduced by a factor of $\sim$6 when the metal abundance increases by a factor of 3. This result can equivalently be expressed in terms of the effective temperature for which the approximate gradient is

$$\frac{\Delta \log T_{eff}}{\Delta \log Z} \simeq -0.2$$

As mentioned before, the metal abundance gradient in the Galaxy has been determined with different methods and for different classes of objects. In the solar neighborhood all the results agree rather well with each other (cf. Panagia and Tosi, 1979) and indicate a value for the local gradient of about

$$\left(\frac{\Delta \log Z}{\Delta r}\right)_{solar\ vicinity} \simeq -(0.08 \div 0.10)\ \mathrm{Kpc}^{-1}$$

On a larger scale, however, the gradient appears to be shallower (cf. Janes, 1979; Mezger et al., 1979) and values around -0.05 Kpc^{-1} may be more appropriate for describing the average metal abundance variation across the Galaxy, i.e. extending from the galactic center to several kiloparsecs past the solar circle (cf. Panagia, 1979).

Within this framework, calculations have been performed of model HII regions with the following assumptions:

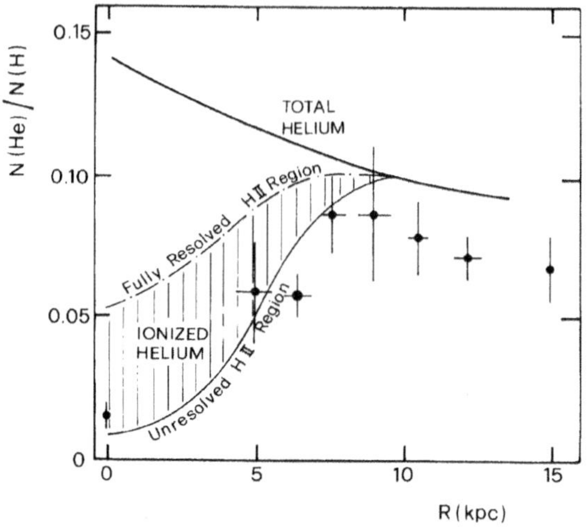

Fig. 1. The behaviour of He abundance relative to H is presented as a function of the galactic radius. The various curves represent the adopted total helium abundance (heavy solid line), the He^+/H^+ ratio for unresolved HII regions (light solid line) and the He^+/H^+ ratio for fully resolved HII regions (dotted-dashed line), respectively. The dots represent averages of the observed values for several groups of giant HII regions. The bars give the corresponding dispersions around the mean values.

1) The metal abundance is assumed to vary as

$$Z = Z_{G.C.} \exp(-r/10 \text{ Kpc})$$

which corresponds to $\Delta\log Z/\Delta r = -0.043 \text{ Kpc}^{-1}$.

2) The helium abundance by number relative to hydrogen varies as

$$y = 0.08 + 0.054 \exp(-r/10 \text{ Kpc})$$

which corresponds to $\Delta Y/\Delta Z \simeq 2$ and implies $y(r_\odot) = 0.10$ and $y(G.C.)=0.134$ at the solar circle and in the galactic center, respectively.

3) The average effective temperature which characterize the ultraviolet radiation field in a giant HII region varies as

$$\overline{T}_{eff} = 3.1 \times 10^4 \exp(r/50 \text{ Kpc}) \text{ K}$$

Thus, at the solar circle $\overline{T}_{eff}(r_\odot) = 3.8 \times 10^4$ K and $\xi(r_\odot) = 0.14$.

4) The opacity of internal dust in the Lyman continuum is assumed to be independent of frequency (i.e. no selective absorption) and to increase with the metal abundance. The computed variation of the ratio He^+/H^+ as a function of the galactic radius is presented in Fig. 1. Two extreme configurations are considered, namely: a) the HII region is completely unresolved, i.e. the whole source is enclosed within the observing beam; b) the HII region is fully resolved and thus the observed He^+/H^+ ratio refers in fact to a line of sight along the main diameter of the source. For comparison, the observational results of Churchwell et al. (1978) are also shown. By inspecting the figure it is apparent that, although the calculations were meant to be merely illustrative, the theoretical curves can satisfactorily represent the observations. A better fit to the experimental data may be obtained by slightly modifying the assumptions but it is clear that the adopted parameters already provide such a close representation of observations that no big change should be needed for any of them. In particular, the large scale metal abundance gradient cannot be much greater than the adopted value because, otherwise, a too low ionization of He in the galactic center would be predicted, in contradiction with the observational evidence.

It is of interest to note that the same model which can explain the behaviour of the He ionization can account as well for the observed correlation (Churchwell et al., 1974; Emerson and Jennings, 1978) between the ratio He^+/H^+ and the infrared excess, defined as

$$IRE = \frac{L(IR)}{N_L\ \varepsilon(Ly\text{-}\alpha)}$$

where L (IR) is the infrared luminosity of the HII region and associated matter, N_L is the Lyman continuum photon flux absorbed by the gas of the HII region and $\varepsilon(Ly\text{-}\alpha)$ is the energy of a Lyman-alpha photon. Values of the IRE have been computed with the same assumptions as listed above and using the theory developed by Panagia and Smith (1978) for two extreme cases, namely: i) there is no dust to absorb radiation outside the HII region; ii) the HII region is surrounded by a dust envelope which is infinitely opaque to UV and optical radiation but transparent to IR radiation. The resulting curves for unresolved HII regions are presented in Fig. 2. The observational data reported by Emerson and Jennings (1978) are also displayed. It is evident that the two curves envelope very nicely the experimental points. Also, it must be noted that the fit is definitely better than anyone which is possible to obtain with the hypothesis of selective absorption (see e.g. Panagia and Smith, 1978). This result confirms that a systematic variation of the effective temperature of O-type stars with the galactic radius can explain very well all observed phenomena connected with the ionization of helium.

Also, it is important to realize that this explanation follows quite naturally from well established facts and easily predictable

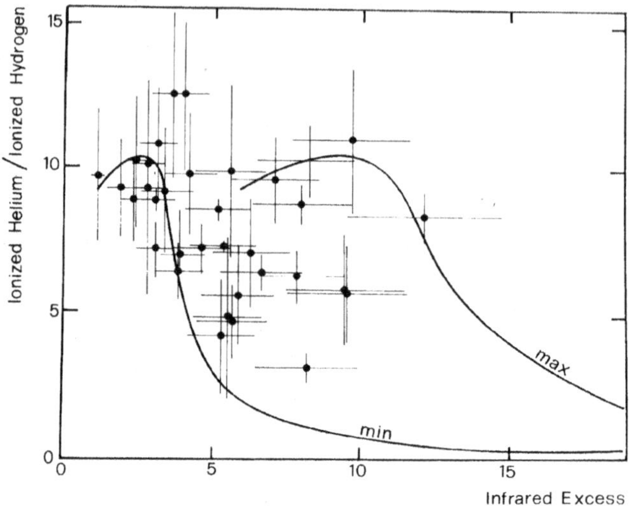

Fig. 2. The ratio He^+/H^+ (in percentage) is shown as a function of the infrared excess (IRE). Each dot corresponds to a giant HII region; the error bars are 1σ. The theoretical curves refer to the extreme cases in which either i)no dust absorption occurs outside the HII region (min) or ii) the HII region is surrounded by a dust envelope opaque to UV and optical radiation (max).

phenomena. Therefore, the success in accounting for all observational aspects indicates that no additional or competing process is needed for explaining the behaviour of the He ionization in the Galaxy. It is possible that other factors, such as high rates of mass-loss, local abundance fluctuations and some selective absorption, either "direct" or "inverse" (i.e. with average cross section which is either increasing or decreasing with frequency) contribute to alter the He ionization, thus making the situation more complicated than envisaged here. However, we believe that these additional processes may, at most, modify marginally the overall behaviour and/or add some spread superimposed on the general trend but cannot deeply affect the main characteristics of the phenomenon.

ACKNOWLEDGEMENTS. I wish to thank Ms. Rhona Kamer for typing the manuscript and Mr. L. Baldeschi for drawing the figures.

REFERENCES

Churchwell, E., Mezger, P.G. and Huchtmeier, W., 1974, Astron. Astro-

physics. 32, 283.
Churchwell, E., Smith, L.F., Mathis, J.S., Mezger, P.G. and Huchtmeier, W., 1978, Astron. Astrophys. 70, 719.
Emerson, J.P. and Jennings, R.E., 1978, Astron. Astrophys. 69, 129.
Janes, K.A., 1979, Astrophys. J. Suppl. Ser. 39, 135.
Kahn, F.D., 1974, Astron. Astrophys. 37, 149.
Mezger, P.G., Smith, L.F. and Churchwell, E., 1974, Astron. Astrophysics. 32, 269.
Mezger, P.G., Pankonin, V., Schmid-Burgk, J., Thum, C. and Wink, J., 1979, Astron. Astrophys., in press.
Panagia, N., 1979, Mem. Soc. Astron. Ital. 50, 79.
Panagia, N. and Smith, L.F., 1978, Astron. Astrophys. 62, 277.
Panagia, N. and Tosi, M., 1979, Astron. Astrophys., in press.
Stothers, R., 1972, Astrophys. J. 175, 431.
Stothers, R., 1976, Astrophys. J. 209, 800.
Torres-Peimbert, S. and Peimbert, M., 1977, Rev. Mex. Astron. Astrofis. 2, 81.

PRESSURE BROADENING EFFECTS ON THE EMISSION OF He RECOMBINATION LINES

Diego A. Cesarsky, DERAD, Observatoire de Meudon, France

(paper presented by E. Falgarone)

The first rigorous treatement of pressure broadening of hydrogen-like emitters of radio recombination lines is due to Griem (1967). The calculations showed that most of the radio observations that were available at that time (mostly 109 and lower frequency transitions) must have been affected by pressure broadening. This conclusion was ignored or contested by most of the people involved with the observations, on the ground that there was no evidence showing any observed amount of broadening. Indeed the widths of recombination lines are remarkably constant over several frequency octaves. Brocklehurst and Seaton (1972) were the first to show that theory and observations were not necessarily incompatible. They considered a model source of variable density. At a given frequency there are two regions of interest in the source: the dense region that emits broadened lines, and the less dense regions, that emit unbroadened, gaussian looking lines. The combined spectrum would be an almost gaussian line, sitting on the extended wings of the broadened line. The usual expedient of fitting baselines to any odd looking spectra removed the extended wings and left just the gaussian looking profile. Obviously, this profile carries a small fraction of the total line radiation emited by the source. The relevance of this fact in the understanding of LINE/CONT or ALFA/BETA ratios will be certainly discussed in this workshop. I want to call your attention to the consequences of pressure broadening on the interpretation of He/H ratios.

Consider a spectrum containing two emission lines. Assume that both have comparable peak intensities and are equally affected by pressure broadening. The baseline removal procedure eliminates the extended wings associated to each line. What remains of the original spectrum is two lines, each with a fraction of the total energy carried by the original emission but the ratio of line energies has not changed. On other words, the ratios LINE/CONT are wrong and the ratio LINE/LINE is correct. It is important to realize that all ratios will be wrong if one of the lines was more broadened than the other and this will be the case for He and H lines, at least if the frequency is not very high. At a given density, the damping parameter for He is 20 - 30% higher than that for H because of the smaller doppler width of the He

P. A. Shaver (ed.), Radio Recombination Lines, 107–108.

lines. Consider now a low excitation H II region or a region where dust effectively removes a fair number of the He ionizing radiation. He II would be concentrated to the central parts of the region. Assume further that the density increases towards the center. It follows that He will be more affected by pressure broadening than H will. It is clear that the ratio He/H will be observed to vary as both a function of the resolving power of the telescope (the "geometrical effect",Cesarsky,1971; Mezger et al., 1974) and as a function of observing frequency because of broadening. This is precisely the behaviour observed in Sgr B2: superimposed to the smooth variation due to the geometrical effect one observes large fluctuations that we interpret as being due to pressure broadening effects. To test this hypothesis we have constructed a variable density model of Sgr B2 (Pitault and Cesarsky, 1979). The model predicts, with a reasonable accuracy, the observed continuum fluxes, hydrogen line intensities and He/H line ratios. An interesting feature of the model is that a very small amount of neutral helium is required, thus easing the requirements on the mechanism responsible for keeping helium partially neutral. A full description of the model and the numerical results can be founf in the reference cited above.

REFERENCES

Brocklehurst,M. and Seaton, M. 1972, M.N.R.A.S., 157, 179
Cesarsky, Diego A. 1971, unpublished Ph.D. thesis, Harvard University
Griem, H.R. 1967, Ap.J. 148, 547
Mezger,P.G., Smith,L.F., Churchwell,E. 1974, Astron. Astrophys, 32, 269
Pitault,A. and Cesarsky, Diego A. 1979, Astron. Astrophys. (to be published in late 1979 or early 1980)

PART III

RADIO RECOMBINATION LINES FROM CII REGIONS

THE PARTIALLY IONIZED MEDIUM ADJACENT TO HII REGIONS

V. Pankonin
Max-Planck-Institut für Radioastronomie
Auf dem Hügel 69
5300 Bonn 1
Federal Republic of Germany

INTRODUCTION

Recombination line emission from a partially ionized medium has been detected toward approximately fifteen galactic HII regions. The carbon line is the strongest and is the best studied, but lines from elements heavier than carbon and from hydrogen have also been observed. Therefore, I want to include in my review the characteristics of the more general partially ionized medium (PIM) which is adjacent to HII regions, and I will concentrate on developments after about 1974 when Dupree (1974) published a compilation of carbon line data and presented an interpretation. The components of the PIM are not only the C^+ region, but also what I will term Z^+ regions where elements heavier than carbon are ionized, and what I will call an H^o region where hydrogen is partially ionized (i.e. relative ionization of 10^{-4} - 10^{-3}). Many properties are shared by the three types of region, and there is probably some spatial coincidence.

It is fairly well established that the partially ionized medium forms the interface between the fully ionized gas (i.e. the HII region) and a neutral (molecular) cloud. The carbon and heavier elements are ionized by the star which excites the HII region. Stellar radiation with a wavelength longer than 912 Å escapes from the nebula and ionizes all elements with ionization potentials less than 13.6 eV in the contiguous, dense neutral material. Observations of radio recombination lines provide the most straight forward probe of the properties of this partially ionized medium.

It is now clear that HII regions are intimately related to molecular clouds, and the evolution can be studied only by examing the complex as a whole. Although the partially ionized medium may not occupy a very large fraction of the volume of the HII region-molecular cloud complex, information from this interface improves our understanding of the coupling of the HII region to the molecular cloud. Thus we can assess the influence of the HII region on the energetics and evolution of the cloud.

P.A. Shaver (ed.), Radio Recombination Lines, 111–125.

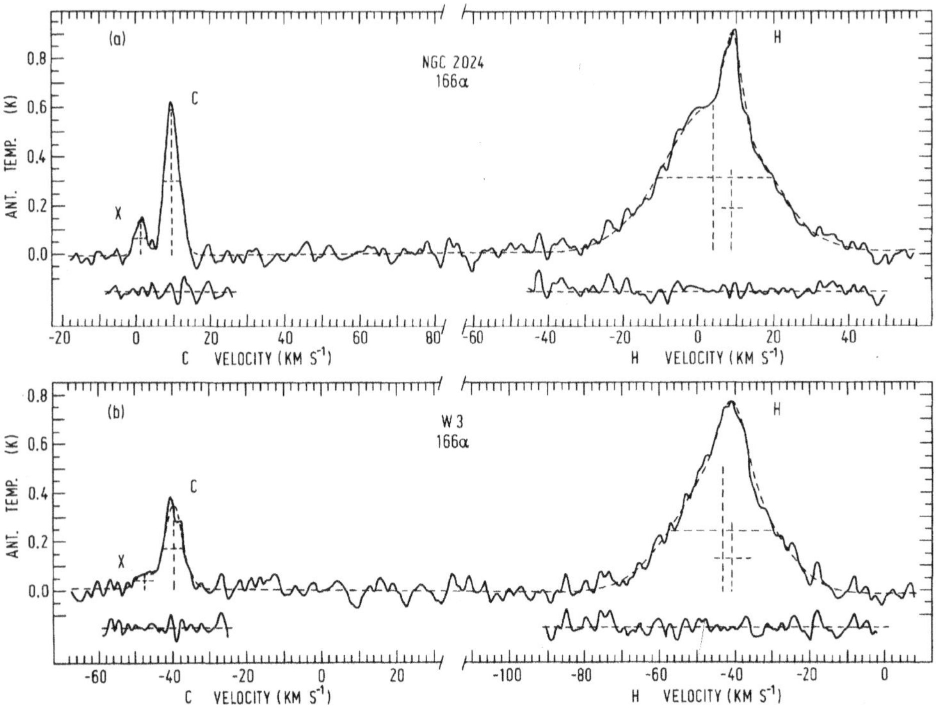

Fig. 1: The 166α spectra of a) NGC2024 and b) W3. The Gaussian fits are indicated by the dashed lines. The residuals are shown below each profile (from Pankonin et al. 1977b).

Figure 1 illustrates the observational data available to recombination line observers. This shows the 166α spectrum of NGC2024 at the top and W3 at the bottom which were obtained with the 100-m telescope of the MPIfR. The spectrum of NGC2024 is the best example of emission from all of the partially ionized regions which I will discuss. The carbon line is prominent in this spectrum (there is no helium emission from the HII region observed at this frequency). The heavier element line, which in this case is labelled X signifying unknown, is at a lower velocity on the carbon velocity scale. The hydrogen line from the H^o region is the narrow feature sitting atop the broader emission from the HII region. This in fact brings up the important point of the observational characteristics of lines from the partially ionized medium which distinguish them from emission from the HII region:

i) The lines from the PIM have a different frequency dependence than lines from the HII region. In particular, the lines from the PIM become stronger relative to the HII lines at lower radio frequencies.

ii) The velocity of the PIM often differs from the velocity of the HII region by a few km s^{-1}. The velocity of the PIM is in much closer agreement with the velocity of the adjacent molecular cloud.

iii) The widths of the lines from the PIM are usually significantly smaller than the widths of the HII region lines.

These three characteristics are particularly important for distinguishing between hydrogen lines from the H^o region and lines from the HII region, since these lines are usually blended. The carbon and heavier element lines are offset from hydrogen in frequency, although carbon emission is often blended with helium lines from the HII region.

I will now discuss each component of the PIM and describe what we know of the characteristics of each, primarily by giving examples.

C^+ Regions. The early history of carbon recombination line research was reviewed by Dupree and Goldberg (1970) in Annual Reviews of Astronomy and Astrophysics. I will not repeat that here other than to state a few established facts about the nature of CII regions. The line, which was first detected by Palmer et al. (1967), is firmly identified as due to electronic recombination onto singly ionized carbon atoms. It is also clear that the line emission originates in a cold gas (i.e. $T_e \lesssim 200$ K) rather than in the ionized gas in the HII region. It is now well accepted that the CII region forms a shell around the HII region, and the exciting star of the HII region is ionizing the carbon. However, this shell may be incomplete, and it is surely not uniform and homogeneous. The observations have been concerned primarily with locating the C^+ relative to the HII region, and in determining the physical conditions within the CII.

Carbon recombination lines have now been detected throughout the frequency range from 328 MHz (C271α) to 23.4 GHz (C65α). It has been especially interesting to follow the frequency dependence of the emission to the low end of the range because the carbon line is about the same strength as the hydrogen line emitted from the HII region at low frequencies (Pankonin et al. 1974; Pedlar et al. 1978). Despite the large frequency range available, those key observations which have revealed the physical and structural properties of the CII regions have been limited to ∿1.4 to 15 GHz. This is due to a combination of the sensitivity and angular resolution of the observing instruments. I believe it is accurate to say that, since it became operational in ∿1972, the 100-m telescope of the MPIfR has been the primary instrument for this research.

Two recent surveys detected several new CII sources (Pankonin et al. 1977a; Silverglate and Terzian 1979), and today CII regions have been observed toward approximately 15 HII regions. The surveys are valuable for improving the statistical correlations of the properties of CII regions and HII region-molecular cloud complexes, but one conclusion I draw from the surveys is that there is quite a large variation in the relative observable properties of carbon lines, hydrogen lines from the HII regions, and molecular lines, and it is not even possible to predict which HII regions are most likely to have observable CII regions. For example, the lack of a CII region associated with NGC7538 is still puzzling, since

NGC7538 seems to have the right prerequisites-- an HII region adjacent to a molecular cloud.

It is particularly difficult to deduce accurate physical parameters of the C^+ from the line observations because the line emission depends on a number of factors, several of which are external to the CII region itself. Not only is the geometry and distribution of the CII important, but also its location relative to the HII region and the observer. Dust and element abundances in the CII region and the spectral type of the exciting star play important roles in the ionization structure and thermal equilibrium (cf. Pankonin and Walmsley 1976).

The data are not, and I believe never will be, complete enough to support a detailed model of a CII region. The observations have shown CII regions to be very complex structures and a simple model with a single set of physical parameters and a simple geometry will not suffice. In spite of this pessimistic introduction, as long as one realizes the limitations, very useful information can be extracted from the recombination lines:

i) We can determine the spatial distribution of the CII relative to the HII region and the molecular cloud.

ii) We can examine the dynamic properties of the CII.

iii) We can determine whether the dominant process for line formation is stimulated or spontaneous emission.

iv) We can establish ranges for the electron temperature and electron density in the CII region.

The first two points require reasonable angular resolution for a map of the source (the lines are stronger at lower frequencies, e.g. $\sim$166α at 1.4 GHz, but the telescope beams are also larger). The second two points above require observations over a range of frequencies or quantum levels. Very little can be learned with observations of only one line.

The most detailed map of a CII region is that of Orion A made by Jaffe and Pankonin (1978) at 5 GHz. The quality of these data is aided considerably by the nearness of Orion which gives the source an adequate angular size. These data give a fairly detailed picture of the dynamic properties of the CII. They are best described with the aid of Figure 2, which is a schematic of the Orion HII region-molecular cloud complex due to Balick et al. (1974). The CII map data show three velocity componets in the ionized carbon which are each identified with a specific location relative to the Nebula. The highest CII velocity is concentrated in the direction of the core of the Nebula where there is little obscuration. We interpret this velocity component as originating in the CII shell at the back of the ionized gas. The C^+ is moving into the molecular cloud with a velocity of $\sim$3 km s^{-1} (in the rest frame of the molecular cloud). The middle velocity component in the CII is inter-

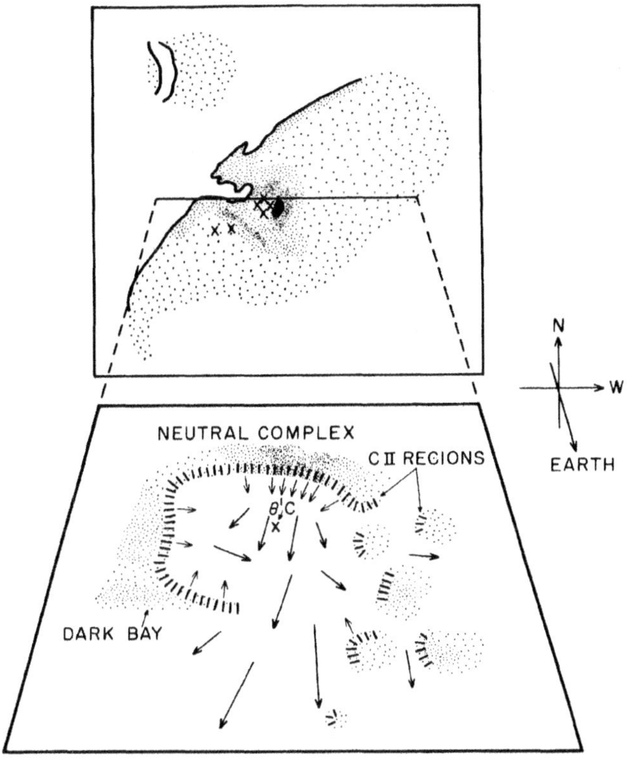

Fig. 2: A schematic diagram of the Orion A HII region-molecular cloud complex. The lower portion of the figure represents a cross section through the nebular center. The arrows designate the flow structure of the ionized gas of the HII region. (From Balick et al. 1974).

preted as originating at the edges of the Nebula where the ionization fronts are seen edge on. Any motion of the C^+ relative to the molecular gas is then transverse to the line of sight. The velocity of this C^+ component agrees very well with the velocity of the molecular gas. The lower velocity CII component is concentrated toward the dark bay which appears to be a foreground cloud. The C^+ velocity agrees with the velocity of molecules which are known to be in front of the HII region, e.g. 18 cm OH absorption lines. This velocity component of the CII emission dominates low frequency observations because the line is dominated by stimulated emission as would be expected from CII which is located between the HII region and the observer. It has been known for some time that the velocity of the carbon line in Orion depended on the frequency of the observation, and the 5 GHz map has now clarified the dynamic picture. The carbon lines observed toward NGC2024 and W3 also show similar systematic velocity structure (Pankonin et al. 1977b). The interpretation is assumed to be similar to that adopted for Orion A, although it is not as easy to correlate radio with optical properties in NGC2024 and W3. The W3 source may contain two discrete CII clouds which differ in velocity by ~ 3 km s^{-1} (Jaffe and Wilson 1979).

Early interpretation of carbon line observations were primarily concerned with determining whether the line emission was stimulated or spontaneous. The answer was required to help locate the source relative to the HII region and to derive the physical conditions in the CII. Information on the line formation process and the physical conditions in the CII comes from observing lines over a range of quantum levels. One can use nα lines observed over a range of frequencies (cf. Dupree 1974; Hoang-Binh and Walmsley 1974). This has the disadvantage of requiring an assumed geometry (a source model) for the CII region which must be convolved with a variety of telescope beams. A more straightforward method of distinguishing between stimulated and spontaneous emission and estimating T_e and n_e is to observe nα and nβ transitions which occur at nearly the same frequency. This has the advantage of covering a large range of quantum levels with a constant telescope beam size. In the ratio of the nα/nβ line strengths, the geometry becomes a secondary correction and at some frequencies, the correction is negligible (cf. the discussion of the Rho Oph model in Pankonin and Walmsley 1978). The disadvantage in this method is that nβ lines are very weak. We have pursued the nα/nβ method using the 100-m telescope to examine the CII regions of NGC2024 and W3 (Pankonin et al. 1977b). The results are shown in Figure 3. Figure 3a shows the theoretical ratio of T_L (157α)/T_L (197β) at 1.6 GHz

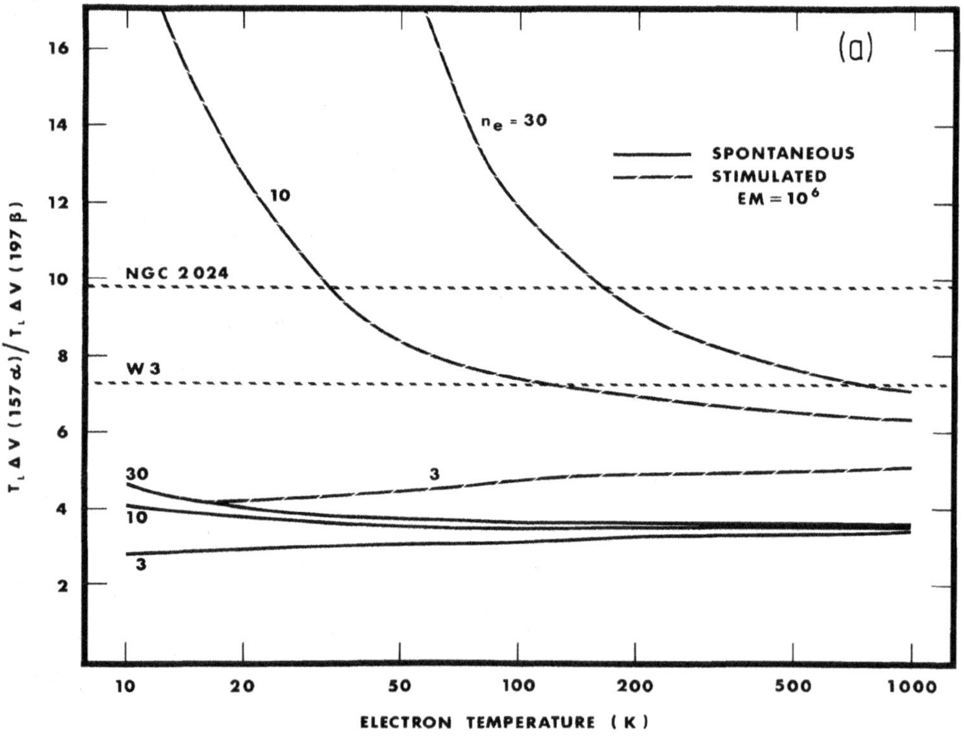

Fig. 3: a) The calculated ratio of the peak line temperatures of 157α/197β as a function of temperature for several electron densities. Solid lines show the ratio for spontaneous emission only; dashed lines show the ratio for stimulated emission plus spontaneous emission for a background HII region with emission measure 10^6 cm^{-6} pc. Also shown (dotted lines) are the measured ratio for C157α/C197β from NGC2024 and W3.

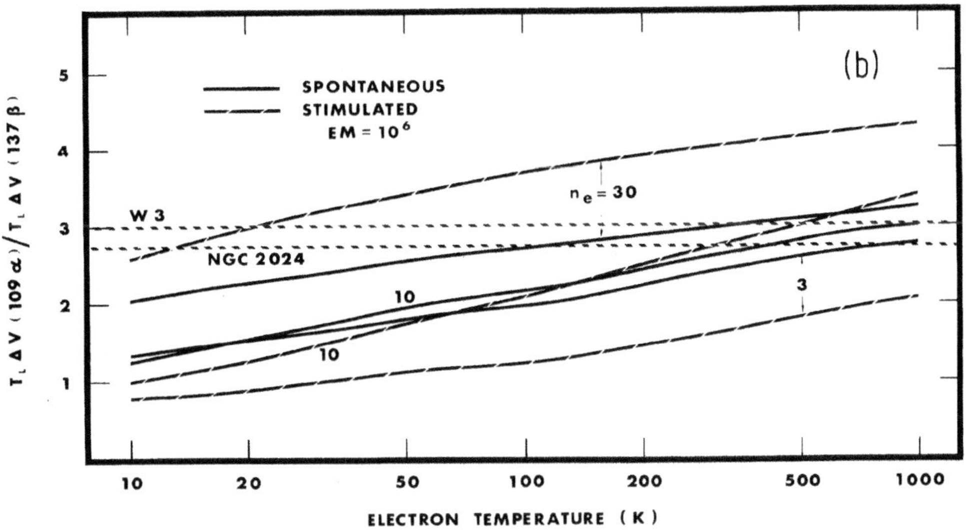

Fig. 3: b) The same as a) except the calculations and measurements apply to the 109α/137β ratio. (From Pankonin et al. 1977b).

versus T_e for three electron densities for simple sources which are either entirely in front of the HII region, i.e. stimulated emission, or are to the side of or behind the HII region, i.e. spontaneous emission. The dashed straight lines are the observed values. A characteristic of stimulated emission models is that the ratio of the nα/nβ line intensities can reach large values at low frequencies. From these data we conclude that stimulated emission is the dominant process of line formation at 1.6 GHz; hence this emission originates primarily in a foreground CII region. The 5 GHz data which are shown in Fig. 3b (109α/137β), are not very decisive in differentiating spontaneous from stimulated emission, but the intersection of the results from the nα/nβ ratios at 1.6 and 5 GHz for NGC2024 and W3 yields $T_e \sim 100$ K and $n_e \sim 20$ cm^{-3}. Assuming a cosmic abundance for carbon and that all carbon is atomic and singly ionized, this implies a total hydrogen density of $\sim 10^5$ cm^{-3}. We also conclude that there is a combination of stimulated and spontaneous emission and the relative contribution of each to the observed line profile is frequency dependent. Similar conclusions apply to Orion A, although this particular type of data is not so good in Orion A. I emphasize that this procedure gives ranges of physical parameters; it is impossible to derive a unique set of parameters.

From the theoretical side, the quantum levels of the carbon atoms are not populated according to LTE even for spontaneous emission. The level populations depend on the intensity of radiation incident on the CII region, hence on the relative geometries of the CII and HII regions. The derivation of the electron temperature and density in the C^+ region depends on the LTE departure coefficients, b_n. The calculations of the b_n have been refined over the years; improvements have been achieved primarily by using more recent collision cross sections, although these cross sections remain uncertain for low temperature regions, i.e. $T_e \lesssim 50$ K The latest b_n calculations to be published are those of Salem and Brocklehurst (1979), and they are probably the most accurate. The most significant change in the CII region parameters which results from using the newer departure coefficients is an increase in the electron density from ~ 5 to 50 cm^{-3}. The density was a factor of 5 to 10 smaller with the earliest b_n.

Z^+ Regions. The frequencies of recombination lines of all singly ionized atomic elements heavier than carbon are usually included in the carbon line spectra, and recombination line emission from heavy elements was first detected toward NGC2024 and W3 by Chaisson et al. (1972). Magnesium, silicon, sulfur, and iron are all relatively abundant elements with first ionization potentials less than 13.6 eV which should be singly ionized in the CII region. In fact, since their ionization potentials are less than that of carbon, the ionized heavy element shell may be larger than the C^+ shell. The emission measure in the line and structural properties of the Z^+ region determine whether the emission is strong enough to be observed.

Heavy element recombination lines have definitely been observed toward only three HII regions -- NGC2024, W3, and Orion A. Since the heavy element lines are generally $\sim 1/3$ the strength of the carbon lines, the small number of sources is probably due primarily to observational sensitivity, although there are exceptions to this, i.e. W48, DR21 (Pankonin et al. 1977a).

The line emission attributed to heavy elements is generally too wide (2 to 5 km s^{-1}) to identify it with any one species of emitter. This is not an observational problem, but rather the fact that the rest frequencies of the lines from these elements are too close together. However, this emission has recently been observed toward the dust clouds Rho Oph and L1630 (Chaisson 1975; Pankonin and Walmsley 1978). These lines are narrow ($\lesssim 1$ km s^{-1}), and they have been attributed to sulfur; Mg, Si, and Fe appear to be depleted relative to S. With the aid of the conclusions drawn from the dust cloud data, Pankonin et al. (1977b) identied the heavy element line in NGC2024 as due primarily to sulfur also. For W3 and Orion A, it is still not possible to say if only one element is contributing to the observed emission. I therefore use the collective term "Z-line" to designate emission from the ionized heavy element (Z^+) region.

There is both experimental evidence and theoretical expectation that the C^+ and Z^+ regions overlap to a large extent but that they are not entirely coextensive. The Z^+ region has never been fully mapped in any of the HII regions; the relative characteristics of the C^+ and Z^+ regions are examined only by direct comparison of the line parameters over a large frequency range. The W3 and NGC2024 sources show that the

C/Z line intensity ratio is not constant with frequency. The Z-lines become relatively stronger at lower frequencies (Pankonin et al. 1977b; Jaffe and Wilson 1979). This is probably because the Z^+ region is larger than the C^+ region and the physical conditions differ somewhat in the noncoincident portions. At any rate the different frequency dependence implies that the intensity ratio of the lines is not a reliable estimate of elemental depletion. It is also true that different sources have different C/Z line ratios at the same frequency (cf the data for W48 and DR21 in Pankonin et al. 1977a compared with the data for NGC2024 in Pankonin et al. 1977b). However, this may also not reflect depletion factors since the ratio is sensitive to the physical conditions and the geometry of the sources.

H^o Regions. Hydrogen recombination line emission from a cold, partially ionized medium was first discovered by Ball et al. (1970). The "narrow" hydrogen line as it is often called is blended with hydrogen line emission from the HII region. This blending is the basic cause of the difficulty in obtaining high quality observational data on which to base an interpretation. A gaussian decomposition is required to determine the parameters of the lines from each region. Jaffe and I (unpublished) have made tests of the nonlinear gaussian fitting routine on known line profiles with variable signal-to-noise ratios. The program, which is similar to that that used by most observers, is very proficient at deconvolving the blended gaussians. The derived line parameters are reliable to plus or minus twice the calculated error in the fitted parameters. Therefore, any fault lies not with the fitting program but with the far more basic assumption that the observed profile is composed of gaussian functions. Data available on DR21 and W33 show how erroneous conclusions can be reached by blindly fitting multiple gaussians to a line profile of a single transition (Pankonin et al. 1977a; Bieging et al. 1978). There is often velocity structure in the gas in the HII region, and the HII lines may be pressure broadened. This may lead to asymmetric line profiles or Voigt profiles which cannot be fitted with a single gaussian. It is my opinion that the burden of proving that an H^o region has been observed lies with the observer, and it must be proven for each source observed. One cannot automatically fit multiple gaussians to profiles and claim that emission from an H^o region has been observed if the decomposition yields a "narrow" line.

The presence of an H^o region can best be proven by observations which cover a large range of quantum levels. Thus, one either observes $n\alpha$ lines at several frequencies or observes an $n\alpha$ and an $n\beta$ line at nearly the same frequency. Pankonin et al. (1977b) used the $n\alpha$ - $n\beta$ method at 1.6 GHz to prove that the H^o region exists in NGC2024. That result is shown in Figure 4. The H157α line is at the top; the H197β line is at the bottom. The shape of these two lines is very different. The narrow feature in the H157α line is not visible at all in H197β, i.e. the narrow feature in the H157α profile has become relatively much weaker at 197β. This proves that the observed line profile is composed of emission from two regions which have significantly different physical conditions. The quantum level dependence of the relative line strengths is in

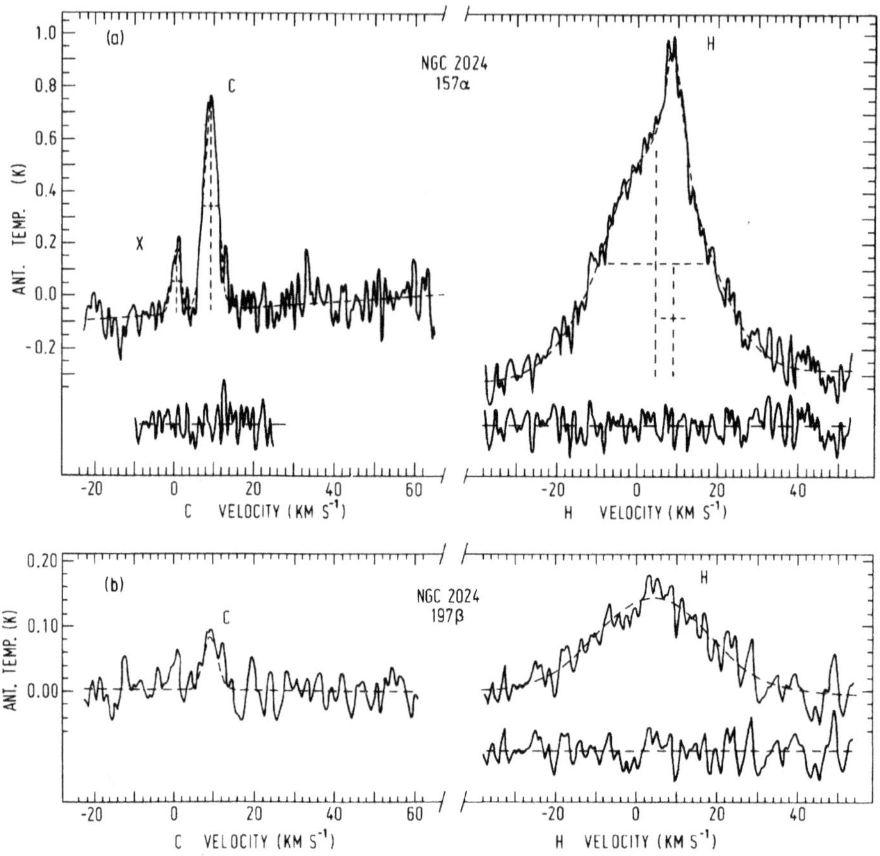

Fig. 4: The 157α spectrum a) and the 197β spectrum b) of NGC2024. Gaussian fits are indicated by the dashed lines, and the residuals are shown below each profile. (From Pankonin et al. 1977b).

the direction predicted if the narrow line is from a cold, partially ionized medium with physical conditions similar to those in the CII region. This same nα/nβ technique was also applied to W3 (Pankonin and Walmsley 1979). The data lead to the same conclusion as for NGC2024, although it is not as firmly based because of a poorer signal-to-noise ratio.

Both the frequency range and nα/nβ techniques were used on Mon R2 (Pankonin and Walmsley 1979), and in this case the frequency range provided a more convincing argument. The data are shown in Figure 5. The upper portion of the figure shows spectra at ∿4.6 GHz -- C114α, H112α, and H143β. The signal-to-noise ratio of the H143β line is not really good enough to accurately define the profile shape. However, the spec-

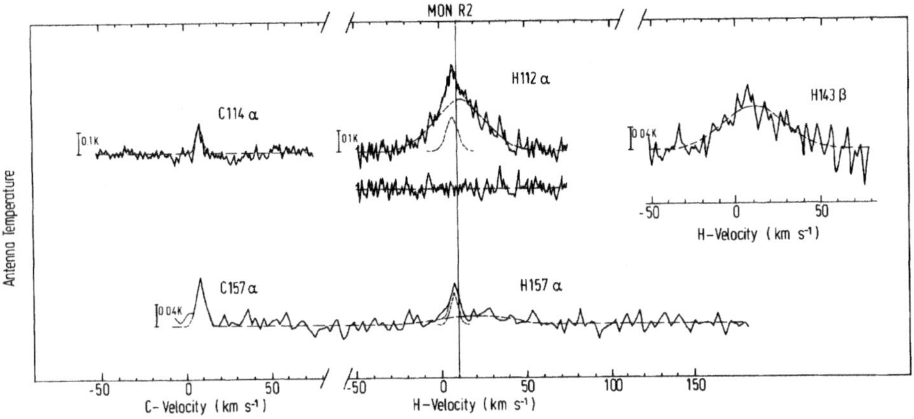

Fig. 5: C114α, H112α and H143β spectra (upper segment) and the 157α spectrum (lower segment) of Mon R2. Gaussian fits are indicated by dashed lines. (From Pankonin and Walmsley 1979).

trum at 157α, shown in the lower portion of the figure, is decisive. In this case, we are fortunate that the emission from the HII region has disappeared, leaving only the narrow line, H^{o} emission. This rarely happens, and it means that the Mon R2 HII region has a fairly well defined electron density which is large (rms $n_e \sim 10^4$ cm^{-3}). The HII region is then optically thick in the radio continuum at 1.6 GHz, and any line emission from the HII region is pressure broadened. From these data, we conclude that Mon R2 is also a source of H^{o} line emission.

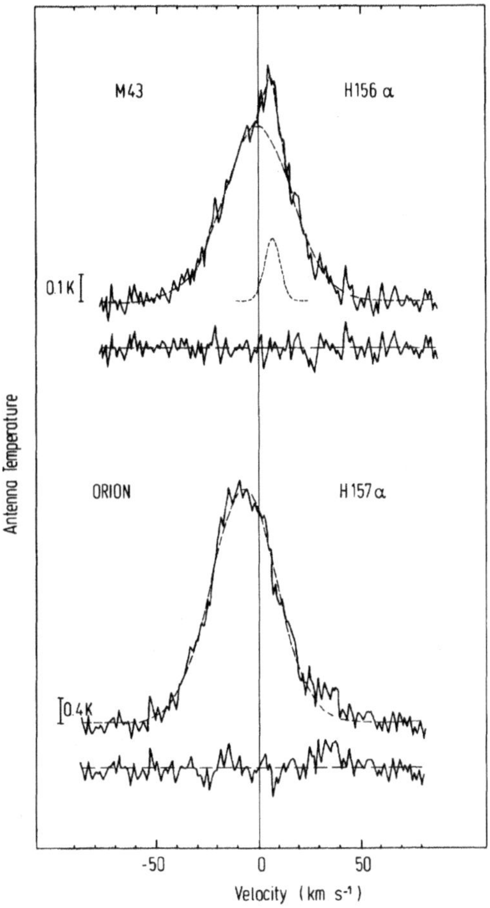

Fig. 6: H156α spectrum of M43 (upper segment) and H157α spectrum of Orion A (lower segment). Dashed lines indicate Gauss fits. (From Pankonin and Walmsley 1979).

Orion A is a more difficult source. We do not have a good nβ profile of Orion at low frequencies. We did observe H157α toward the continuum peak of Orion A and toward M43, a smaller HII region 7.5' (1 beam width for the 157α observations) northeast of Orion (Pankonin and Walmsley 1979). The spectra are shown in Figure 6. The spectrum of M43 is in the upper portion, and that of Orion A is in the lower portion. There is no evidence in the Orion A spectrum for a narrow component; a single gaussian fits the observed profile very well. The M43 profile is asymmetric and a two gaussian decomposition is shown, which is composed of one wide gaussian and one narrower one (13 km s^{-1}). The wide line is emission from the Orion Nebula. The narrower line could be interpreted as H^{o} emission. However, if that interpretation is made, then there is no evidence for the line emission from the M43 HII region. Conversely, we know the velocity and continuum flux density of the M43 HII region. The narrower line in the decomposition has this velocity, and its strength is apparently that expected from the continuum flux density of M43. I tend to believe that this narrower line is the emission from the M43 HII region, and there is no conclusive evidence for an H^{o} region in Orion A.

That leaves only two definite sources of emission from H^{o} regions -- NGC2024 and Mon R2 -- and a third very likely candidate -- W3. This is an unfortunately small sample. Still a few general remarks are appropriate. There have been observations of hydrogen recombination lines for several years and still there are only three sources of H^{o} lines. One could argue that the H^{o} emission is present in a larger number of HII regions, but the confusion with the HII region emission is too great to distinguish it. This may be partially true. Still there have been many observations with very good signal-to-noise ratios, and if the phenomenon were more common I believe several more sources should have been observed. Thus I must conclude that the limited set of conditions which leads to observable lines from H^{o} regions occurs only rarely. It is true that two of the three H^{o} sources (in fact the best two) are associated with HII regions which are ionized by early B stars. It is tempting to conclude that the spectral type of the exciting star is inherently connected to the properties of the H^{o} region.

The H^{o} lines appear to have nearly the same frequency dependence as the carbon lines, and there is a CII region associated with each of the three H^{o} sources. Therefore, the physical conditions in the H^{o} regions are probably not too different from the conditions in the C^{+} region. On the other hand, there seems to be a small but significant shift in the velocity between the H^{o} and CII (Pankonin et al. 1977b). Thus the two regions may overlap to a certain extent, but they are probably not entirely coextensive.

In addition to a lack of a really good set of data, we are also faced with an observed phenomenon which does not yet have a good theoretical explanation. In particular the source of the partial hydrogen ionization in the H^{o} region is not known. Narrow hydrogen recombination lines have not been detected toward reflection nebula and dust clouds where carbon lines are observed. Thus the source of ionization is probably not cosmic rays or ambient ultraviolet radiation external to the HII region-molecular cloud complex, but it is probably associated with the exciting star of the HII region. Zuckerman and Ball (1974) suggested that the H^o lines may originate in the outermost portion of the ionization front which bounds the HII region. Following this hypothesis, Hill (1977) showed that H^o lines of approximately the observed intensity can be formed in the outer layers of a weak D-type ionization front. Pankonin et al. (1977b) speculated that the partial ionization of the H^o region may be due to soft (0.5-1.0 keV) X-rays which escape the HII region. Krügel and Tenorio-Tagle (1978) elaborated on this hypothesis by examining the consequences of mass loss by the exciting star of the HII region. The resulting stellar wind leads to a soft X-ray flux which traverses the HII region and partially ionizes the hydrogen in the neutral cloud. The X-rays probably do not escape the neutral cloud and therefore would not be observable. Although 1 - 8 keV X-rays have been detected toward the Trapezium stars in Orion A (den Boggende et al. 1978), Orion is not at the moment a good candidate for H^o line emission. Unfortunately, we simply know too little about the observed H^o regions to distinguish between the two proposed explanations or to offer any better alternative.

OVERVIEW

Despite the obstacles -- very weak lines; the complex dependence of the line emission on many factors -- we have learned a lot about the structure of the interface between HII regions and the (parent) neutral cloud. The observational facts, i.e. the frequency dependence of the line emission, the geometrical distribution and the dynamics, are understood in the context of source models where the partially ionized medium forms a shell (perhaps incomplete) around the HII region.

For the future, infrared spectroscopy offers exciting prospects for directly measuring the principal coolants of the PIM, e.g. the 156μ line of CII and the 35μ line of SiII. Additional radio spectroscopy will be important for adding the details of the structure of the PIM. For example, line maps of nearby CII regions with an angular resolution of 1' or better would be useful. We also need $n\alpha$, $n\beta$, $n\gamma$ etc. line measurements at nearly the same frequency with the same telescope. High frequency, i.e. 9 - 15 GHz, observations of these higher order lines offer the best means of distinguishing among source models to specify T_e and n_e (cf. Pankonin and Walmsley 1978). For the Z^+ and H^o regions, we need to establish the frequency dependence of the line emission relative to the carbon lines and to sort out the geometrical distributions of the various species of atom in the PIM.

The most useful sources are those where the HII region is optically visible. This provides one additional means of correlating the various properties of the HII region - molecular cloud complex. In this context a recent development in PIM research should be pursued. In Orion, the neutral cloud more or less surrounds the HII region. It would be much easier to locate the constituents of the complex if the neutral cloud were confined to one edge of the HII region, preferably a side so that the transition zone could be mapped edge on. NGC1977, an HII region which lies ∿30' north of Orion A, appears to provide the desired configuration. The molecular cloud seems to come to an abrupt end at the southern boundary of the HII region. Kutner et al. (1979) mapped the carbon and hydrogen lines across the ionization front and found the CII was fairly well restricted to the vicinity of the ionization front. A similar experiment of our own (Pankonin and Walmsley 1979) produced concurring results. Unfortunately the lines are weak even by carbon line standards, and the mapping is not complete. We did not detect an obvious hydrogen line from an H^{o} region although we did detect the line from the HII region. We also selected three other HII regions which appeared to have similar relationships with molecular clouds (S255-S257, S252, the HII associated with the B35 cloud) and observed at the frequency of the C166α line with the Arecibo telescope. No lines were detected. This latter experience simply emphazises one of the most frustrating aspects of recombination line observations of the partially ionized medium -- we cannot predict which HII region - molecular cloud complex will have observable C^{+}, Z^{+}, or H^{o} regions. Actually performing the observations is the only way to find out.

REFERENCES

Balick, B., Gammon, R.H., Hjellming, R.M.: 1974, Pub. Astron. Soc. Pacific 86, 616
Ball, J.A., Cesarsky, D., Dupree, A.K., Goldberg, L., Lilley, A.E.: 1970, Astrophys. J. 162, L25
Bieging, J.H., Pankonin, V., Smith, L.F.: 1978, Astron. Astrophys. 64, 341
den Boggende, A.J.F., Mewe, R., Gronenschild, E.H.B., Heise, J., Grindlay, J.E.: 1978, Astron. Astrophys. 62, 1
Chaisson, E.J., Black, J.M., Dupree, A.K., Cesarsky, D.A.: 1972, Astrophys. J. 173, L131
Chaisson, E.J.: 1975, Astrophys. J. 197, L65
Dupree, A.K.: 1974, Astrophys. J. 187, 25
Dupree, A.K., Goldberg, L.: 1970, Ann. Rev. Astron. Astrophys. 8, 231
Hill, J.K.: 1977, Astrophys. J. 212, 692
Hoang-Binh, D., Walmsley, C.M.: 1974, Astron. Astrophys. 35, 49
Jaffe, D.T., Pankonin, V.: 1978, Astrophys. J. 226, 869
Jaffe, D.T., Wilson, T.L.: 1979, in prep.
Krügel, E., Tenorio-Tagle, G.: 1978, Astron. Astrophys. 70, 51
Kutner, M.L., Guelin, M., Evans, N.J. III, Tucker, K.D., Miller, S.C.: 1979, Astrophys. J. 227, 121
Palmer, P., Zuckerman, B., Penfield, H., Lilley, A.E., Mezger, P.G.: 1967, Nature 215, 40
Pankonin, V., Parrish, A., Terzian, Y.: 1974, Astron. Astrophys. 37, 411
Pankonin, V., Thomasson, P., Bahrsuhn, J.: 1977a, Astron. Astrophys. 54, 335
Pankonin, V., Walmsley, C.M.: 1976, Astron. Astrophys. 48, 341
Pankonin, V., Walmsley, C.M., Wilson, T.L., Thomasson, P.: 1977b, Astron. Astrophys. 57, 341
Pankonin, V., Walmsley, C.M.: 1978, Astron. Astrophys. 64, 333
Pankonin, V., Walmsley, C.M.: 1979, Astron. Astrophys., submitted

Pedlar, A., Davies, R.D., Hart, L., Shaver, P.A.: 1978, Mon. Not. Royal. Astron. Soc. 182, 473
Salem, M., Brocklehurst, M.: 1979, Astrophys. J. Suppl. 39, 633
Silverglate, P.R., Terzian, Y.: 1978, Astrophys. J. 224, 437
Zuckerman, B., Ball, J.: 1974, Astrophys. J. 190, 35

CARBON RADIO RECOMBINATION LINE EMISSION FROM DARK CLOUDS

Robert L. Brown
National Radio Astronomy Observatory
Green Bank, West Virginia 24944

Abstract. We review the observations of carbon and sulfur radio recombination line emission from CII regions that are associated with dark interstellar clouds. Such CII regions detectable by their radio recombination line radiation share several characteristics: they are all excited by an early B-type star that is embedded within or closely associated with a dense ($n \gtrsim 10^4$ cm^{-3}) neutral cloud; the exciting star is identifiable both as a near infrared point source and as an extended source of far infrared thermal dust emission; the CII region coincides with a localized region of enhanced molecular line emission. We summarize the physical properties (temperature and density) of the CII regions as derived from the recombination line observations and comment on prospects for further research.

I. INTRODUCTION

All of the papers in this volume are addressed to a description and interpretation of the radio recombination line emission from ionized gas within or adjacent to bright HII regions that are excited by extremely luminous O stars--this paper is the single exception. Here I review the expectations and the observations of carbon and sulfur recombination line emission from dense, nearby molecular clouds that lack a bright source of excitation. For convenience I will refer to such regions collectively as "CII regions" but mean by this term not just regions of ionized carbon but also ionized regions of all elements having a first ionization potential less than 13.6 eV.

That regions of partially ionized gas exist coincident with, or perhaps on the periphery of, dense molecular clouds is not wholly unexpected. If the flux of ionizing protons is sufficiently great or if the ambient interstellar radiation field λ < 1100 A is sufficiently intense, we expect a shell of partially ionized material to surround dense clouds; for nearby clouds the recombination line emission from that shell should be detectable. That such lines are, however, not detected (Gordon 1973) means that one or both of these conditions is in general not satisfied.

P.A. Shaver (ed.), Radio Recombination Lines, 127–140.

The carbon radio recombination lines that are detected in a very specific subset of dense interstellar molecular clouds--the Rho Ophiuchi cloud, L1630 (NGC 2023, NGC 2068, NGC 2071), and the clouds adjacent to NGC 7023 and S140--all have associated with them bright infrared sources indicative of the presence of embedded, newly formed stars. It is these stars, almost certainly, that excite the CII regions and provide the conditions necessary for the production of detectable recombination line emission. Hence by studying the line emission it is possible to investigate physical conditions in the dense material most closely associated with newly formed stars and to determine the manner in which these stars interact with their placental cloud.

The outline of this article is the following: In section II, I examine first the physical conditions expected in the CII regions that surround B-type stars which are embedded in dense gas; in section III, I summarize the conditions as deduced from the observations; section IV is a comparison of the expectations with the observations; and finally in section V, I briefly address the future of this research.

II. DESCRIPTION OF AN IDEALIZED CII REGION IN A DENSE INTERSTELLAR CLOUD

An early B-type star found embedded within or immediately adjacent to a dense cloud will have at least the following 3 effects on that cloud:

1. It will dissociate the molecular cloud material closest to the star;

2. It will ionize all those atomic species having ionization potentials less than 13.6 eV;

3. It will heat the gas and dust in its vicinity.

A simple example will illustrate each of these effects. Consider a B1V star embedded in a homogeneous cloud of density 10^4 cm^{-3}. If we adopt the Van Citters and Morton (1970) model atmosphere for this star, then it is a straightforward matter to balance the ionization and recombination rates in the gas as a function of distance from the star (e.g. Viala and Walmsley 1976), and determine the extent of the CII, SII etc. regions. The results of this calculation are illustrated in Figure 1. In performing this calculation we must, of course, allow for the very rapid conversion of atomic C to CO when τ(1100 A) $\sim$ 2-3 (Glassgold and Langer 1975).

Here it can be seen that in this particular example carbon is fully ionized to a distance $\sim$ 0.25 pc from the exciting star beyond which it is associated into the molecular form CO. The other trace elements, magnesium, iron, sulfur and silicon which have much lower ionization potentials are, as a consequence, fully ionized to greater distances than is carbon. Hence we expect, for example, the size of the SII region to be larger than the CII region while the size of the SiII region should be even larger still.

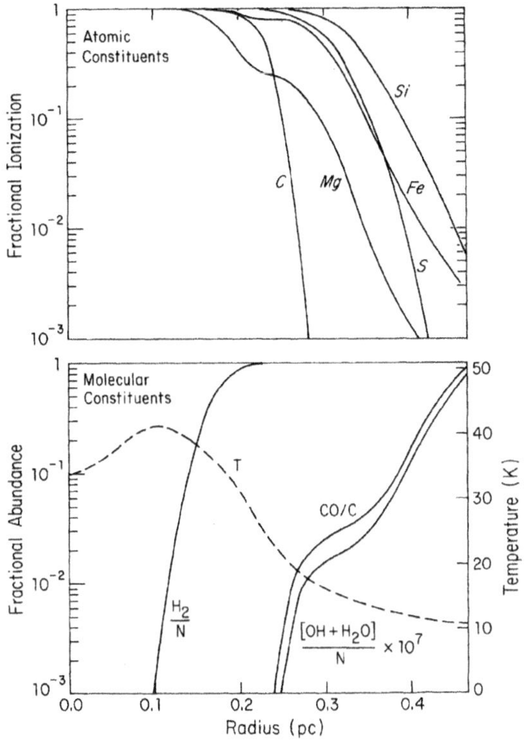

Fig. 1. The variation of fractional atomic ionization (upper panel) and molecular abundance (lower panel) as a function of distance from a B1V star embedded in a cloud of density 10^4 cm^{-3}. The resultant gas temperature is indicated by a dashed line in the lower panel.

At the same time that the ionization equilibrium calculation is carried out we must, of course, also account for thermal equilibrium. Here we consider the embedded star to be the sole heat source. With this constraint there are 4 principal heating mechanisms: atomic photoemission, grain photoemission, H_2 dissociation and collisional heating between hot grains and the cooler gas (e.g. Dalgarno and McCray 1972). In this example , atomic photoemission dominates the heating both very near the star, $R < .06$ pc, and far from the star, $R > 0.14$ pc, while at intermediate distances photoemission from grains is by far the dominant heat source. This curious behavior obtains because the grains are so highly charged near the star, I > 13.6 eV, that the stellar radiation field cannot ionize them. At intermediate distances the charge on the grains drops below 13.6 eV and photoemission becomes possible; when this occurs it quickly dominates all other heating mechanisms. The resultant temperature profile, illustrated in the lower panel of Figure 1, reflects this effect.

The cooling of the CII region occurs principally by means of collisional excitation of the 156μ transition of C^+, and/or by the 610μ and 369μ transitions of neutral atomic carbon. All these infrared lines should be very intense in CII regions.

Throughout these calculations of the thermal and ionization structure of CII regions it is important to emphasize the overriding role that dust plays. Not only does dust photoemission dominate the heating, but more importantly as Pankonin and Walmsley (1978a) stressed, dust determines the ionization and the size of CII regions; fewer than 1 in 100 of the stellar photons λλ912-1100 A is absorbed by the gas, dust grains absorb all the (vast majority) of the remaining photons.

III. DESCRIPTION OF THE OBSERVED CII REGIONS IN DENSE INTERSTELLAR CLOUDS

(a) The Rho Ophiuchi Dark Cloud

The dark cloud near the bright star Rho Ophiuchi (Heiles' Cloud 4) has been intensely studied over the past few years owing to the observation that an entire cluster of stars appears to have formed very recently from this cloud. The newly-formed stars are seen not as visible objects, but they are identifiable as point sources at 2μ in the near infrared (Grasdalen et al. 1973; Vrba et al. 1975). Two of the infrared sources appear to be obscured stars of spectral type sufficiently early as to have compact HII regions around them that are detectable as sources of radio continuum emission (Brown and Zuckerman 1975). And one of these latter 2 stars is sufficiently luminous as to heat the dust in its vicinity enough to render this part of the cloud detectable as a source of far infrared radiation (Fazio et al. 1976). It is at this point in the cloud--the point where a near infrared source, a far infrared source and a source of radio continuum emission coincide--that carbon radio recombination lines were first detected (Brown and Knapp 1974). Moreover, this spatial coincidence strongly suggests that the embedded star responsible for the infrared and radio continuum emission is also the source of excitation of the CII region (Brown et al. 1974).

Subsequent to these early observations a great deal of observational work has been devoted both to the study of the radio recombination line emission and to the radio continuum structure of this cloud. Figure 2, which is a 9 cm continuum map of the ρ Oph cloud will serve to orient the discussion of these observations.

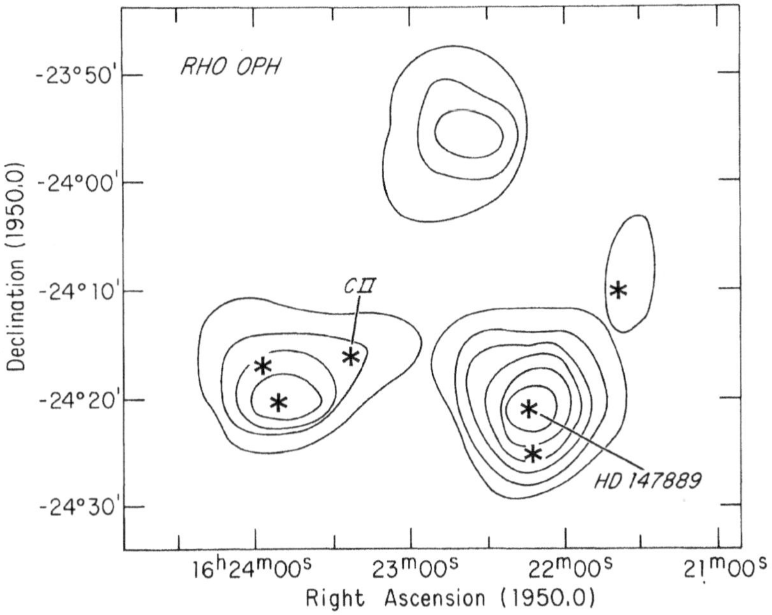

Fig. 2. The radio continuum structure of the ρ Oph cloud at 9 cm made with the 140-foot telescope. The contours are antenna temperature, the lowest being 6 mK; the contour interval is 4 mK. The crosses mark the position of the 6 compact radio continuum sources identified by Brown and Zuckerman (1975).

As can be seen from Figure 2, there are at least 4 readily identifiable radio continuum sources seen toward the ρ Oph cloud. The continuum properties of these 4 sources are summarized in Table 1. With the single exception of Oph-West (an extremely compact source with an inverted spectrum) carbon radio recombination lines have been detected at the positions of each of the radio continuum peaks (Brown and Knapp 1974; Pankonin and Walmsley 1978a; Falgarone et al. 1978). Since each of these 3 sources has a thermal bremsstrahlung spectrum, we conclude that they are in fact weak HII regions excited within, or on the face of, the ρ Oph dark cloud by early B-type stars that are spatially associated with the molecular cloud. Moreover, as a direct corollary, we note that since the recombination lines are found at the positions of the radio continuum sources and not elsewhere in the cloud, there exists a one-to-one causal relationship between the existence of embedded B stars and the emission of detectable carbon radio recombination line radiation.

Table 1. The Principal Radio Continuum Sources in the ρ Oph Dark Cloud

	α(1950)	δ(1950)	3250 MHz Flux Density (mJy)
Ophiuchus -			
East	$16^h23^m58^s$	-24°21'	62
Center	16 23 33	-24 16	26
North	16 22 42	-23 55	54
West	16 22 23	-24 21	110

Although most of the observations and physical interpretation of the carbon radio recombination line emission seen in the ρ Oph cloud have concentrated on the region of strongest line emission, Oph-Center, it is nevertheless clear that the physical conditions, especially the density, differ from one CII region to the other even within the same cloud. Specifically, in Oph-C, Pankonin and Walmsley (1978a) note that the very intense high frequency, ν = 5-10 GHz, recombination lines require a 2 component model of the CII region: they suggest for this CII region a "core-halo" description with the core having a density $n(C^+)$ = 15 cm^{-3} and a diameter d = 0.1 pc while the halo has a much lower density, $n(C^+)$ = 1 cm^{-3} and a larger diameter, d = 0.5 pc. In the CII regions in the remainder of the cloud, the line emission appears to arise wholly in a lower density medium, $n(C^+) \simeq 1$ cm^{-3} (Falgarone et al. 1970).

The sulfur radio recombination has also been detected toward ρ Oph, but only at the position of Oph-C, not in the other CII regions (Chaisson 1975; Knapp et al. 1976; Pankonin and Walmsley 1978a; Falgarone et al. 1978). The latter authors, in an extensive set of recombination line observations at 18 cm, demonstrate not only that the sulfur line emission is restricted to the region of Oph-C but also that the extent of the SII region is nearly twice as large as the extent of the CII region. No line emission from Mg, Si or Fe has been detected in any of the observations of ρ Oph.

(b) The Reflection Nebula NGC 2023

NGC 2023 is a very compact, ~6', reflection nebula found immediately south of Orion B (NGC 2024) and just east of the conspicuous "Horsehead" nebula. The reflection nebula itself is found on the near side of the giant dust cloud L1630; NGC 2023 is illuminated by a B1.5V star HD37903.

NGC 2023 is a very attractive region for the study of all aspects of the interaction of B-type stars with dense molecular material because the region is compact, simple and the source of excitation is optically visible. Moreover, the effect that HD37903 has on heating the adjacent dust cloud is evident both from an enhancement in the brightness temperature of the CO lines toward NGC 2023 (Milman et al. 1975) and in the far infrared dust emission seen at this position (Emerson et al. 1975). Curiously, neither the peak CO brightness nor the peak far IR emission coincides precisely with the star; rather both are most intense ∿1' south of HD37903. A weak HII region of approximately the flux density expected to be produced by a B1.5V star is also evident toward NGC 2023, but this again is displaced ∿1' southeast of the exciting star. A map of the CO brightness and the radio continuum emission of NGC 2023 is presented in Figure 3.

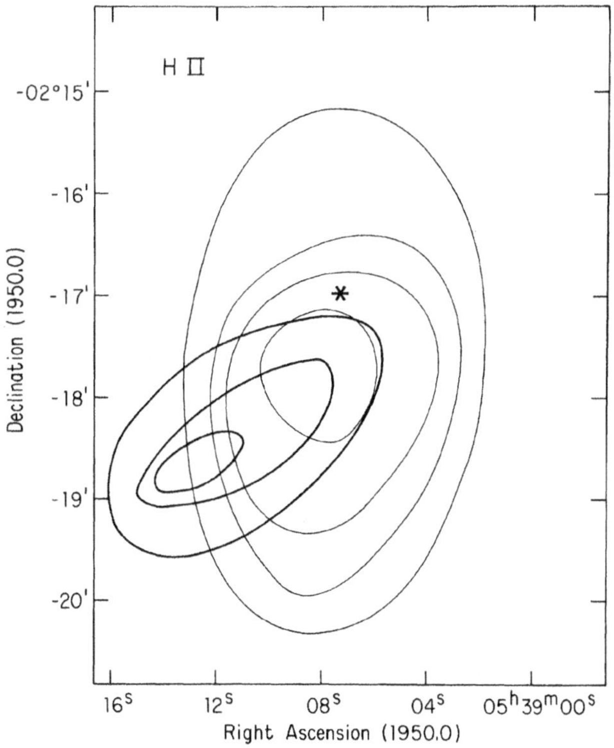

Fig. 3. The 11 cm radio continuum emission (heavy contours) and the CO brightness (light contours--the contours are 32, 34, 36 K) of NGC 2023.

Carbon radio recombination line emission was first detected toward NGC 2023 by Knapp et al. (1975) and subsequently studied in much greater detail by Pankonin and Walmsley (1976). From maps of the carbon radio recombination line emission from NGC 2023 made with high spatial and angular resolution, Pankonin and Walmsley (1976) find evidence for 2 distinct spatial and velocity features. As in the case of Oph-C, they argue that their observations require a "core-halo" description for the density structure of the line emitting gas in NGC 2023. The core, whose centroid is again displaced $\sim 1'$ south of HD37903, has a density of $n(C^+) \sim 10\ cm^{-3}$ and a size of 0.05-0.1 pc. Since the lower density halo, $n(C^+) \sim 1\ cm^{-3}$, is most evident at low frequencies, its centroid is more difficult to establish precisely but its size is certainly ~ 0.5 pc.

Although at first glance the similarities of the CII regions in Oph-C and NGC 2023 are striking--both have a core-halo density structure, the sizes and densities of the cores and halos are remarkably similar--it is nevertheless clear that these regions are in fact quite distinct. The most direct way to appreciate this point is to note that whereas in Oph-C the ratio of the sulfur to carbon lines is $\sim 1/3$, in NGC 2023 on the other hand the sulfur line has not yet been detected, $S/C < 1/4.5$.

(c) S140-IR in the Dark Cloud L1204

S140 is an emission nebulosity seen interacting with the dark cloud L1204--a bright rim ionization front on the edge of L1204 facing S140 is a conspicuous feature of this region. Immediately to the east of the ionization front, and apparently within L1204, is an intense point near-infrared source, S140-IR (Blair 1976), which has been interpreted as a newly formed early B or late O type star. This interpretation is strengthened by the detection of far infrared radiation from heated dust in the vicinity of S140-IR (Rouan et al. 1977). Furthermore, at this same position a local maximum in the CO brightness is evident.

The 5 GHz radio continuum structure of the S140/L1204 region is shown in Figure 4. The source of ionization of S140 is commonly assumed to be the B0.5V star HD211880 which is noted on this figure. Curiously, S140-IR appears in this map much closer to the centroid of the radio brightness thereby begging the question as to whether S140-IR does not contribute significantly to the ionization of the larger S140 region. This question should be examined in greater detail.

Carbon recombination lines have been detected toward S140-IR by Knapp et al. (1976) at 13 and 21 cm (C166α and C141α) and at 9 cm (C125α) by Brown (unpublished). These observations, taken together, suggest that the CII region toward S140-IR is of much lower density than that found in either ρ Oph or in NGC 2023; specifically $n(C^+) \lesssim 1\ cm^{-3}$. Unfortunately, owing to the large telescope beams used to study this particular region, it has not been possible to obtain a reliable estimate of the size of the CII region. Nor has it been possible to identify

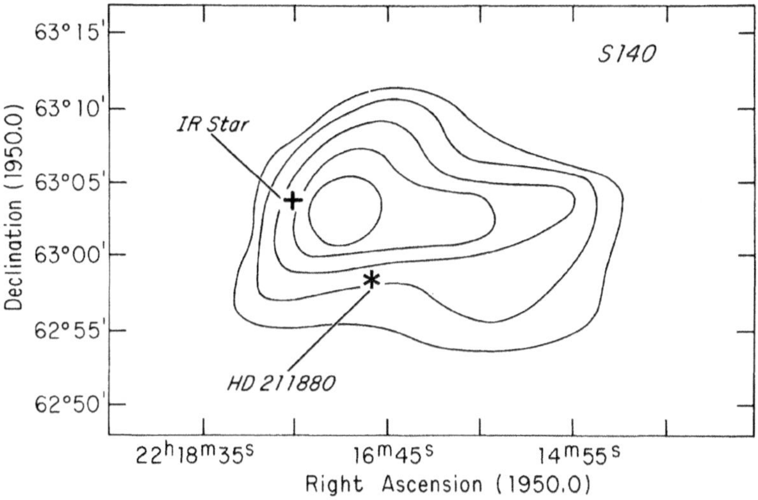

Fig. 4. The 5 GHz radio continuum brightness of S140 made with the 140-foot telescope. The contours show antenna temperature. The lowest contour is 15 mK; the contour interval is 15 mK.

with certainty whether the line emitting region is related to S140-IR or to the very nearby S140/L1204 ionization front. One suspects from the similarity between the infrared and molecular emission properties of S140-IR and Oph-C or NGC 2023 that the CII region is indeed excited by S140-IR, but this cannot presently be proved.

(d) The M78 Reflection Nebulae NGC 2068 and NGC 2071

NGC 2068 and NGC 2071 are two reflection nebulae separated by only ∿ 12' that are seen projected against the same dark cloud, L1630, with which NGC 2023 is associated. The M78 nebulae lie ∿ 2°.5 northeast of NGC 2023. As with all the regions discussed above, a number of point near infrared sources are found near and around M78 (Strom et al. 1975). And again, the interpretation of these infrared sources is that they are highly obscured stars that are embedded within or behind L1630. Similarly, once again the reflection nebulae NGC 2068 and NGC 2071 are evident as local maxima in the CO brightness of L1630 indicating that the stars which excite these nebulae also heat the molecular gas in their vicinity.

Carbon radio recombination lines were first detected from M78 at 1.4 and 2.4 GHz by Brown et al. (1975). Subsequent observations of these lines at 1.6 GHz by Pankonin and Walmsley (1978b) confirmed that the lines arise in a medium whose density, $n(C^+) \lesssim 1$ cm^{-3}, is considerably less than the peak density found either in NGC 2023 or ρ Oph. In the case of M78, however, the extent of the ionized region must be quite a bit larger than found in other CII regions, ~ 3 pc. In addition to this large size, the M78 CII region is remarkable in one other respect: viz. Pankonin and Walmsley (1978b) identify SII emission as well as CII emission. This latter result is particularly surprising in view of the fact that no SII line emission was seen in the same cloud, L1630, toward NGC 2023. Naively, this suggests that the relative depletion of abundant elements varies by factors at least as large as ~ 2 within a single interstellar cloud.

IV. PHYSICAL CONDITIONS IN THE CII REGIONS ASSOCIATED WITH DARK CLOUDS

The radio recombination line observations described above, while often made with limited angular resolution, are nevertheless very sensitive probes of the density of CII regions in dark clouds. Using these tools we have seen that the density is typically $n(C^+) \simeq n_e \simeq 1$ cm^{-3} with two exceptional cases where $n(C^+) \gtrsim 10$ cm^{-3}. The observations are much less sensitive probes of the temperature of the CII regions; most of the observations can be fit by temperatures anywhere between 20-200 K (e.g. Pankonin and Walmsley 1978a). If we are to make further progress in understanding either the temperature or the total gas density [i.e. not $n(C^+)$ but $n(H + H_2)$], then we must rely on theoretical models of CII regions.

Consider first the temperature. Referring to Figure 1 we see that even in the presence of grain photoemission heating it is difficult to increase the temperature of a CII region above approximately 50 K. Most modifications one would propose to make the models more realistic work to <u>lower</u>, not raise, the temperature (for example, if the CII region is on the edge of a molecular cloud rather than fully within it as I've considered here, then the optical depth of the principal cooling lines decreases and hence the cooling efficiency increases). Moreover, even if we arbitrarily postulate an unknown heating mechanism whose energy input is equivalent to that of grain photoemission, we raise the temperature very little because, again, the cooling processes are very efficient. It seems most prudent, therefore, to regard the temperature in the CII region to lie somewhere between the temperature of the molecular material adjacent to the CII region (as measured by CO for instance) and 50 K. That is, $25 \lesssim T(CII) \lesssim 50$ K.

The total density in the CII regions depends wholly on the extent to which carbon is depleted into grains within the CII region. From the fact that we see sulfur radio recombination lines and not lines from the more cosmically abundant species Mg, Fe and especially Si, we know that these latter species are indisputably heavily depleted (factors of 5-10) within the CII region. Is carbon also depleted? To answer this question we construct a series of model CII regions, assume that carbon is not depleted, and compute the expected line strengths. An example of a calculation such as this is shown in Figure 5.

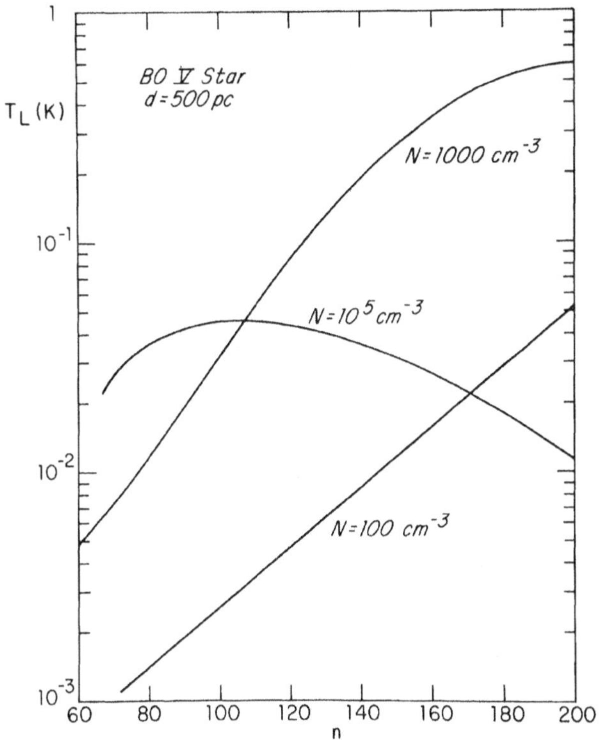

Fig. 5. The expected carbon recombination line strengths as a function of principal quantum number from a homogeneous CII region ionized by a BOV star if that star is embedded in a cloud of density $n(H_2)$ = 100, 1000, or 10^5 cm^{-3} (as indicated) at 500 pc distance and observed with a 100-m telescope.

Two features of this figure are worthy of note. First, in low density gas the line strength increases with n (the lines are stronger at low frequencies) owing to the variation of departure coefficient with n. For CII regions in very dense gas the variation of line strength with n is much flatter owing to the increasing importance of beam dilution at low frequencies. Secondly, for clouds of density $1000 \lesssim n \lesssim 10^5$ cm^{-3} the expected line strengths at the frequencies most often observed (n = 110 - 160) are quite substantial.

To determine whether or not carbon is depleted, we need to compare these predictions on a case-by-case basis with the observations. This, unfortunately, has not yet been done carefully. However, if we plot the observations in the same manner as we plotted the expectations, Figure 6, we see immediately the differences mentioned above between the high density CII regions ρ Oph and NGC 2023 and the low density CII regions S140 or M78. The latter are much steeper functions of n at low frequencies as we expect (Fig. 5). It is important to emphasize here that in all these cases, even if carbon is wholly undepleted, when we consider $n(C^+) = 1 - 10$ cm^{-3} we are explicitly considering very dense clouds, $n(H + H_2) = 3000 - 3 \times 10^4$ cm^{-3}.

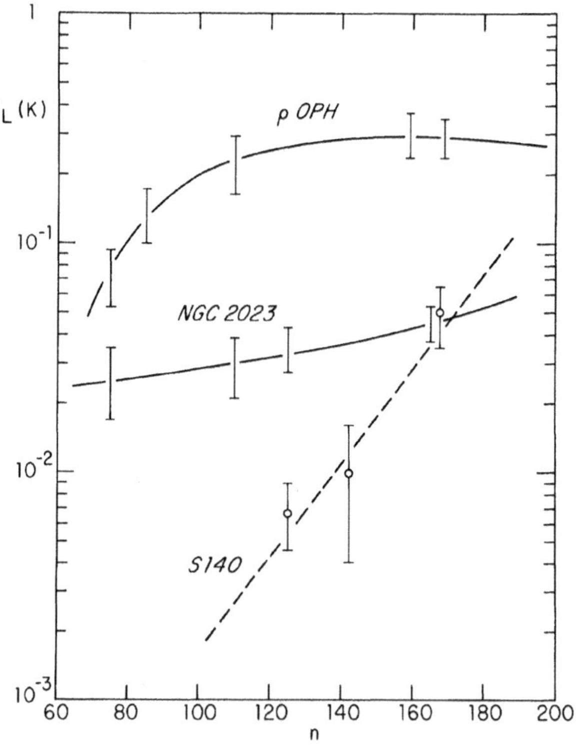

Fig. 6. The observed variation of carbon recombination line strength with principal quantum number for ρ Oph, NGC 2023 and S140-IR.

V. CONCLUDING REMARKS: FUTURE RESEARCH

The relation between embedded early B-type stars in dark clouds and their resultant CII region, regions of intense infrared radiation, molecular hot spots and compact HII regions is an extremely interesting study to which radio recombination line observations can contribute in a fundamental way. Most particularly, these observations provide an estimate of the properties of the cloud material most closely associated with nearly formed stars. The unfortunate reality is, however, that carbon radio recombination lines are detectable from a very few embedded B-type stars. From the approximately 50 candidate regions (reflection nebulae) that have been searched thoroughly for carbon recombination lines, such lines have been detected in fewer than a half dozen. The precise reason why the detection rate is so low is puzzling; carbon radio recombination line emission from an embedded early B-type star will be suppressed if any of the following obtain:

1. Carbon is heavily depleted into grains;

2. The density in the cloud surrounding the star is very high ($\gtrsim 10^6$ cm $^{-3}$). In this case beam dilution would be insurmountable.

3. The temperature in the CII region is high, T $\gtrsim$ 300 K, as a result of an extremely efficient, but unknown, heating mechanism.

4. The distance between the exciting star and the "face" of the molecular cloud is sufficiently large ($\gtrsim$ 0.3 pc) that the stellar radiation field $\lambda < 1100$ A is too dilute to efficiently excite a CII region.

Any, or all, of these possibilities could account for the null results.

In the future it is certainly possible and desirable to search farther for new CII regions in dark clouds particularly if one's search is guided by detections made in the far infrared. However, equally important, or perhaps more so, are more precise, detailed observations of the CII regions which we know exist. It is readily evident that much remains to be learned from comprehensive radio, infrared and (not least) theoretical studies of CII regions in dark clouds.

References

Blair, G. (1976) Thesis, Univ. Texas (unpublished)

Brown, R. L., Gammon, R. H., Knapp, G. R., Balick, B. (1974) Astrophys. J. 189, 253.

Brown, R. L., Knapp, G. R. (1974) Astrophys. J. 189, 253
Brown, R. L., Knapp, G. R., Kuiper, T. B. H., Rodriguez-Kuiper, E. N. (1975) Astrophys. J. 195, L23
Brown, R. L., Zuckerman, B. (1975) Astrophys. J. 202, L125
Chaisson, E. J. (1975) Astrophys. J. 197, L65
Dalgarno, A., and McCray, R. A. (1972) Ann. Rev. Astron. Ap. 10, 375.
Emerson, J. P., Furniss, I., Jennings, R. E. (1975) Mon. Not. Roy. Astron. Soc. 172, 411
Falgarone, E., Cesarsky, D. A., Encrenaz, P. J., Lucas, R. (1978) Astron. Astrophys. 65, L13.
Fazio, G. G., Wright, E. L., Zeilik, M., Low, F. J. (1976) Astrophys. J. 206, L165
Glassgold, A. E., and Langer, W. D. (1975) Astrophys. J. 197, 347
Gordon, M. A. (1973) Astrophys. J. 184, 77
Grasdalen, G. L., Strom, K. M., Strom, S. E. (1973) Astrophys. J. 184, L53
Knapp, G. R., Brown, R. L., and Kuiper, T. B. H. (1975) Astrophys. J. 196, 167
Knapp, G. R., Brown, R. L., Kuiper, T. B. H., and Kakar, R. K. (1976) Astrophys. J. 104, 781
Knapp, G. R., Kuiper, T. B. H., Brown, R. L. (1976) Astrophys. J. 206, 109
Milman, A. S., Knapp, G. R., Kerr, F. J., Knapp, S. L., Wilson, A. (1975), Astron. J. 80, 93.
Pankonin, V., Walmsley, C. M. (1976) Astron. Astrophys. 48, 341
Pankonin, V. and Walmsley, C. M. (1978a) Astron. Astrophys. 48, 341
Pankonin, V. and Walmsley, C. M. (1978b) Astron. Astrophys. 67, 129
Rouan, D., Lena, P. T., Puget, T. L., de Boer, K. S., Wijnbergen, J. J. (1977) Astrophys. J. 213, 235
Strom, K. M., Strom, S. E., Carrasco, L., Vrba, F. J. (1975) Astrophys. J. 196, 484
Van Citters, G. W. and Morton, D. C. (1970) Astrophys. J. 161, 695
Viala, Y. P. and Walmsley, C. M. (1976) Astron. Astrophys. 50, 1
Vrba, F. J., Strom, K. M., Strom, S. E., Grasdalen, G. L. (1975) Astrophys. J. 197, 77

CARBON AND SULPHUR IONIZED REGIONS IN DARK CLOUDS

Edith Falgarone,
DERAD, Observatoire de Meudon, France

Observations of carbon and sulphur recombination line (CRL and SRL) emission from dark clouds provide means to detect and quantitatively study the sites of current massive star formation.

Unfortunately, two CRL surveys of such sites (van Gorkom et al. 1979, Cesarsky and Falgarone, unpublished results) gave mostly negative results. The reasons invoked to explain the usually low luminosity of that kind of CII regions are mainly: carbon depletion and absorption of the carbon-ionizing photons by the dust.

At present, CRL sources have been detected in only four dark clouds : ρ Ophiuchus, Monoceros R2, NGC 2023, M 78 and NGC 7023 (see references in van Gorkom et al. 1979). The C166α and C141α detections reported by Knapp et al. (1976), in the S140 dark cloud, have not been confirmed by the detection of the C158α line (Pankonin and Walmsley 1978, Cesarsky and Falgarone, idem). The latter have mapped all the area covered by the beam of Knapp et al. (1976) and obtained, by integration of the data, an upper limit lower than the antenna temperature of the C166α line. In ρ Ophiuchi, three embedded stars, of approximately the same spectral type, excite three CII regions which emit C158α fluxes differing by a factor of eight (Falgarone 1980).

Calculations of heavy elements (C, S, Si, Fe, Mg) ionized regions have been carried out, within a wide range of parameters, to cast some light on this puzzling behaviour. The outline of the sensitivity of CRL and SRL fluxes and radii of CII and SII regions, with respect to physical conditions in the cloud, are presented here.

I. CHARACTERISTICS OF THE MODEL

I.1. The choice of the parameters

Idealized observations of CII and SII regions supply typically the six following quantities:

i) $S_c\,\delta\nu$ and $S_s\,\delta\nu$, the total fluxes, proportional to the numbers of stellar photons effectively absorbed by the gas of each species, if the ionized

P.A. Shaver (ed.), Radio Recombination Lines, 141–147.

regions are optically thin,

ii) R_c and R_s, the radii of the ionized regions defined as the distance to the star where the line brightness temperature has decreased by a factor of two,

iii) $T_c\,\delta\nu$ and $T_s\,\delta\nu$, the line brightness temperatures through the center of the regions. Even if they are not completely independent (like the flux and the central brightness temperature), all of them provide interesting informations on the structure of the ionized regions.

Six important free parameters have been selected to enter the model:

i) A_C and A_s, the carbon and sulphur abundances,

ii) k_d^{abs}, the dust absorption coefficient, or ω, the dust albedo, related by : $k_d^{abs} = k_d^{ext}\,(1-\omega)$, k_d^{ext} being considered as an observational data derived from the extinction curve of Bless and Savage (1972) and the extinction laws of Jenkins and Savage (1974) and Bohlin et al. (1978),

iii) two densities, n_m and n_p, describing a two components distribution:

$$n(r)=n_m\,\exp(-r/r_1)^2+n_p\;;$$

the radius r_1 turns out to be unimportant, except in the case of sharply peaked distribution ($n_m \gg 10^5$ cm^{-3} and $r_1<0.05$ pc),

iv) T_m, the maximal electronic temperature. The adopted distribution:

$$T_e(r)=T_m\,\exp(-r/r_2)+T_o$$

is a rough analytical fit to the results of Pankonin and Walmsley (1976).

I.2. Basic assumptions

The transfer problem is considerably simplified by the assumption (justified a posteriori) that CII and SII regions are optically thin. Brown et al. (1978) had already suggested that, if it were not the case, the contribution to the line of the emission stimulated by the thermal 2.7K background ($\sim 2.7\,\tau_c$) would be much larger than any observed line.

UV stellar spectra are computed from the model atmospheres of Kurucz (1974). The spectral types of the exciting stars (characterized by T_{eff}) are not considered as free parameters since they may be directly determined by the continuum flux of the associated compact HII regions, or by near infrared photometry.

The wavelength dependence of the heavy elements photoionization cross-sections is derived from the atomic data referenced in Walmsley (1973). The small difference between the C and S cross sections near 1100Å and the fact that $A_c > A_s$ justify the "on the spot" approximation for ~1100Å photons emitted by an electron captured in the ground state.

Scattering of UV photons is neglected in the transfer treatment: for $717\text{Å} < \lambda < 1620\text{Å}$, the phase function is sharply forward throwing (Lillie and Witt 1976). The dust albedo, ω, is kept independent of λ. In the case of ρ Ophiuchi, the value $R=A_v/E(B-V)=4$ is in better agreement with near infrared photometry of a few stars associated with the cloud (Chini *et al.* 1977) than is R=3.

The b_n coefficients, characterizing the departure of the atomic level populations from LTE, are those of Brocklehurst and Salem (1977).

II. RESULTS

II.1. Dust plays a leading part: typically, 10^{-2} to 10^{-3} of the C stellar ionizing photons are absorbed by the gas. This ratio is about 10 times smaller for sulphur and it decreases noticeably as T_{eff} increases: if the number of C ionizing photons is multiplied by a factor of 200 (as it is the case between a late B3V star and a B0V star), the CRL flux increases only by a factor of ~10.

Consequently, the rapid decrease of the flux with the distance of the CII region cannot be balanced by large increases of CRL luminosity with the temperature of the exciting star.

II.2. The RL flux is strongly dependent on the gas density <u>and</u> density distribution around the star. For the same mass of gas, the more the distribution of the density is peaked, the more the emitted flux and the radii decrease.

In Table I, are given observational quantities computed for four gas density distributions, all the other parameters are constant and equal to the reference set: $\sigma_d^{ext}=3.7\ 10^{-22}\ cm^2$, $\omega=0.67$, $A_c=7\ 10^{-5}$, $A_s=1.3\ 10^{-5}$, $T_m=25K$ and $T_{eff}=15000K$.

TABLE I

n_m cm^{-3}	n_p cm^{-3}	r_1 pc	$(n_m-n_p)/r_1$ $cm^{-3}\ pc^{-1}$	$S_c\delta\nu$ Jy KHz (*)	R_c pc
$5.6\ 10^5$	$5\ 10^3$	0.07	$8\ 10^6$	1.2	0.01
$1.9\ 10^5$	"	0.1	$2\ 10^6$	1.7	0.04
$2.4\ 10^4$	"	0.2	10^5	3.4	0.15
0	$2\ 10^4$	-	0	5.2	0.22

(*) Fluxes are computed for d=160pc, the distance of the ρ Oph cloud, at ν=1652.4 MHz

The reduction factor of the CRL flux between the first and the last model is considerable and is, however, a lower limit: all the fluxes of Table I have been computed with the same distribution of T_e. It is a crude approximation, since density peaks (as those created by collapsing gas) are known to be associated with peaks of kinetic temperature.

ρ Oph is again an example of the statement. Around two of the three embedded B stars found in the densest part of this cloud, the gas is still possibly collapsing, as suggested by the shape of the CO and ^{13}CO profiles observed at these locations (Falgarone, 1980). The CRL fluxes detected there are eight times weaker than that detected, by the same antenna, at the location of OPH 4 (Falgarone et al. 1978).

A pessimistic remark should be noted here: as soon as a compact HII region is sufficiently evolved that the dense post-shock layer, preceding the ionization front, has accumulated enough matter to become optically thick to C-ionizing photons, the CII region is mainly confined within that shell (see the model 1 of Table I) and emit a very low flux. The common existence of such dense layers surrounding compact HII regions might explain the large amount of previous non detections. Typically, the layer pushed, by the ionization front of a B2 star, into a $3\ 10^4 cm^{-3}$ gas is transparent to C-ionizing photons only for $t<5\ 10^3$ yrs, the origin of time is the beginning of the HII region expansion phase.

II.3. A quantity:

$$\rho(\sigma_d^{abs}, n(r), T_e(r), A_c, A_s, T_{eff}) = \frac{S_c\delta\nu \,/\, S_s\delta\nu}{A_c \,/\, A_s}$$

is plotted in Figure I, for two different values of A_c/A_s. Within a large set of physically meaningful values of σ_d^{abs} (materialized on Figure I by the two arrows), the ratio ρ is essentially independent of A_c/A_s. Furthermore, the calculations show that:

a) as σ_d^{abs} varies, by a factor of 10, around the mean value $\sigma_d^{abs} = 7\ 10^{-23} cm^{-2}$, corresponding to $\omega=0.8$ and $\sigma_d^{ext}=3.7\ 10^{-22} cm^{-2}$, the ratio ρ varies by only ~30%,

b) ρ is surprisingly insensitive to the spectral type of the exciting star (Table II),

c) ρ is a slowly varying function of T_e: if T_e is multiplied by a factor of two, ρ decreases by only 6%,

d) ρ is also a slowly varying function of the adopted density law. If the peak density (or the halo density) is multiplied by 5, ρ increases by only 3%.

There is, of course, an immediate and useful consequence to these properties: it is possible to derive the ratio A_c/A_s from the observed ratio $S_c\delta\nu/S_s\delta\nu$, and therefore to determine A_c, since sulphur is known to be weakly depleted (Morton 1974).

The accuracy of this determination is only 30% if no assumption is made on the dust properties and if the stellar spectral

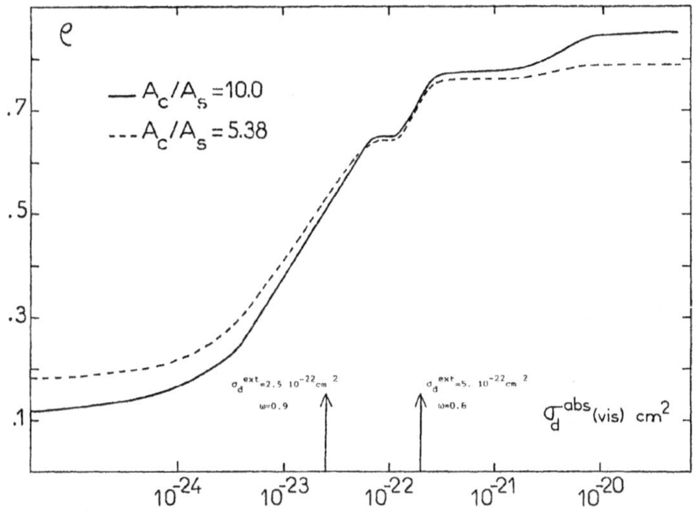

Figure I : Plot of the computed values of ρ vs. σ_d^{abs} in the visible range, all the non specified other parameters being equal to the reference set defined above.

type is unknown. In the best case, it is smaller than 10%. In ρ Oph, for example, the fluxes emitted by the CII and SII regions centered on OPH 4 are respectively 2JyKHz and 0.6JyKHz (Falgarone et al. 1978). Since the star is known to be a late B3V star, A_c/A_s=5.1. If sulphur is assumed not to be depleted, A_c=8 10^{-5}. If the depleted value A_s=8.6 10^{-6} is adopted A_c=4.3 10^{-5}. In both cases, carbon appears to be depleted in the gas phase by a factor ≥4 from its solar value.

TABLE II

T_{eff} (K)	Sp. type	N^*_c/N^*_s	ρ			
			A_c/A_s=5.38		A_c/A_s=10.	
			ω=0.67	ω=0.8	ω=0.67	ω=0.8
15000	B 3.5	0.39	0.66	0.64	0.67	0.65
20000	B 2	0.52	0.77	0.74	0.74	0.72
30000	B 0	0.63	0.79	0.74	0.79	0.78

N^*_c and N^*_s are respectively the numbers of stellar carbon and sulphur ionizing photons.

The behaviour of ρ is difficult to justify theoretically, because of the large number of interacting parameters. When no dust is present, the limit value:

$$\rho_o = \frac{N^*_c}{N^*_{c-s}} \left(1 - \frac{A_c}{\Sigma A} \right)$$

where N^*_{c-s} is the stellar flux of photons in the range 1100Å < λ < 1250Å, and ΣA the sum of the abundances of the elements considered, is a function of both the exciting star and the abundances. As the dust absorption increases, $S_s \delta\nu$ decreases faster than $S_c \delta\nu$ because sulphur is less abundant, which explains the general increase of ρ. The fact that the two curves cross each other and remain almost confounded for a large set of σ_d^{abs} values, independent of T_{eff}, might be due to the decrease of $\sigma_d^{abs}(\lambda)^d$ towards high λ.

Let us come back to the value of A_c determined above. If the high carbon depletion, observed in the ρ Oph cloud, is a common property of CII regions in dark clouds, calculations show that the CRL optical thickness of those regions remains ≤0.05, even for low electronic temperatures ($T_e \sim 25K$).

II.4. The last result is that the distribution of CRL and SRL brightness temperatures are good indicators of the gas density distribution around the star:

i) . the limb-brightening of the RL emission disappears as soon as a peak of density exists at the transition between the HII and the CII region;

ii) . if there is an extended low density halo, an important fraction of the RL emission arises in the low density medium, under the condition that the core is not opaque to the carbon and sulphur ionizing photons. This fraction, more important for sulphur than for carbon, increases as the halo density decreases;

iii) . the radius of the CII region depends only (for a given star and fixed dust properties) on the central density ($\sim n_m + n_p$). It must be recalled that this radius, as defined in §I.1, is rather the core radius of the CII region. An analytical law has been fitted to the computed values, in the case of a slowly varying density distribution, all the other parameters being equal to the previously defined reference set:

$$R_c \text{ (pc)} = 593 \; n \; (\text{cm}^{-3})^{-0.79} \; .$$

Within the explored range of density ($10^3 \text{cm}^{-3} < n < 10^5 \text{cm}^{-3}$), the quality of the fit is surprising. But no attempt has been made to take into account gas density fluctuations and possible large scale asymmetry in the temperature and density distributions.

III. CONCLUDING REMARKS

Even if excited by bright stars, compact CII regions are never very luminous sources. However, they are probably quite common objects and the purpose of the present model, which is still too coarse, was:

i) to explain the previous mostly negative results,
ii) to determine to which extent high spatial resolution observations of carbon and sulphur recombination lines in dark clouds might allow the study of several physical parameters, in a wide variety of clouds.

To sum up:
- the abundance ratio A_c/A_s may be directly deduced from $S_c\delta\nu/S_s\delta\nu$,
- in the case of a smooth density distribution, suggested by the shape of the $T_c\delta\nu$ distribution, the maximal density is derived from the measured radius of the CII region,
- at last, once the maximal density is known, the constraint on the line optical thickness sets a lower limit on the mean electronic temperature in the densest parts of the CII region.

REFERENCES

Bless, R.C., Savage, B.D. 1972 Astrophys.J. 171, 293
Bohlin, R.C., Savage, B.D., Drake, J.F. 1978, Astrophys.J. 224, 132
Brown, R.L., Lockman, F.J., Knapp, G.R. 1978, Ann.Rev.Astron.Astrophys. 16, 445
Chini, R., Elsässer, H., Hefele, H., Weinberger, R. 1977, Astron.Astrophys. 56, 323
Falgarone, E. 1980, in preparation
Falgarone, E., Cesarsky, D.A., Encrenaz, P.J., Lucas, R. 1978, Astron.Astrophys. Letters 65, L13
van Gorkom, J.H., Shaver, P.A., Goss, W.M. 1979, Astron.Astrophys. 76, 1
Jenkins, E.B., Savage, B.D. 1974, Astrophys.J. 187, 243
Knapp, G.R., Brown, R.L., Kuiper, T.B., Kakar, R.K. 1976 Astrophys.J. 204, 781
Kurucz, R.L., Peytremann, E., Avrett, E.H. 1974, Blanketed Model Atmospheres for Early Type Stars (Smithsonian Astrophysical Observatory)
Lillie, C.F., and Witt, A.N. 1976 Astrophys.J. 208, 64
Morton, D.C. 1974, Astrophys.J.Letters 193, L35
Pankonin, V., and Walmsley, C.M. 1976, Astron.Astrophys. 48, 341
" " 1978, Astron.Astrophys. 67, 129
Walmsley, C.M. 1973, Astron.Astrophys. 25, 129

CARBON (?) RECOMBINATION LINES IN W49A (G43.2-0.0)

M.R. Viner and V.A. Hughes
Queen's University, Kingston, Ontario, Canada

1. INTRODUCTION

In 1971, Chaisson and Ball (henceforth referred to as CB) reported detection of an anomalous recombination line in the low-frequency wings of He94α and He85α in W49A. The strength of the line was about 3% of the nearby hydrogen alpha line, and the velocity with respect to the local standard of rest (LSR) was 60-70 km s^{-1}, assuming the emitting species is carbon.

The reality of the anomalous emission in W49A was well established by CB through the absence of the line when observing blank sky, non-thermal sources, and other thermal sources at 94α. Subsequently, a narrow recombination line at a similar LSR velocity was detected in the 247α and 248α line sets (Pankonin et al. 1973, Parrish 1975), and in the 86α line set (Lichten et al. 1979).

However, the anomalous line was apprently not detected in the somewhat lower sensitivity measurements of Churchwell et al. (1974) at 85α and 109α, and is clearly not seen in published observations of 166α emission (Parrish 1975, Payne and Terzian 1976, Pankonin et al. 1977).

We have made high-sensitivity recombination line measurements of W49A in several α-sets in order to follow up on the anomalous emission line at high frequencies. Our earliest observations failed to detect the line at the levels reported by CB, but our later data shows that the emission is more anomalous than previously thought. This paper presents our observations and some speculation on their implications.

2. OBSERVATIONS

Most of our data was obtained with the 46-m telescope of the Algonquin Radio Observatory (ARO), using a 100-channel filter spectrometer with a resolution of 87 kHz in a bandpass of 10 MHz. We centred the bandpass so that both hydrogen and helium lines were measured simultaneously in all line sets.

P.A. Shaver (ed.), Radio Recombination Lines, 149–155.

Table 1: Summary of Observations

Epoch	α-set	Integration Time (minutes)	Beam Size	$\frac{T_{anom}}{T_H}$ (%)	Reference
1970.46	94α	-	4!2	3.4 ± 0.8	Chaisson & Ball 1971
1971.04	85α	-	3.0	2.5 ± 0.4	Chaisson & Ball 1971
1972.36	85α	375*	2.7	<1.8	Present results
1973.18	85α	30			
1973.35	85α	80			
1973.42	85α	100			
1974.66	85α	360	2.7	<1.0	Present results
1975.044	101α	330	4.6	<1.5	
1975.068	92α	355	4.0	1.7 ± 0.5	
1975.76	100α	460	4.5	<1.8	
1975.87	85α	327	2.7	4.9 ± 0.6	
1978.0	86α	100	3.5	4.9 ± 0.7†	Lichten et al. 1979

*In Dicke-switching mode; all others in Total Power mode

†May be confused with nearby H162η

We measured the 85α line set at several epochs as indicated in Table 1, but only made one measurement at each of the 100α and 101α line sets. Our first set of observations at 85α was taken in the Dicke-switching mode, using a cold load as the reference. For this epoch, the system temperature was about 240 K. For the later epochs, we used the spectrometer in the total power made with a system temperature of about 130 K at 85α and 80 K at both 100α and 101α.

We also obtained a measurement of the 92α line set using the 36.6-m telescope of the Haystack Observatory. A 100-channel autocorrelation spectrometer was used in the total power mode with a 6.67 MHz bandpass, giving a resolution of 167 kHz (60 km s^{-1}) after Hanning smoothing. The bandpass was positioned separately for the hydrogen and helium lines in order to obtain sufficient baseline for reliable fitting. The system temperature for the Haystack measurements was about 80 K.

Our observing procedure at Haystack was similar to that used at ARO, in which a series of 5 or 10-minute integrations on and off the source were interrupted every 2 to 4 hours to measure the continuum radiation from the source by means of orthogonal scans and/or on-off measurements. For comparison with Table 1, the angular size of W49A is about 3 arcmin.

3. RESULTS

Figure 1 shows the averaged data for each epoch, with the adopted baseline fit superimposed. The peak of the hydrogen recombination line emission has been blanked out for those spectra in which hydrogen and helium both appear. The velocity scale is relative to the LSR velocity of carbon line emission for each transition, and the intensity scale is in mJy per beam area. For most of the ARO data, the baseline consisted of a straight line plus the standing wave sinusoid of period 8.2 MHz, plus the second harmonic of the standing wave when necessary ($\geq 3\sigma$). For the Haystack data, a quadratic fit was sufficient to remove the baseline.

We examined each spectrum for evidence of the anomalous emission reported by CB at a carbon line velocity of 60 to 70 km s^{-1}. The results are indicated in Table 1 as a percentage of the hydrogen line

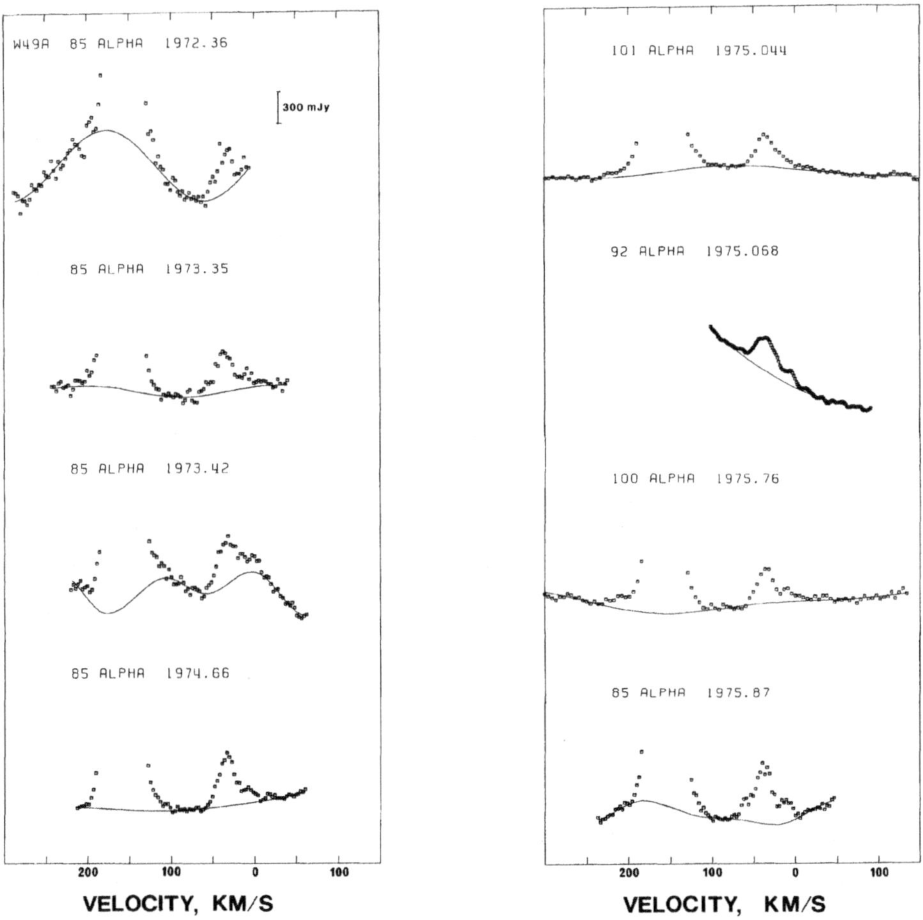

Fig. 1. Observed recombination-line spectra from the direction of W49A, with baseline fits superimposed.

strength, and show that the emission was considerably weaker than reported by CB except for one epoch (1975.87), where it appears with about twice the originally reported intensity. The upper limits in Table 1 are three times the rms scatter within each spectrum.

During the observations of 1975.87, two different calibration algorithms were used with the spectrometer: both sets of data are substantially identical. The anomalous emission shoulders on both sides of the helium line could be seen in most of the individual 10-minute spectra, and show up clearly in daily averages over the 4 days of observing. They could not be removed by any standard baseline function known to us. No instrumental or interference problems were encountered during that period, and the anomalous emission does not appear on similar spectra of W3 obtained on the same days. We are convinced that the observed line profile represents emission from W49A at that epoch rather than some terrestrial effect.

In order to study the distribution of the anomalous emission in velocity and in time, we decided to subtract the "expected" helium recombination line profile from each spectrum. The "expected" profile was derived from Table 2, which summarizes all published values of the relative line strengths, velocities, and widths of the helium and hydrogen recombination lines in W49A. Included in Table 2 is our data from epochs 1973.35 and 1974.66 (85α), 1975.76 (100α) and 1975.044 (101α), when contamination by anomalous emission appears to be smallest and the baselines are most reasonable.

Table 2: Comparison of Hydrogen and Helium Recombination Lines

α-set	Beam Size	$T_L(He)/T_L(H)$ %	$V_{LSR}(He)-V_{LSR}(H)$ km s^{-1}	$\Delta V(He)/\Delta V(H)$*	Reference
85	3'	9.2 ± 0.4	0.6 ± 0.6	0.72 ± 0.04	1
85	3	8.7 ± 0.8	1.1 ± 1.1	0.60 ± 0.03	2
85	2.7	10.1 ± 0.3	-1.1 ± 0.5	0.77 ± 0.03	3
86	3.5	10.2 ± 0.5	2.2 ± 0.5	0.80 ± 0.05	4
94	4.2	10.6 ± 0.9	0.0 ± 0.8	0.67 ± 0.05	1
100	4.5	10.6 ± 0.5	0.7 ± 0.5	0.70 ± 0.05	3
101	4.6	10.2 ± 0.3	-0.1 ± 0.4	0.72 ± 0.04	3
109	6	8.9 ± 3.6	3.0 ± 1.2	0.63 ± 0.06	5
109	4	10.6 ± 1.5	-	0.81 ± 0.11	6
109	6	8.9 ± 2.0	-2.6 ± 1.8	0.83 ± 0.09	2
166	8.5	8.2 ± 1.4	2.2 ± 1.3	0.72 ± 0.11	7

*Corrected for instrumental broadening

References: 1. Chaisson & Ball 1971; 2. Churchwell et al. 1974; 3. Present results; 4. Lichten et al. 1979; 5. Reifenstein et al. 1970; 6. Mezger et al. 1970; 7. Pankonin et al. 1977

Keeping in mind that some of the published data may be distorted by unrecognized anomalous emission, we adopt the values: $T_L(He) = 0.100\ T_L(H)$, $V_{LSR}(He) = V_{LSR}(H)$, and $\Delta V(He) = 0.72\ \Delta V(H)$. Figure 2a shows the baseline-corrected spectra for each epoch at 85α with the expected helium profile superimposed, and Figure 2b shows the residuals after removal of the helium emission. Figures 3a and 3b present the 101α, 92α and 100α data in the same way. In most cases, the helium profile is directly related to the simultaneously observed hydrogen profile. For the Haystack data, the hydrogen profile was observed at the same average epoch as the helium profile, but poor weather

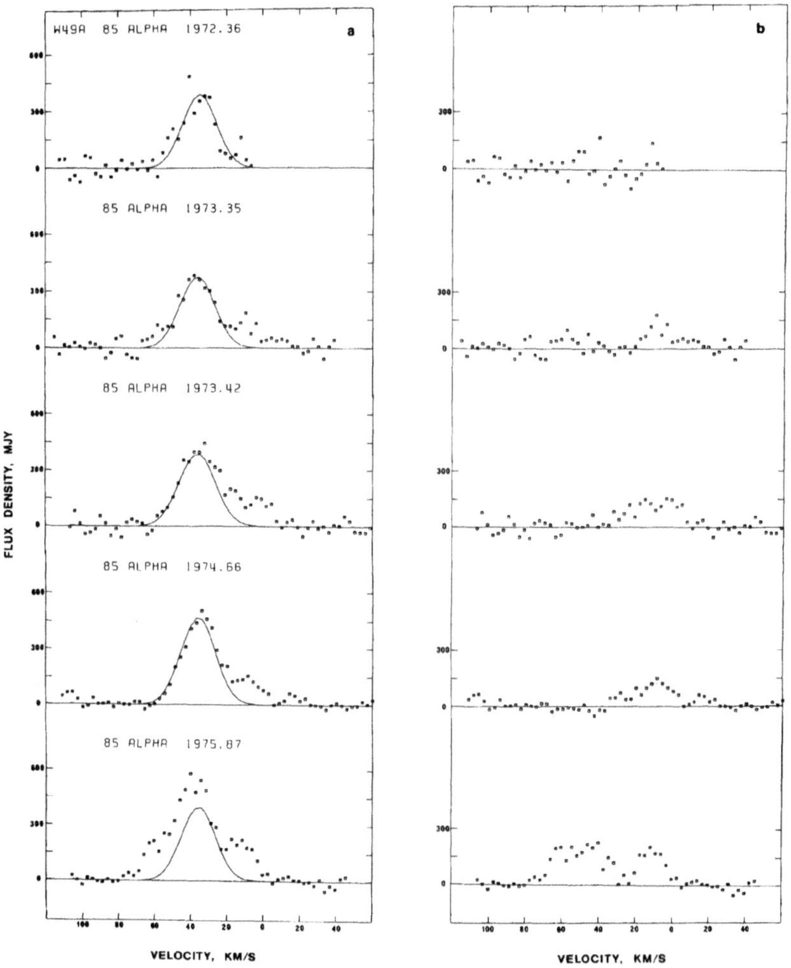

Fig. 2. a) Baseline-corrected spectra at 85α, with the expected helium line profile superimposed. b) Residual emission after removal of baseline and helium emission.

conditions add a 5% uncertainty to the comparison of line strengths.

In Figures 2 and 3 we see three examples of possible changes in the residual emission with time. Between 1973.35 and 1973.42 there appears to be a broadening of the emission near 10 km s^{-1}. Since the lines are only at the 3.5 X rms level, they might be explained as baseline effects.

Over a period of 9 days in January 1975 (1975.044 to 1975.068), we see significant changes in the emission from 0 to 30 km s^{-1} at a level well above 4 X rms, and evidence of emission at 47 and 62 km s^{-1} at the 3.5 X rms level. These changes are difficult to explain as baseline effects, since the baselines for these two epochs are simple and well defined (see Figure 1). An alternative explanation would have the He92α line wider than the adopted profile by some 36%, a 4.2σ effect.

Finally, between 1975.76 and 1975.87, the 10 km s^{-1} emission became considerably stronger and more broad, and very strong emission appeared at 44 and 62 km s^{-1} at a level 9 X rms. The resolution of our data is not sufficient to determine whether we are seeing three broad components or several narrow, partially overlapping lines at 1975.87. However, as discussed above, we believe these components to be real.

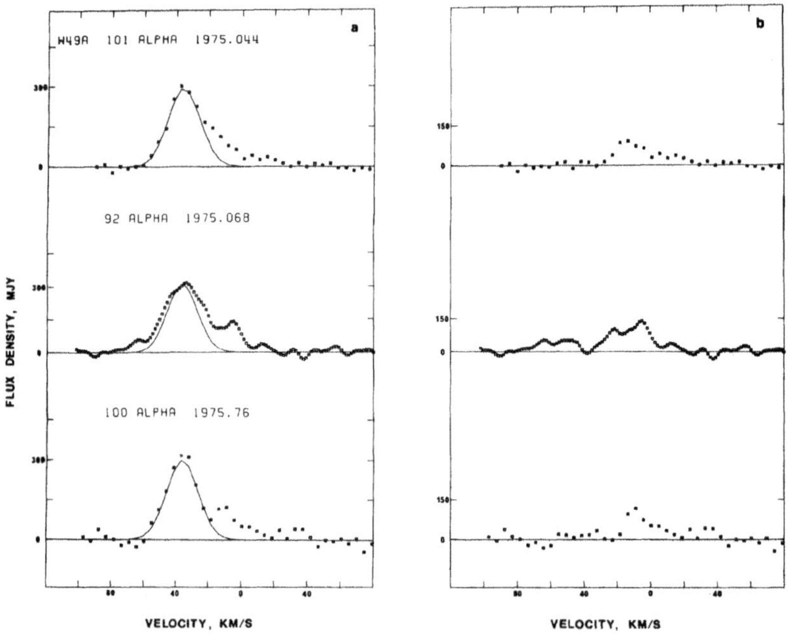

Fig. 3. a) Baseline-corrected spectra at 101α, 92α and 100α, with the expected helium profile superimposed. b) Residual emission after removal of baseline and helium emission.

4. DISCUSSION

From Figure 2 alone, we see evidence for time variations in the 85α spectrum, which we believe to be intrinsic to W49A. This by itself could be due to an unknown molecular maser line. However, emission at a carbon velocity of about 60 km s^{-1} has been seen at 85α, 86α, 92α, 94α, 247α, and 248α, but not seen at 100α, 101α, 109α and 166α. Hence we would need up to 6 previously unknown molecule masers, emitting by coincidence near six specific helium alpha lines.

On the other hand, if the anomalous emission really is due to electronic recombination, the time scales of the apparent variations are difficult to understand. Significant changes in a 9-day period correspond to a source size $< 2 \times 10^{16}$ cm and a brightness temperature $> 10^5$ K for a source of emission at a distance of 14 kpc. These limits assume that the line emission is triggered instantly by a disturbance travelling at the speed of light. However, even if the exciting UV flux was turned on instantaneously, it would take a long time to ionize enough material to produce significant line emission.

Clearly, more measurements are necessary to confirm and further define the time variations. We need better time resolution in a given α-set to study the time scales involved, and simultaneous observations in at least two α-sets to confirm that the anomalous emission is due to electronic recombination and to determine the wavelength dependence.

REFERENCES

Chaisson, E.J. and Ball, J.A.: 1971, Astrophys. J. 109, pp. 495-501.
Churchwell, E., Mezger, P.G., and Hutchmeier, W.: 1974, Astron. Astrophys. 32, pp. 283-308.
Lichten, S.M., Rodriguez, L.F., and Chaisson, E.J.: 1979, Astrophys. J. 229, pp. 524-532.
Mezger, P.G., Wilson, T.L., Gardner, F.F., and Milne, D.K.: 1970, Astrophys. Letters 6, pp. 35-37.
Pankonin, V., Parrish, A., and Terzian, Y.: 1973, Astrophys. J. 180, pp. L113-L116.
Pankonin, V., Thomasson, P., and Barsuhn, J.: 1977, Astron. Astrophys. 54, pp. 335-344.
Parrish, A.: 1975, Astrophys. J. 200, pp. 594-597.
Payne, H. and Terzian, Y.: 1976, Publ. Astron. Soc. Pacific 88, pp. 888-890.
Reifenstein, E.C., Wilson, T.L., Burke, B.F., Mezger, P.G., and Altenhoff, W.J.: 1970, Astron. Astrophys. 4, pp. 357-377.

PART IV

LARGE-SCALE PROPERTIES OF THE GALAXY

THE GALACTIC CENTER: RESULTS OF RADIO RECOMBINATION LINE SURVEYS

T. Pauls *
Max-Planck-Institut für Radioastronomie,
Auf dem Hügel 69, 5300 Bonn 1, Federal Republic of Germany

The detection of a radio recombination line from an astronomical radio source immediately tells us: (i) that the source contains ionized hydrogen, i.e. whether the source is thermal or nonthermal; (ii) the radial velocity of the source relative to some standard frame of reference; and (iii), when combined with a measurement of the continuum intensity, an estimate of the electron temperature of the source. In this chapter we will primarily be concerned with applying points (i) and (ii) to radio sources in the Galactic center.

Figure 1 shows the distribution of continuum intensity seen toward the Galactic center at 5 GHz, with a resolution of 2.6'. In this figure we see a number of discrete sources superimposed on a more extended background, and we realize that the detection of radio recombination lines will help us distinguish between supernova remnants (SNRs, nonthermal sources) and H^+ regions (thermal sources), and perhaps yield information about source kinematics and distribution along the line of sight.

Mezger and Höglund (1967) were the first to detect radio recombination lines from sources in the Galactic center, and the subsequent research is summarized in the Table 1. Some of this data has been discussed previously in reviews by Mezger (1974), Oort (1977), and Mezger and Pauls (1979).

RESULTS OF LARGE-SCALE SURVEYS

A. The Line-of-Sight Problem

Figure 2 shows the radio continuum from the Galactic center at 5 GHz superimposed on a photograph of the region taken in Hα (from Pauls and Mezger 1975). We see that there are several areas where the radio contours seem to be associated with sources of Hα emission, and this raises the

---

* Present address: I. Physikalisches Institut der Universität zu Köln, Universitätsstraße 14, 5000 Köln 41, F.R.G.

P.A. Shaver (ed.), Radio Recombination Lines, 159–170.

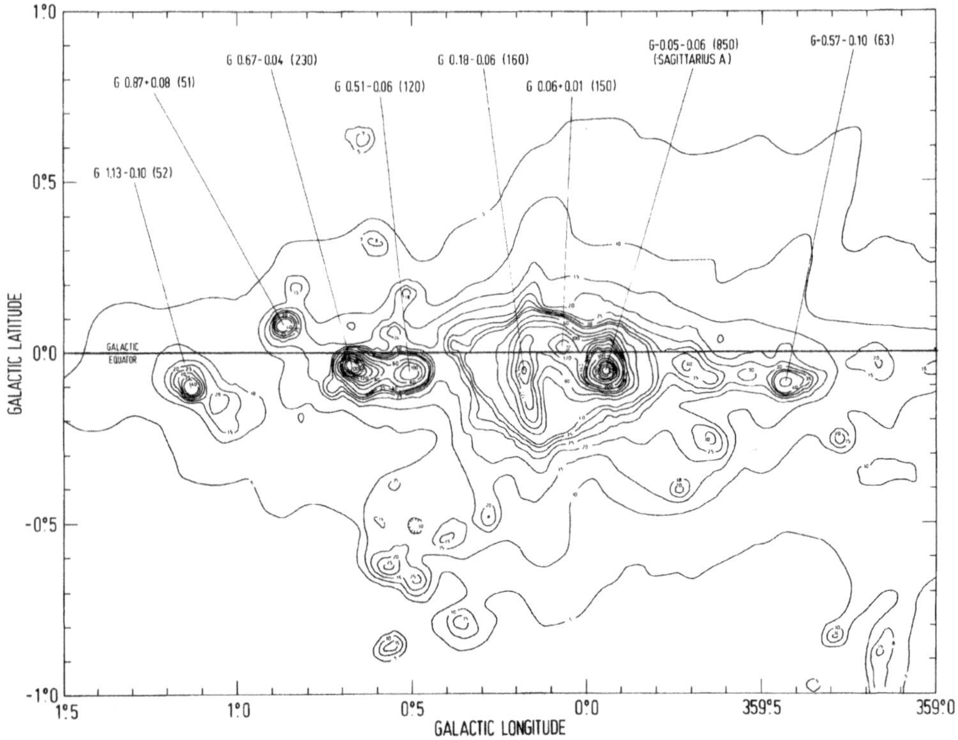

Fig. 1. *A map of the continuum radiation distribution at 5 GHz made with the Effelsberg 100-m telescope (Altenhoff et al. 1979). The half-power beam width is 2.6'.*

Table 1

Frequency (GHz)	Resolution (arc min)	Number of Positions Observed	Authors
5	6	4	Mezger and Höglund 1967
1.6	19	14	Lockman and Gordon 1973
5	4	8	Whiteoak and Gardner 1973
1.4	31 x 33	6	Matthews et al. 1973
5	6	32	Pauls and Mezger 1975
10.5	3	22	Pauls et al. 1976
5	4	44	Gardner and Whiteoak 1977
1.4	12	21	Kesteven and Pedlar 1977
5	2.6	32	Pauls and Mezger 1980

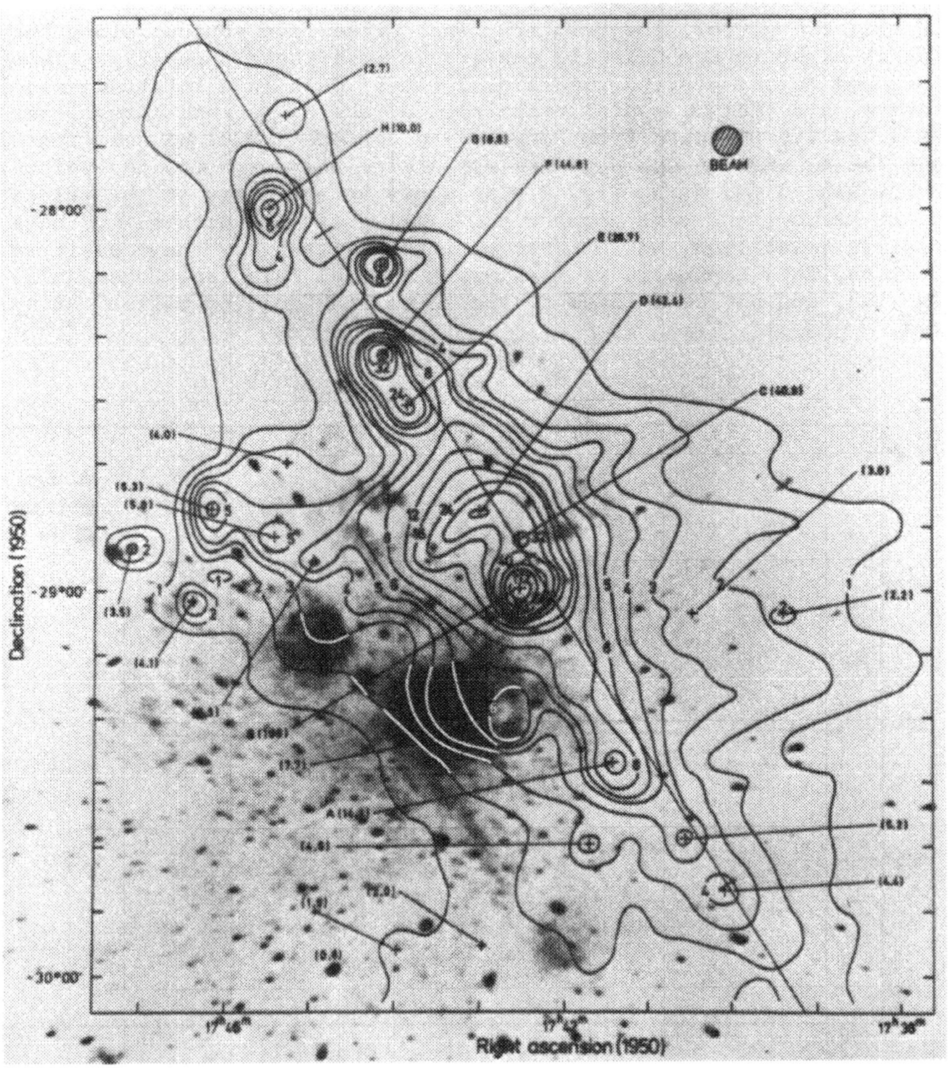

Fig. 2. The 5 GHz contour map of Whiteoak and Gardner (1973) combined with the Hα photograph of Cruvellier (1967), from Pauls and Mezger (1975).

question of whether some of the radio sources lie along the line of sight to the Galactic center. Pauls and Mezger (1975) found that their 5 GHz recombination line data showed two types of line emitting regions: one region emits broad lines ($\Delta v \sim 80$ km s^{-1}) with large radial velocities ($v_{LSR} = 50 \rightarrow 100$ km s^{-1}), the second type shows narrow lines ($\Delta v \sim 20$ km s^{-1})

and radial velocities near zero. Pauls and Mezger suggested that the broad, high-velocity lines originate from the Galactic center sources and that the narrow, low-velocity lines arise from sources along the line of sight to the Galactic center. To make their suggestion plausible, Pauls and Mezger plotted (see Fig. 3) their data in a latitude-velocity diagram, and include radial velocities of the Hα regions (Cruvellier 1967) and 1.6 GHz data from Lockman and Gordon (1973). We see from Fig. 3 that the Hα and 1.6 GHz data are associated only with the low-velocity, narrow-line 5 GHz data. Fig. 3 also shows an asymmetry in the velocity of the narrow-line data about b = 0°. All positions with b > 0° have negative velocities, while those positions with b < 0° have positive velocities. This asymmetry is also present in HI data (see Cohen 1975, Fig. 2k), and may be related to the so-called "rolling motion" of spiral arms.

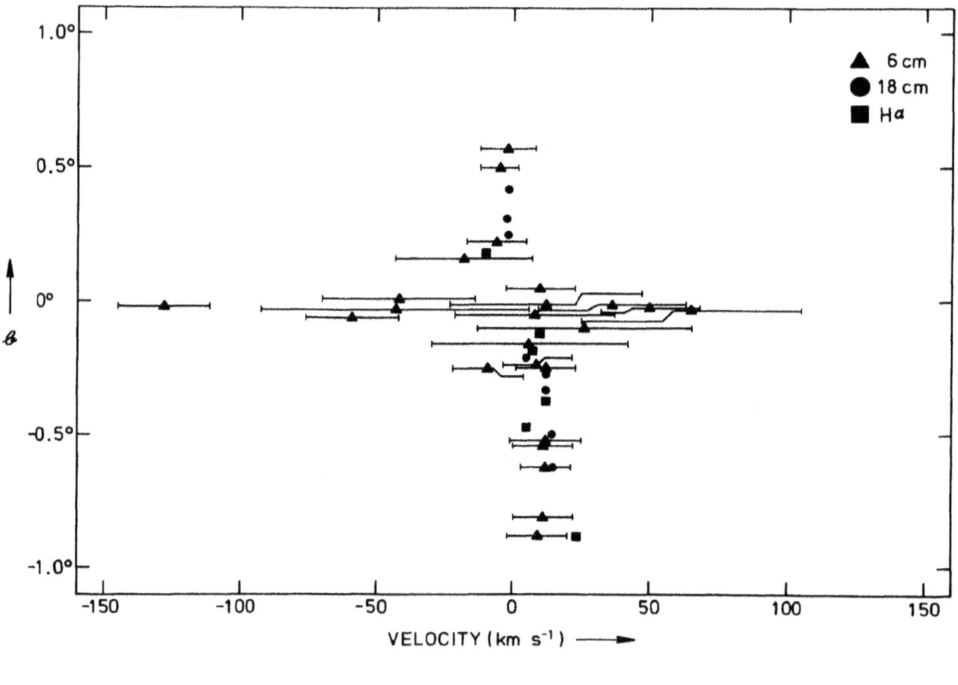

Fig. 3. Latitude-velocity diagram showing 5 GHz and 1.6 GHz recombination line data and Hα velocities. The horizontal bars are the full widths of the 5 GHz lines at half maximum intensity (Pauls and Mezger 1975).

B. Extended Sources of Emission

Surveys of the continuum radiation from the Galactic center show, in addition to compact sources, two regions of extended emission which are sketched in Figure 4. The larger source (labeled EI) is 90' x 36' ($\Delta l \times \Delta b$) at the half-power points, has a flux, $S_5 \sim 300$ Jy at 5 GHz, and an emission measure, EM $\sim 10^4$ cm^{-6} pc (Downes and Maxwell 1966). The second source (EII) appears on high resolution maps; it has a size of 38' x 22', a flux of $S_5 \sim 180$ Jy, and EM $\sim 3 \times 10^4$ cm^{-6} pc (Mezger 1974, Schmidt 1978).

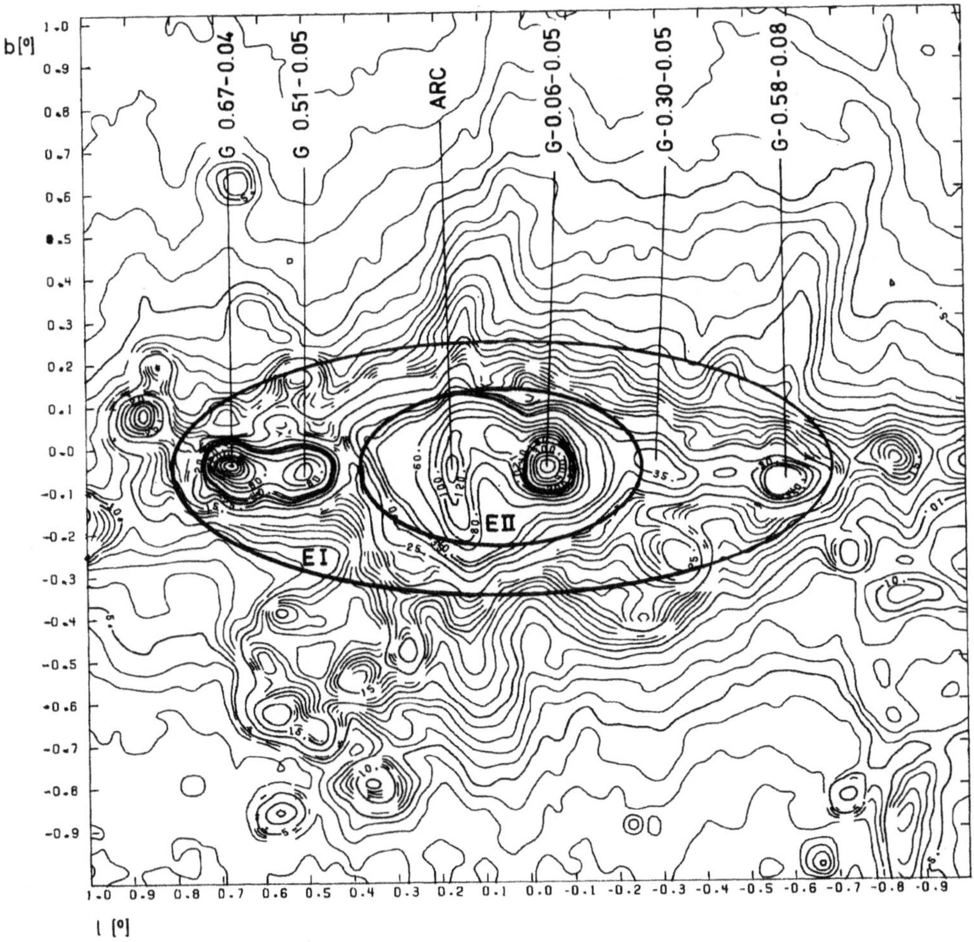

Fig. 4. The extended Galactic center sources EI and EII superimposed on 5 GHz, 100-m telescope contours (Schmidt 1978).

Radio recombination line emission attributed to the source EI has been reported by Mezger et al. (1974), Pauls and Mezger (1975), Kesteven and Pedlar (1977). This source shows broad, $\Delta v \sim 50 \rightarrow 100$ km s^{-1}, lines with a systematic longitude variation shown in Figure 5. This figure also shows the rotation curve of the HI nuclear disk, and we see that the ionized gas rotates at a lower velocity than that needed to be in dynamical equilibrium with the gravitational field of the stars near the nucleus.

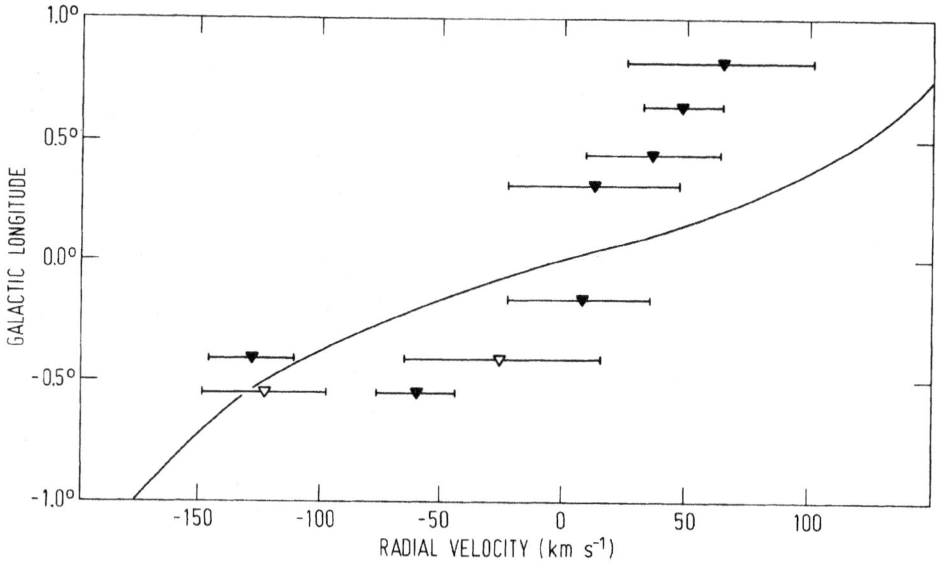

Fig. 5. Recombination line data from the extended thermal source (EI); horizontal bars are the full widths of the lines at half maximum intensity. The solid curve represents the HI rotation curve (extrapolated to positive longitudes).

The radio recombination line emission from EII was probably first detected by Roberts and Lockman (1970), although its nature was not realized at the time. Kesteven and Pedlar (1977) and Rodriguez and Chaisson (1978, 1979b) report recent observations of broad-line emission from this source, and the latter authors find evidence that this emission occurs under conditions near local thermodynamic equilibrium (LTE).

C. The Arc

Within ~ 15' of Sgr A is a group of sources which form an arc-shaped distribution on continuum maps made with a resolution of a few arc minutes or better (cf. Altenhoff et al. 1979). Figure 6 shows the highest resolution single-dish map, made at 10.7 GHz with a resolution of 77", and we see that there are a number of compact sources superimposed on more extend emission. Radio recombination line studies (Pauls et al. 1976, Gardner and Whiteoak 1977, Pauls and Mezger 1980) show that most emission comes from around b = 0, l = 6' and at negative radial velocities ($V_{LSR} \sim -40$ km s^{-1}), with some systematic motion around l = 0.1^o, b > 0. A comparison of the total integrated line intensity and the radio continuum is given in Figure 7. If the electron temperature is fairly uniform, then the solid contours in Fig. 7 are proportional to the emission measure, and therefore, to the thermal emission; while the dashed contours denote the sum of thermal and nonthermal emission. Thus, Fig. 7 shows that within the Arc the spatial distribution of thermal and nonthermal radiation is

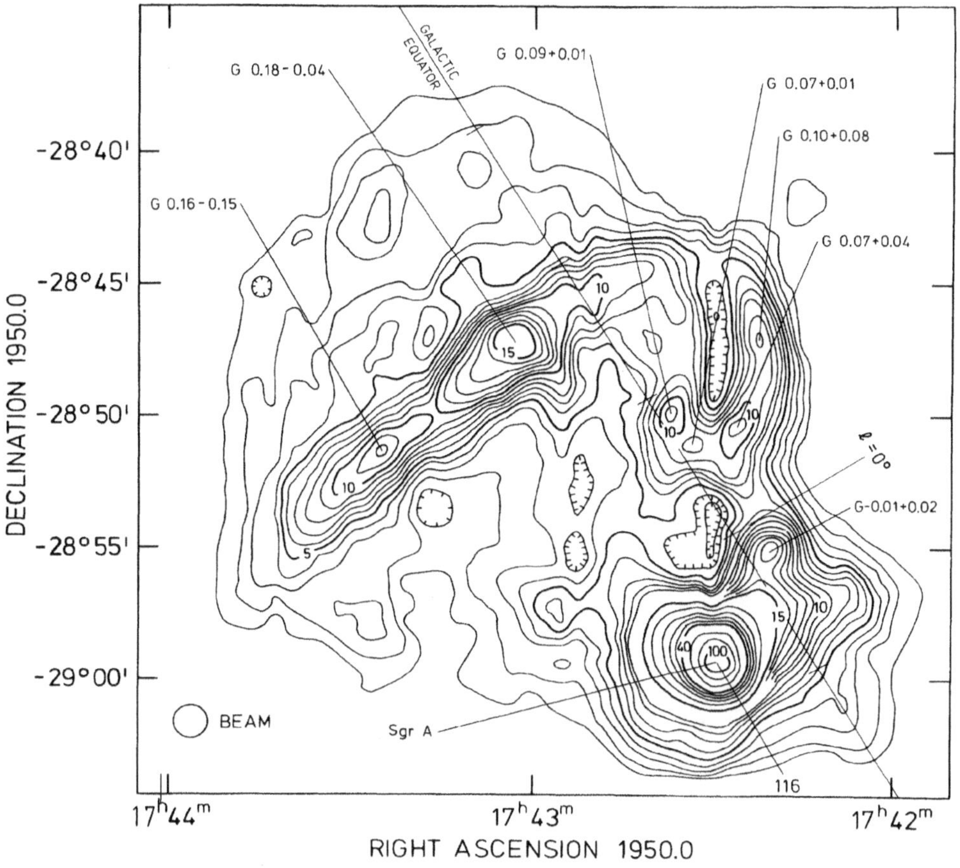

Fig. 6. Contour map of the continuum emission in the region near Sgr A, made with the 100-m telescope at 10.7 GHz (Pauls et al. 1976).

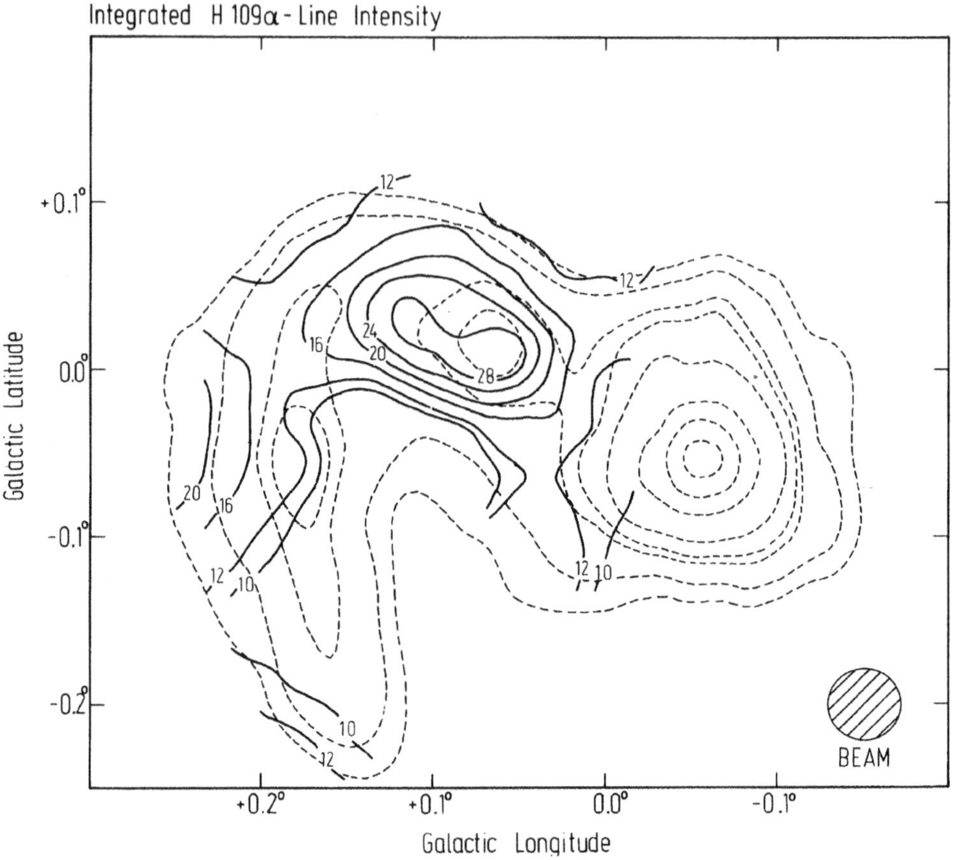

Fig. 7. The total integrated radio recombination-line intensity at 5 GHz (solid contours) superimposed on the 5 GHz continuum distribution (dashed contours). Observations with the 100-m telescope (Pauls and Mezger 1980).

different. In fact, the source G 0.16 - 0.15 in Fig. 6 is believed to be a supernova remnant, because no recombination lines have been detected toward it (Pauls et al. 1976) and appears stronger than G 0.18 - 0.04 on the 0.408 GHz map of Little (1974).

RADIO RECOMBINATION LINE EMISSION FROM SAGITTARIUS A

Radio recombination line emission from the direction of Sagittarius A (Sgr A), the radio source associated with the Galactic nucleus, was first detected at 1.4 GHz with a resolution of ∿ 20' (Riegel and Kilston 1970, Roberts and Lockman 1970), after attempts at 5 GHz with a resolution of ∿ 6' (Mezger and Höglund 1967, Reifenstein et al. 1970) were unsuccessful. These spectra show a relatively narrow line ($\Delta v \sim 30$ km s^{-1}) at $V_{LSR} \sim 0$ km s^{-1} superimposed on a much broader feature ($\Delta v \sim 100$ km s^{-1}), whose velocity centroid is about $V_{LSR} = -20 \rightarrow -40$ km s^{-1} (see Roberts and Lockman 1970, Lockman and Gordon 1973, Brown and Balick 1973 and Kesteven and Pedlar 1977). It is now felt that the narrow line arises from stimulated emission of the line-of-sight gas by the strong continuum radiation of Sgr A (see Pedlar and Davies, this volume); and that the broad line comes from the Arc, which makes a significant contribution at resolutions of ∿ 20'.

Spectra showing radio recombination line emission thought to arise from within Sgr A itself were first reported by Pauls et al. (1974). The line emission is believed to originate in an H^+ region surrounding the nucleus of the Galaxy. The suggestion that there was an H^+ region at the Galactic center was made by Downes and Martin (1971) based on broad-band continuum interferometric measurements. In addition to the thermal source (called Sgr A West), Downes and Martin also identified a nonthermal source (Sgr A East) about 2' away, which is probably a supernova remnant. These sources are shown in Figure 8, in which the 5 GHz data from several interferometers has been combined to give an effective resolution of 6.3" x 34" (Ekers et al. 1975).

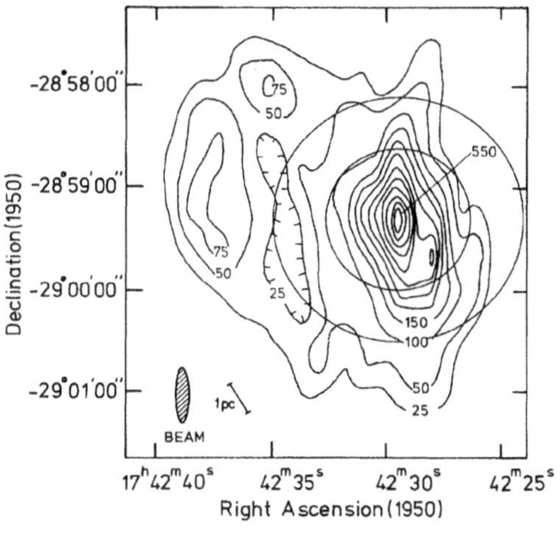

Fig. 8. The Galactic nucleus at 5 GHz observed with a resolution of 6.3" x 34" (Ekers et al. 1975). Circles represent the beam of the 100-m telescope at 4.8 GHz (2.6') and 8.6 GHz (1.5').

The spectrum of Pauls et al. (1974) showed a very broad line ($\Delta v \sim 200$ km s^{-1}) centered near $V_{LSR} = 0$ km s^{-1}, and was interpreted as a blend of He109α, H109α and H137β. However, subsequent observations have shown that the line must be a blend of three hydrogen lines at different velocities, instead. Figure 9 shows spectra of H110α and H91α taken with resolutions of 2.6' and 1.5', respectively (shown as circles centered on Sgr A West in Fig. 8. The H110α spectrum clearly shows a blended line, but

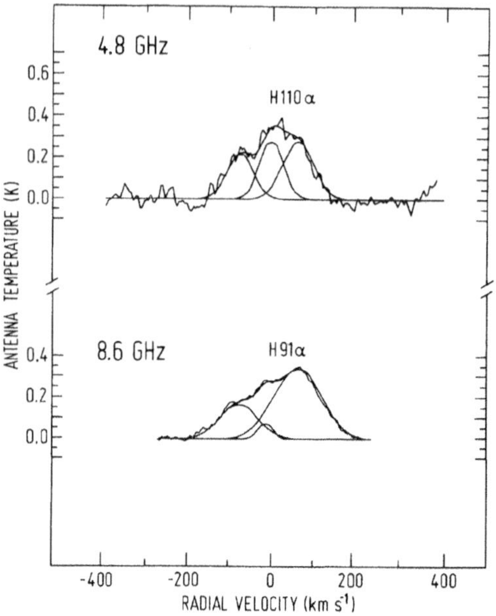

Fig. 9. The H110α and H91α radio recombination line spectra of Sgr A made with the 100-m telescope.

there is no other strong recombination line near H110α (as H137β is near H109α), so the blend must be due to different velocity components. This is confirmed by the H91α observations as can be seen by the comparison in Table 2. H110α and H91α show three velocity components: one near zero velocity and two symmetrically placed about zero at ∿ ± 70 km s⁻¹. No He^+ line is present (this line would appear at -120 km s⁻¹ on both spectra); but, for a normal helium-to-hydrogen abundance ratio of 10% by number, He^+ would be below our detection limit. The line near 0 km s⁻¹, which is about three times stronger at 4.8 GHz than at 8.6 GHz, probably arises from simulated emission of the thermal part of the continuum halo in which Sgr A West and Sgr A East are imbedded (see Pauls et al. 1976 and Schmidt 1978, for a discussion of this source).

Table 2 [a)]

Line	Velocity Component	Line Antenna Temperature	Radial Velocity (km s⁻¹) [b)]	Line width (km s⁻¹) [c)]
H91α	1	0.18 ± 0.01	-76 ± 10	110 ± 15
	2	0.08 ± 0.03	-10 ± 10	43 ± 15
	3	0.38 ± 0.01	+68 ± 10	140 ± 10
H110α	1	0.22 ± 0.10	-73 ± 20	74 ± 30
	2	0.30 ± 0.10	+ 2 ± 20	65 ± 35
	3	0.30 ± 0.10	+67 ± 30	85 ± 30

a) *all errors are 3σ (Pauls and Mezger 1977).*
b) *radial velocity relative to the Local Standard of Rest (LSR).*
c) *full width at half maximum intensity for a Gaussian line-shape.*

Recent spectra with the Effelsberg 100-m telescope at 15 GHz and 23 GHz (resolutions of 55" and 43", respectively) show that the + 70 km s^{-1} line becomes dominant and that its width increases with increasing frequency (Pauls and Mezger, unpublished). This seems to conflict with the spectra of Rodriguez and Chaisson (1979a) which peak near 0 km s^{-1}; but the difference may simply be due to the lower angular resolution used by these authors.

The discovery of radio recombination lines from Sgr A West has lead to searches for line emission from other ionized elements. The most useful of these has proved to be the 12.8 μm fine-structure line of Ne^+, first detected by Aitken et al. (1974, 1976). With high spectral and spatial resolution, Ne^+ can be used to probe the dynamics of the central parsec of the Galaxy (Wollman et al. (1976, 1977), Lacy et al. 1979; see also the review by Geballe, 1979). These infrared data, combined with the radio data, should increase our understanding of the entire Galactic center region.

REFERENCES

Aitken, D.K., Jones, B., Penman, J.M.: 1974, Monthly Notices Roy. Astron. Soc. 169, 35 P
Aitken, D.K., Griffiths, J., Jones, B., Penman, J.M.: 1976, ibid 174, 41 P
Altenhoff, W.J., Downes, D., Pauls, T., Schraml, J.: 1979, Astron. Astrophys. Supp. 35, 23
Brown, R.L., Balick, B.: 1973, Astrophys. J. 185, 843
Cohen, R.J.: 1975, Monthly Notices Roy. Astron. Soc. 171, 659
Cruvellier, P.: 1967, Ann. Astrophys. 30, 1059
Downes, D., Maxwell, A.: 1966, Astrophys. J. 146, 653
Downes, D., Martin, A.H.M.: 1971, Nature 233, 112
Ekers, R.D., Goss, W.M., Schwarz, U.J., Downes, D., Rogstad, D.H.: 1975, Astron. Astrophys. 43, 159
Gardner, F.F., Whiteoak, J.B.: 1977, Proc. Astron. Soc. Aust. 3, 150
Geballe, T.R.: 1979, Sci. Am. 241, 60
Kesteven, M.J., Pedlar, A.: 1977, Monthly Notices Roy. Astron. Soc. 180, 731
Lacy, J.H., Baas, F., Townes, C.H., Geballe, T.R.: 1979, Astrophys. J. Lett. 227, L17
Little, A.G.: 1974, in Galactic Radio Astronomy, Proc. IAU Symp. No. 60, eds. F.J. Kerr and S.C. Simonson III, p. 491
Lockman, F.J., Gordon, M.A.: 1973, Astrophys. J. 182, 25
Matthews, H.E., Davies, R.D., Pedlar, A.: 1973, Monthly Notices Roy. Astron. Soc. 165, 173
Mezger, P.G., Höglund, B.: 1967, Astrophys. J. 147, 490
Mezger, P.G.: 1974, Proc. ESO/SRC/CERN Conf. Prog. New Large Telescopes, p. 79
Mezger, P.G., Churchwell, E.B., Pauls, T.A.: 1974, in Stars and the Milky Way System, ed. L.N. Mavridis, p. 140
Mezger, P.G., Pauls, T.: 1979, in The Large-Scale Structure of the Galaxy, Proc. IAU Symp. No. 84, ed. W.B. Burton, p. 357

Oort, J.H.: 1977, Ann. Rev. Astron. Astrophys. 15, 295
Pauls, T. Mezger, P.G., Churchwell, E.: 1974, Astron. Astrophys. 34, 327
Pauls, T., Mezger, P.G.: 1975, Astron. Astrophys. 44, 259
Pauls, T. Downes, D., Mezger, P.G., Churchwell, E.: 1976, Astron. Astrophys. 46, 407
Pauls, T., Mezger, P.G.: 1977, Bull. Am. Astron. Soc. 9, 286
Pauls, T., Mezger, P.G.: 1980, Astron. Astrophys., in press
Reifenstein III, E.C., Wilson, T.L., Burke, B.F., Mezger, P.G., Altenhoff W.J.: 1970, Astron. Astrophys. 4, 357
Riegel, K.W., Kilston, S.D.: 1970, Astrophys. J. 159, L155
Roberts, M.J., Lockman, F.J.: 1970, Astrophys. J. 161, 877
Rodriguez, L.F., Chaisson, E.: 1978, Monthly Notices Roy. Astron. Soc. 184, 145
__________, 1979a, Astrophys. J. 228, 734
__________, 1979b, ibid 231, 697
Schmidt, J.: 1978, unpublished Ph.D. thesis, Bonn University
Whiteoak, J.B., Gardner, F.F.: 1973, Astrophys. Lett. 13, 205
Wollman, E., Geballe, T.R., Lacy, J.H., Townes, C.H., Rank, D.M.: 1976, Astrophys. J. Lett. 205, L5
__________, 1977, ibid 218, L103

LOW FREQUENCY RECOMBINATION LINES

A. Pedlar and R. D. Davies
Nuffield Radio Astronomy Laboratories
Jodrell Bank, Macclesfield, Cheshire, England

1. INTRODUCTION

Low frequency recombination lines will be defined for the purpose of this paper, to be those transitions with frequencies significantly less than 1 GHz. Most radio recombination line observations have been carried out at high frequencies, as not only can higher angular resolution be obtained, but also the signal-to-noise ratio is higher for small diameter, high emission measure objects and there are less problems with the optical depth of the free-free emission.

Low frequency radio recombination lines are useful for investigating the properties of extended, low density ionised gas. They are also useful for investigating deviations from local thermodynamic equilibrium (LTE) of the upper levels of the emitting atoms. This can result in a weak maser action which can significantly affect the intensities of the lines. This effect is particularly interesting when the line-emitting gas is in front of a strong continuum source, and can result in strong amplification of the line intensity. Theoretical calculations have suggested that very large enhancements may occur at low frequencies if the lines are emitted from cool, partially ionised gas. However, the present observational evidence for the presence of such gas rests largely on lines from elements such as carbon, and it appears that such gas may only be present in localised regions at the interface between ionised and neutral gas.

In Table 1 we have summarised a selection of observations of low frequency hydrogen recombination lines. Most of these have been observed on the Arecibo 305m or the Jodrell Bank 76m radio telescopes. The main difficulty in observing these lines is the presence of man-made interference which has to be eliminated by careful filtering in the radio receiving system and also by selecting periods when man-made interference is at a minimum (e.g. Christmas Day!). At these low frequencies a number of α transitions may be available in the pass-band of a given receiver, and hence, unlike other spectral

P. A. Shaver (ed.), Radio Recombination Lines, 171–183.

lines, interference can often be avoided by observing an adjacent line. The narrow weak lines from carbon are particularly difficult to distinguish from interference, and only after careful analysis can their profile parameters be established.

Table 1 A SELECTION OF LOW FREQUENCY HYDROGEN RADIO RECOMBINATION LINES

Transition	Frequency (MHz)	Sources	References
352α	150	Cas A, W80, Sgr A (upper limits)	1,2
300α	242	SgrA, W43	3,4
274α/275α	314/318	W49, W51	5
271α	328	Sgr A, W35, W43	3
257α	386	M17, W43, W51, W16, W80	6
253α	404	W80	7
252α	408	Sgr A, W35, W43	3,8
248α	428	3C291, W49B, G44.2, W51	9,10
221α	606	W51	10
220α	613	W1, W3, W10, W12, W16, W29, W35, W37, W43, W51, W80.	11

1. Shaver, Pedlar & Davies (1976)
2. Hart et al. (1980)
3. Pedlar et al. (1978)
4. Casse & Shaver (1978)
5. Parrish et al. (1977)
6. Gordon et al. (1974)
7. Penfield et al. (1967)
8. Batty (1974)
9. Pankonin (1975)
10. Parrish et al. (1972)
11. Pedlar & Davies (1972)

2. OBSERVATIONS OF EXTENDED, LOW BRIGHTNESS HII REGIONS

The ratio of the recombination line temperature to free-free continuum temperature for an optically thin HII region emitting lines in LTE is given by

$$\Delta\nu \frac{T_L}{T_C} = \frac{2.33 \times 10^4}{a(\nu,\ T_e)} \times \frac{1}{(1 + N(H_e^+)/N(H^+))} \left[\frac{\nu_L}{\text{GHz}}\right]^{2.1} \times T_e^{-1.15} \qquad 1.$$

where $a(\nu, T_e)$ is a factor tabulated by Mezger and Henderson (1967), $\frac{N(H_e^+)}{N(H)}$ is the number ratio of helium ions to hydrogen ions, and $\Delta\nu$ is the half-power width in kHz. If an HII region is extended and larger than the telescope beam then $T_C \propto \nu^{-2.1}$ hence from equation (1) $T_L \propto \nu^{-1}$.

Consequently the line temperature observed from extended, low emission measure HII regions increases at low frequencies. Recombination lines from extended regions with $E \lesssim 10^4$ pc cm $^{-6}$ can be measured easily at frequencies $\lesssim$ 1.5 GHz, whereas they are very difficult to detect above 5 GHz. As many of these regions have diameters of several degrees, the modest beamwidths attainable at low frequencies are adequate to determine their properties. Parameters such as velocity, electron temperature, turbulent velocity and electron density can be deduced from the line centroids, intensities and widths.

Low frequency lines can also be used to determine the effect of stimulated emission on the derived electron temperature. As n increases the level population become closer to LTE, but the increased free-free optical depth at low frequencies results in more continuum emission and hence more stimulated emission. For an isolated, optically thin, uniform density, uniform temperature HII region the line brightness temperature is given by

$$T_L = T_e \left(b_m \tau_L^* - \frac{\tau_L \tau_c}{2}\right) \qquad 2.$$

where τ_c is the optical depth of the free free continuum, τ_L^* is the line optical depth assuming the levels to be in LTE and τ_L is the line optical depth taking into account non-LTE effects. They are related by the expression (Shaver 1976)

$$\tau_L = b_n \left(1 - 20.84 \frac{T_e}{\nu} \frac{\Delta b}{b_n}\right) \tau_L^* \qquad 3.$$

where b_n is the departure coefficient which relates the population of level n, to its value in thermodynamic equilibrium. At low frequencies the second term in parenthesis increases and can result in a negative value of τ_L, which when substituted into equation 2 results in an increased line temperature corresponding to line masering. Consequently when equation 1 is used to derive an electron temperature, the derived electron temperature T_e^* differs from the actual electron temperature and can be calculated for the case of a uniform temperature, uniform density HII region (see Dyson 1967) from

$$T_e^* = T_e \left(b_n \left(1 + \frac{\tau_c}{2} \frac{kT_e}{h\nu} \frac{d \ln b_n}{dn}\right)\right)^{-1/1 \cdot 15} \qquad 4.$$

This equation does not take into account stimulated emission caused by background continuum sources; this will be discussed in Section 4. In Fig. 1 we show the variation of the 'LTE electron temperature' T_e^* as a function of principal quantum number n for an HII region with

$E = 5000$ pc cm^{-6} and $T_e = 5000$ K. It can be seen that if $N_e \lesssim 10\ cm^{-3}$ there is a decrease of the LTE electron temperature at low frequencies which is the result of increasing stimulated emission at high n.

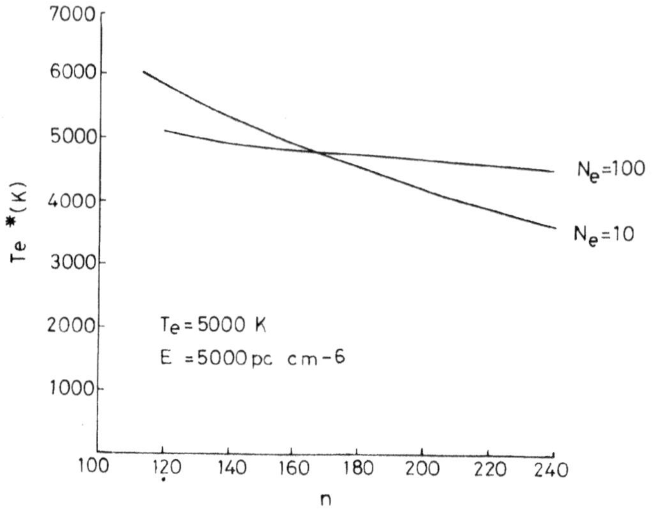

Fig. 1 The "expected" electron temperature (T_e*) derived using the LTE assumption, as a function of principal quantum number, n, if the actual electron temperature (T_e) = 5000 K.

W1 and W80 are examples of low brightness HII regions which give consistent LTE electron temperatures down to n = 220 and n = 257 respectively (Pedlar & Davies 1972, Pedlar & Matthews 1973, Gordon, Gordon & Lockman 1974, Pedlar 1980), and hence we can deduce from Fig. 1 that their LTE electron temperatures are approximately equal to their true electron temperatures and that $N_e \simeq 100\ cm^{-3}$. The 257α observations of W16 of Gordon et al. (1974) show a significant decrease in temperature compared with 166α and 220α observations, indicating that stimulated emission is important at this frequency. This implies that W16 has a lower mean density than W80, which does not show this effect. W35 is surprising in that there appears little evidence for line enhancement by stimulated emission even at frequencies as low as 328 MHz (Pedlar et al. 1978).

The lack of measurable pressure broadening in the low frequency lines can be used to set upper limits on the electron density of $<200\ cm^{-3}$ for W1, W16 and W80 (Pedlar & Davies 1971) and $<50\ cm^{-3}$ for W35 (Pedlar et al. 1978).

Further observations of low frequency recombination lines from low brightness HII regions would be useful in determining the importance of stimulated emission in such regions. So far the indications are that such effects are quite small and hence LTE electron temperatures for low density HII regions deduced from 166α lines (e.g. Pedlar 1980) are close to the actual electron temperatures.

3. OBSERVATIONS OF CONVENTIONAL HII REGIONS

Many of the more well-known HII regions contain small diameter, high emission measure components which emit strong recombination lines at high frequencies. At low frequencies these components become optically thick in the free-free continuum and the lines are weakened by pressure broadening. The recombination line radiation is also diluted by the large beams which are often used for low frequency observations. Consequently the low frequency lines which are detected from the direction of such HII regions are unlikely to originate in the high density components, and more likely represent emission from the low density outer envelopes of the HII regions or from separate extended low density HII regions. The latter situation would appear to apply in the direction of M17 (Gordon et al. 1974) and W49A (Parrish et al. 1977) where the low frequency lines are at considerably different velocities from the high frequency lines. In the case of W49A the foreground emission is centred at 60 km/s whereas the emission from W49A itself is at 10 km/s. The increasing strength of the 60 km/s component at higher n values is illustrated in Fig. 2.

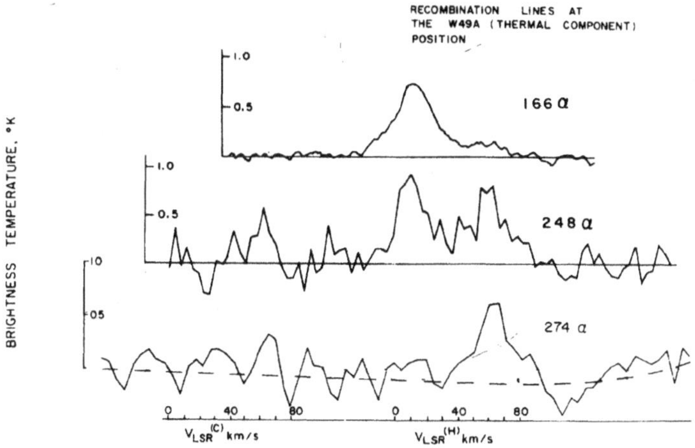

Fig. 2 Spectra obtained from the direction of W49A (from Parrish et al. 1977) observed with comparable angular resolution (10')

For many of the low frequency observations with $|\ell| < 40$ (e.g. Pedlar & Davies 1972, Batty 1974, Gordon et al. 1974) much of the recombination line emission detected could originate in the extended low density HII regions which give rise to the "diffuse" recombination lines at 1.5 GHz (Gordon & Gottesman 1971, Gordon & Cato 1972, Matthews et al. 1973). Clearly a low frequency recombination line survey of diffuse emission from the Galactic plane, which could be compared to the 166α surveys of Hart & Pedlar (1976) and Lockman (1976) would be of value in investigating the properties of the diffuse component.

Although the high density components of HII regions are unlikely to contribute significantly to low frequency recombination lines, their continuum emission can cause the lines from the foreground low density gas to be enhanced by stimulated emission. Thus derivations of the electron temperature from the line-to-continuum ratio are likely to be unreliable and can only be corrected if realistic models of the HII region are constructed. This was first attempted by Brocklehurst & Seaton (1972) to account for the lack of pressure broadening in the low frequency lines from Orion. In many HII regions observed with low angular resolution it is not possible to distinguish an increase in stimulated emission from a real decrease in the electron temperature of the low density gas. However, high resolution measurements of the component G49.1-0.6 in W51 show that the 248α line is definitely enhanced by stimulated emission (Parrish & Pankonin 1975).

Attempts to determine the densities of high density HII regions by measuring pressure broadening in high nα lines are made difficult by the increasing optical depth at low frequencies and the increase in stimulated emission. Consequently in normal HII regions the low frequency lines are emitted from different gas than that emitting at high frequencies. Although it is valid to place upper limits on the density of the tenuous gas emitting the low frequency lines, it is clearly incorrect to place constraints on the densities of high density components which are optically thick at low frequencies. For such objects pressure broadening can best be estimated by comparing the widths of α, β and γ lines which are close together in frequency but have different n values (Churchwell 1971, Lang & Willson 1978).

4. LINE ENHANCEMENT CAUSED BY BACKGROUND CONTINUUM SOURCES

a. Theory

In section 2 we have noted that the line intensity can be enhanced by stimulated emission generated by the free-free radiation from the HII region. It is equally possible for physically unconnected background sources to contribute to the stimulated emission if the low density HII regions lie between the sources and the observer (Dupree & Goldberg 1969). A further interesting possibility is that cold, partially ionised gas, which would be far from LTE, could produce extremely strong recombination lines at low frequencies (Shaver 1975).

To account for the presence of a background source, equation 2 is modified to include an extra term (Shaver 1976) giving

$$T_L = T_e \left(b_m \tau_L^* - \frac{\tau_L \tau_c}{2}\right) - \tau_L T_{BG} \qquad 5.$$

where T_{BG} is the brightness temperature of the background source. As we have discussed in section 2, τ_L can become negative, and hence the recombination line intensity is enhanced. At low frequencies T_{BG} can increase such that the first term can be neglected and hence

$$T_L \approx - \tau_L T_{BG} \qquad 6.$$

The large deviations from LTE predicted for cold, partially ionized gas (Shaver 1975) result in both large negative and positive values of τ_L. These lead to surprisingly intense recombination lines both in emission and absorption at low frequencies. In Fig. 3 we show the peak recombination line temperatures predicted by Shaver (1976) for a typical low density HII region and for cold partially ionized gas.

b. Hydrogen recombination lines from the direction of supernova remnants

In general, recombination lines are not seen in the direction of supernova remnants (Downes & Wilson 1974). However, the supernova remnants 3C391, W49B and W44 show line-of-sight weak recombination line emission at high frequencies (Cesarsky & Cesarsky 1973a,b; Bignell 1973). Cesarsky & Cesarsky, using the intensity of 92α, 157α and 166α lines from the direction of 3C391, deduced that the recombination line emission originated in cool partially ionized gas unrelated to the supernova remnant. Chaisson (1974) and Downes & Wilson (1974) pointed out that this interpretation was not unique, and that the data were more consistent with emission from line-of-sight low density HII regions at normal temperatures. Pankonin (1975) observed 248α lines from the directions of 3C391 and W49A. Although the lines showed evidence of considerable enhancement by stimulated emission, the intensities could not be reconciled with low temperature, partially ionized gas unless the electron density was greater than 10 cm^{-3}. The lines, however, could more readily explained in terms of a foreground HII region with an electron temperature of ≈4000 K, an emission measure of 10^3 pc cm^{-6} and electron densities of ≈10 cm^{-3}, thus removing the need to invoke a cold, partially ionized gas to explain the low frequency lines.

Shaver (1975) predicted that the most intense lines from cold, partially ionized gas would be observable at frequencies near 100 MHz. Accordingly a search was made for the 352α line (150 MHz) in the direction of Cas A (Shaver, Pedlar & Davies 1976), but no recombination lines were detected, implying ionization rates in clouds in front of the strong nonthermal radio source Cas A were as low as 3×10^{-17} s^{-1}.

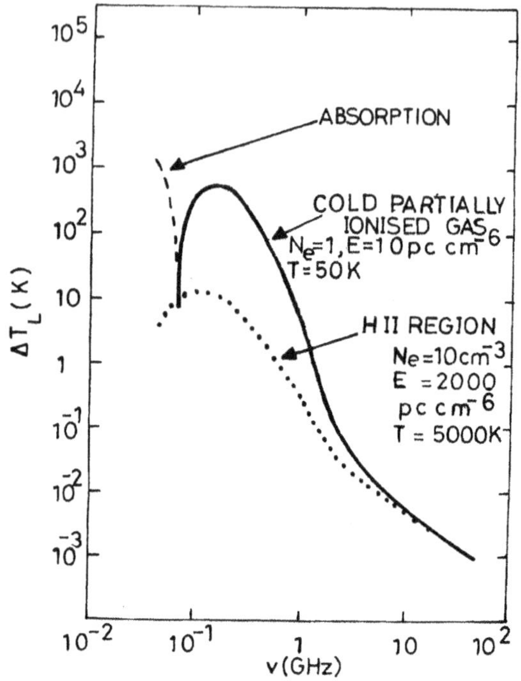

Fig. 3 Theoretical peak brightness temperatures of α lines plotted against frequency for a cold, partially ionized cloud and a typical extended HII region, in front of a background source with T_{BG} = 100 K. at 1 GHz and α = -2.5. (Adapted from Shaver 1976).

c. Low frequency recombination lines from the direction of the Galactic centre

The most intense low frequency lines yet detected are emitted from the direction of the Galactic centre (Pedlar et al. 1978). This is not too surprising as the long path length between the Sun and the Galactic centre results in a high probability that some of the gas will be out of LTE and produce significant stimulated emission in front of the strong continuum background from the Galactic centre. Pedlar et al. have detected 3 hydrogen lines below 500 MHz from this direction: the 252α (408 MHz), 271α (328 MHz) and 300α (240 MHz). The 300α line has been confirmed by Casse & Shaver (1977). Attempts to measure the 352α line (150 MHz) by Hart et al. (1980) have resulted in setting an upper limit of 4 K.

There is little doubt that in the direction of the Galactic centre we are witnessing the amplification of a spectral line by stimulated emission. In Fig. 4 we show the dependence of the line temperature on wavelength.

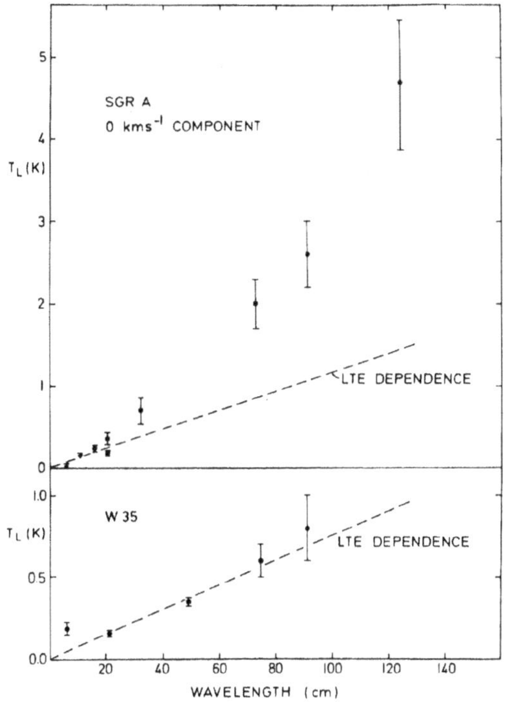

Fig. 4 The variation of line temperature with wavelength for the 0 km/s component of Sgr A and for W35. The dashed line shows the dependence of an optically thin, extended cloud emitting recombination lines in LTE. (From Pedlar et al. 1978).

If the lines were emitted in LTE from an optically thin HII region covering the observing beam, then the line temperature would be proportional to wavelength as explained in Section 2. It should be noted that the 300α observations of Casse & Shaver (1977) show the emission to be considerably more extended than the 70' resolution used by Pedlar et al. (1978). Fig. 4 shows that the line temperature increases more rapidly with wavelength than the LTE case. Even if we assume that stimulated emission is negligible above 1.5 GHz, then the 300α line must be amplified by at least a factor of 3.

This emission from line-of-sight gas in the direction of the Galactic centre was for some years believed to be from cool, partially ionized gas. This was mainly because of upper limits reported to the 200β/159α and 208β/166α line intensity ratios (Lockman & Gordon 1973, Brown & Balick 1973) which indicated extreme departures from LTE and could only be explained by cool, partially ionized gas. Even so this gas had rather extreme properties with electron densities of order 10 cm^{-3} and a path length of order 0.1 pc (Shaver 1976), despite the fact that its lateral extent was several tens of parsecs. Such gas would be expected to produce very strong absorption lines at low frequencies, rather than moderately strong lines in emission. Although some attempts were made to fit the observations by two clouds, one of which cancelled out the low frequency absorption lines from the cold gas, it was felt that such models were physically unrealistic and hence Shaver (1977) reobserved the 200β/159α intensity ratio. The 200β line was easily detected and found to have an intensity of 24% of the 159α line, which is close to the LTE value. Thus we can only conclude that the previous upper limits were in error. The removal of this constraint made it possible to interpret the low frequency lines in terms of conventional, low density HII regions.

As the 300α lines appeared to originate from an area several degrees across (Pedlar et al. 1978, Casse & Shaver 1977), a simple model can be constructed in which the continuum radiation from the direction of the Galactic centre is amplified by a uniform screen which extends over the entire source. Thus if we assume that stimulated emission dominates at frequencies below 1.5 GHz, and that most of the galactic continuum originates behind the HII region, then the line optical depth can easily be calculated from equation 6 to be

$$\tau_L = - T_L / T_C$$

where T_C is the continuum temperature in the direction of the Galactic centre. This ratio was surprisingly constant for lines between 240 MHz and 1700 MHz suggesting that an acceptable fit might be achieved with just one component. In Fig. 5 we show the fit to the data with a component with T_e = 5000 K, N_e = 10 and E = 1500 pc cm^{-6}. This fit is by no means unique and the data can also be fitted by two or more components with slightly different densities and emission measures. The presence of a significant amount of low density ionized gas ($N_e \lesssim 1$) has been ruled out by upper limits on the 352α line (Hart et al. 1980) which suggest that $N_e > 10$ cm^{-3}.

The angular extent of the 252α and 166α emission has been determined by Hart et al. and appears to be approximately 2°.5 x 2°.5. However, significant changes in the cloud properties occur over this region and it is possible that we are observing at least two distinct HII regions whose outer, low density envelopes overlap in front of Sgr A. An absorption feature with a similar angular extent to that of the recombination line emission is present in the 29.9 MHz continuum map of Jones & Finley(1974); it indicates the presence of an HII region with

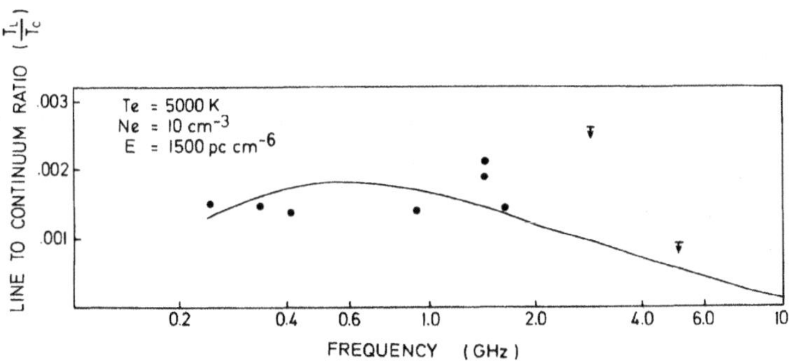

Fig. 5 The observed and predicted line to continuum ratios for a low brightness HII region in front of the Galactic centre (from Pedlar et al. 1978).

an emission measure of approximately 10^3 pc cm^{-6}. Typical low density HII regions have diameters of $\sim$30 pc (Churchwell 1975). Such an HII region would then be approximately 2 → 3 kpc distant placing it in the vicinity of the Sagittarium Arm.

Thus the mystery of the line-of-sight recombination lines from the direction of the Galactic centre is resolved in terms of an extended HII region or regions with conventional properties. The only evidence for cool partially ionized gas is found at the interface between HII and HI regions, as will be discussed below.

5. LOW FREQUENCY CARBON RECOMBINATION LINES

Recombination lines from elements heavier than hydrogen and helium are too weak to detect in normal HII regions because of their much lower abundances. The discovery of carbon recombination lines in such sources as W3, Orion A and NGC 2024 with intensities approaching those of the adjacent hydrogen lines posed a problem of interpretation. However, when a narrow hydrogen line at the corresponding velocity (and not necessarily at the velocity of the identified HII region) was found it became clear that these pairs of carbon and hydrogen lines were emitted in a weakly ionized cool HI cloud. The much higher fractional ionization of the carbon relative to the hydrogen was the result of its lower ionization potential and led to comparable carbon and hydrogen line intensities in these sources. These HI regions may lie near the boundary of normal HII regions from which low energy ionizing radiation enters or alternatively they may lie somewhere on the line-of-sight to the bright HII region. For the reasons given in section 4 such cool gas may show large departures from LTE and could generate strong low frequency recombination lines of any element which was suitably ionized. Table 2 lists observations of carbon recombination lines observed at frequencies below 1 GHz.

Table 2 OBSERVATIONS OF LOW FREQUENCY CARBON RECOMBINATION LINES

Transition	Frequency (MHz)	Sources	References
300α	242	Sgr A	1
274α	318	W49A, W51B (upper limits)	2
271α	327	Sgr A	1
252α	408	Sgr A	1
248α	428	W49A, W49B, W51A, W51B	3,4
221α	605	W49A, W51B (upper limits), W51A	4
220α	613	W10 (Orion A), W12 (NGC 2024)	5

1. Pedlar, Davies, Hart & Shaver (1978)
2. Parrish, Conklin & Pankonin (1977)
3. Pankonin (1975)
4. Parrish & Pankonin (1975)
5. Pedlar & Davies (1972)

The observations of W10 (Orion A) and W12 (NGC 2024) at 220α by Pedlar & Davies (1972) and the detection of 221α and 248α carbon line emission in components of W49 and W51 showed clearly that the carbon lines were increasing in intensity with wavelength in the manner expected for stimulated emission of the radiation from a background whose continuum radiation passed through a low temperature ionized gas. In the case of W10 and W12 the low temperature carbon and hydrogen gas differed in velocity from the background normal HII region by only a few km/s whereas in W49 the velocity difference was ∿50 km/s indicating that the low temperature gas was a few kiloparsecs in front of W49 itself. Parrish et al. (1977) found a maximum in T_L/T_C near the frequency of the 248α line for W49A and W51B and they were able to make a fit to the run of T_L/T_C with quantum number n using a temperature in the range 10 to 100 K and an electron density in the range 0.1 to 1 cm^{-3}.

Pedlar et al. (1978) detected significant carbon line emission from the direction of the Galactic centre in the 252α, 271α and possibly the 300α transitions. The carbon lines appear to have a completely different frequency dependence from the masering hydrogen lines in this direction as described in Section 4(c) and are unlikely to come from the same gas. In addition, the rapid decrease in the carbon line-to-continuum ratio with frequency precludes this emission arising from a high-temperature gas. The available data are compatible with the carbon lines originating in a foreground gas cloud with an electron temperature of about 20 K and electron densities between 0.5 and 1.0 cm^{-3} - conditions very similar to those responsible for the carbon lines in the direction of W49 and W51.

ACKNOWLEDGEMENTS

We wish to thank Dr. L. Hart for assistance with this paper, and Mrs Joan Adams for typing and correcting this manuscript at short notice.

REFERENCES

Batty, M.J., (1974), Mon.Not.R.astr.Soc., 168, 37P.
Bignell, R.C., (1973), Astrophys.J., 186, 889.
Brocklehurst, M. and Seaton, M.J., (1972), Mon.Not.R.astr.Soc. 157, 179.
Brown, R.L. and Balick, B. (1973), Astrophys.J. 185, 843.
Casse, J.L. and Shaver, P.A., (1977), Astron.Astrophys. 61, 805.
Cesarsky, D.A. and Cesarsky, C.J., (1973a), Astrophys.J. 183, L143.
Cesarsky, D.A. and Cesarsky, C.J., (1973b), Astrophys.J. 184, 83.
Chaisson, E.J., (1974), Astrophys.J., 189, 69.
Churchwell, E., (1971), Astron.Astrophys. 15, 90.
Churchwell, E., (1975), Proceedings of Mittelberg Symposium, Lecture Notes in Physics 42, Springer-Verlag, Berlin
Downes, D. and Wilson, T.L. (1974), Astron.Astrophys. 34, 133.
Dupree, A.K. and Goldberg, L., (1969), Astrophys.J. 158, L49.
Dyson, J.E., (1969), Astrophys.Sp.Sci., 4, 401.
Gordon, K.J., Gordon, C.P. and Lockman, F.J.(1974),Astrophys.J. 192, 337.
Gordon, M.A. and Cato, T., (1972), Astrophys.J. 176, 587.
Gordon, M.A. and Gottesman, S.T., (1971), Astrophys.J. 168, 361.
Hart, L., Ediss, G.A. and Pedlar, A. (1980) (in preparation).
Hart, L. and Pedlar, A., (1976), Mon.Not.R.astr.Soc. 176, 547.
Jones, B.B. and Finley, E.A. (1974), Australian J.Phys. 27, 687.
Lang, K.R. and Willson, R.F., (1978), Mon.Not.R.astr.Soc. 183, 5P.
Lockman, F.J., (1976), Astrophys.J. 209, 429.
Lockman, F.J. and Gordon, M.A. (1973), Astrophys.J. 182, 25.
Matthews, H.E., Davies, R.D. and Pedlar, A., (1973), Mon.Not.R.astr.Soc. 165, 173.
Mezger, P.G. and Henderson, A.P., (1967), Astrophys.J. 147, 471.
Pankonin, V., (1975), Astron.Astrophys. 38, 445.
Parrish, A., Conklin, E.K. and Pankonin, V. (1977), Astron.Astrophys., 58, 319.
Parrish, A., Pankonin, V., Heiles, C.E., Rankin, J.M. and Terzian, Y., (1972), Astrophys. J., 178, 673.
Parrish, A., and Pankonin, V., (1975), Astrophys.J., 198, 349.
Pedlar, A. (1980) (Submitted to Mon.Not.R.astr.Soc.)
Pedlar, A. and Davies, R.D., (1971), Nature 231, 49.
Pedlar, A. and Davies, R.D., (1972), Mon.Not.R.astr.Soc., 159, 129.
Pedlar, A., Davies, R.D., Hart, L. and Shaver, P.A., (1978), Mon.Not. R.astr. Soc., 182, 473.
Pedlar, A. and Matthews, H.E. (1973), Mon.Not.R.astr.Soc., 165, 381.
Penfield, H., Palmer, P. and Zuckerman, B. (1967), Astrophys.J.148, L25.
Shaver, P.A., (1975), Pramāna, 5, 1.
Shaver, P.A., (1976), Astron.Astrophys., 49, 1.
Shaver, P.A., (1977), Astron.Astrophys., 59, L31.
Shaver, P.A., Pedlar, A. and Davies, R.D. (1976), Mon.Not.R.astr.Soc., 177, 45.

LOW FREQUENCY RECOMBINATION LINE SURVEYS OF THE GALACTIC PLANE

Felix J. Lockman
National Radio Astronomy Observatory*
Green Bank, West Virginia 24944

Abstract. Weak centimetric recombination lines are observed at almost all locations near the galactic plane where the longitude is less than 50°. Most of the emission probably comes from extended outer parts of normal HII regions whose dense cores are prominent in the radio continuum. At least one component of the emission, however, comes from a more broadly distributed medium that extends over a few hundred parsecs, and that has a density $\sim$1 cm^{-3} and a temperature of a few thousand Kelvins or less. This gas is not associated with sites of current star formation. There is no evidence that the recombination line observations have detected a truly diffuse, low density, electron component of the interstellar medium.

I. INTRODUCTION

Ten years ago, in nearly simultaneous experiments, Gottesman and Gordon (1970) and Jackson and Kerr (1971) detected weak recombination lines from parts of the galactic plane where there were no bright HII regions. The broad and often complex line shapes suggested that the emission might be coming from distributed material rather than from a Stromgren sphere around a massive star. Similar lines were soon observed in the direction of low latitude supernova remnants (e.g. Cesarsky and Cesarsky, 1973) and toward some bright HII regions [the second line component seen by Goss (1968) in the spectrum of W43 probably falls in this class]. The discovery of these lines excited considerable interest, and their existence was taken by some as confirmation of the two-component interstellar medium implied by HI absorption measurements and pulsar dispersion measures. There was, however, no general consensus on which if either component of the ISM was being observed.

* The National Radio Astronomy Observatory is operated by Associated Universities, Inc., under contract with the National Science Foundation.

P.A. Shaver (ed.), Radio Recombination Lines, 185–204.

The lines are very weak (they are among the weakest lines observed in radio astronomy) and observations of them tend to be somewhat difficult and lengthy. Nonetheless, by the mid 1970's research in this topic had become diverse: some investigators were observing the frequency dependence of lines toward supernova remnants in order to model the emission in a few directions (Downes and Wilson 1974; Pankonin 1975). Other groups (Gordon and Cato 1972; Matthews, Pedlar and Davies 1973) made small surveys taking care to avoid obvious continuum sources. Still another approach was taken by Hart and Pedlar (1976b) and by Lockman (1976) who, in two independent surveys in the H166α line at a wavelength near 21 cm (ν=1.4 GHz), aimed for uniform coverage of the galactic plane to determine the overall distribution of these lines at one frequency; there was no bias either toward or away from continuum sources.

The two H166α surveys gave the first adequate description of the galactic distribution of these recombination line sources, and increased the amount of data on them by an order of magnitude. Unfortunately, there was not a corresponding increase in our understanding of the medium that produced the lines. Since the two surveys, Mebold et al. (1976), in an important observation, found a recombination line at a wavelength near 11 cm from the direction of a relative minimum in the galactic continuum radiation. This line can not possibly come from a small angular diameter, dense, HII region, but must originate in moderate or low density gas.

A decade of research in this topic can be summarized:

1) At frequencies of 1-2 GHz, recombination lines with peak antenna temperatures $\sim$0.04 K can be observed at every $\ell \lesssim 50^{\circ}$ when $|b| \lesssim 0\overset{\circ}{.}5$. The sources of these recombination lines are common in the inner parts of the Galaxy (at galactocentric radii R$\lesssim$8 kpc), but are rare or absent in the solar neighborhood and in the anticenter direction in the plane. The emission is more confined to the galactic plane then HI is, and its distribution is qualitatively similar to that of most population I-type objects (molecular clouds, dense HII regions, supernova remnants).

2) The lines often have complex shapes with widths of tens of km/s. The simplest spectra are similar to those observed from bright, isolated HII regions. No narrow ($\Delta v <$ 10 km/s) hydrogen lines have been detected.

3) Over the range of frequencies observed thusfar (0.4 to 5 GHz), the peak antenna temperature of these lines is inversely proportional to the frequency of observation. Line strengths show no strong dependence on the antenna beamwidth.

4) Some of the emission is definitely not associated with bright continuum sources, but some (perhaps much of it) comes from objects that are distinct in the radio continuum.

5) In some directions the line emission has structure on angular scales ≤10 arc-min; in other directions it appears smooth over several degrees.

6) Helium and carbon recombination lines corresponding to the hydrogen lines are not prominent; in a few directions $T_L(He)/T_L(H) \leq 0.03$.

In this review I want to use the basic results listed above, together with data from a new H166α survey (Lockman, 1980) that triples the number of spectra available for study, to discuss the origin of these weak recombination lines. The galactic distribution of the emission will be discussed elsewhere and enters this review only where it provides information on the origin of the emission. Because most of the data have been taken in the H166α line, I define the medium of interest as that seen in H166α surveys of the galactic plane. Observations of other transitions compliment the H166α survey results, and are discussed when they are relevant.

II. SENSITIVITY OF THE SURVEYS: WHAT CAN WE SEE?

H166α observations preferentially detect nebulae of intermediate age or older, and are biased in favor of larger regions. Given the beamwidth and sensitivity of the surveys, an HII region like Orion A will be seen in the H166α recombination line up to a distance ∿5 kpc from the sun; a larger HII region, like W3, will be seen to about 10 kpc from the sun, and objects like the Rosette nebula will be seen out to ∿12.5 kpc. The only sources that will not be detected are the extremely young, small, dense nebulae like W3(OH); pressure broadening of the H166α line is severe in these objects, and the continuum opacity at 1.4 GHz is also very high. But most galactic HII regions are, in principle, accessable to H166α observations even if they are quite far away.

Besides the familiar, isolated HII regions that occupy the frontispiece of many textbooks, we can expect the Milky Way to contain several other forms of ionized gas that are seen in nearby spiral galaxies. In a recent review, Courtes (1977) lists: 1) "conventional HII regions distributed at the front of the spiral arms." These have a "most frequent diameter" of 80 pc (for M31) or 40 pc (for M33) and have emission measures of a few thousand cm^{-6} pc. If they have temperatures ∿8000 K, these conventional, optical HII regions will give a detectable H166α line when they are within about 13 kpc of the sun (M31) or within 6.5 kpc of the sun (M33). A large fraction of the nebulae will be detected at even greater distances. 2) A diffuse nebulosity connecting the larger HII regions. The nebulosity extends over ∿100 pc and has an emission measure ∿200 cm^{-6} pc. If the Galaxy is like M31 or M33, this gas will fill an antenna beam at low longitudes, and it may produce weak but detectable H166α lines over much of the galactic plane. 3) A "diffuse disk emission" that appears in M33 but not in M31. Its properties are not well known, but its ubiquity might make it detectable in H166α observations if the Milky Way contains such gas.

It is clear that there is no lack of ionized gas in spiral galaxies; in fact, it would be surprising if there were no weak recombination lines from the inner parts of the galactic plane. But the abundance and diversity of form of the ionized gas in other galaxies suggests that the goal of early observers, i.e. to look for recombination line emission between the HII regions, might be difficult to accomplish. The problem we face is not in accounting for the existence of weak H166α emission, but in separating it into its various components.

In the following discussion I use the adjectives "isolated" and "discrete" to denote standard HII regions like Orion A and M17 that are relatively "dense" and have definable boundaries. If these also produce enough radio continuum emission to be seen against the galactic background then they are "bright". In contrast, "distributed" material is extensive ($\gtrsim$100 pc), not well defined, and not prominent in the radio continuum. To be "diffuse", a gas must be found over many kiloparsecs. Distributed and diffuse ionized gas roughly correspond to the second and third items, respectively, in Courtes' schema.

III. THE BASIC OBSERVATIONS: WHAT DO WE SEE?

Figure 1 shows the antenna temperature of H166α lines at b=0° between longitudes 33° and 37°. The spectra (from Lockman, 1980) were taken every one-third degree in longitude. The HPBW was 21 arc-min, and each spectrum required ∿5 hours of integration on the NRAO 140-foot telescope. These lines are typical of those observed away from strong continuum sources.

Small scale features and an (apparently) smoother component are visible. The lower velocity lines can change rapidly with position (cf. the spectrum at 33°.67 with that at 34°.0) while the higher velocity component changes more slowly. These are the closest spaced observations made to date, and even here the sampling may be inadequate to reveal the full angular structure of the medium. Inadequate angular sampling tends to create apparent large-scale features where none are present. This shortcoming in the data will not be easy to remedy because of the very long integrative times required to measure a single spectrum.

A more complete view of the H166α emission from the plane is given in Figure 2, where all spectra at b=0° are plotted in velocity-longitude coordinates. Longitudes where observations were made are indicated by marks outside the right border of the figure. The lowest contour is 13 mK of antenna temperature; this is equivalent to about 17 mK of full-beam brightness temperature.

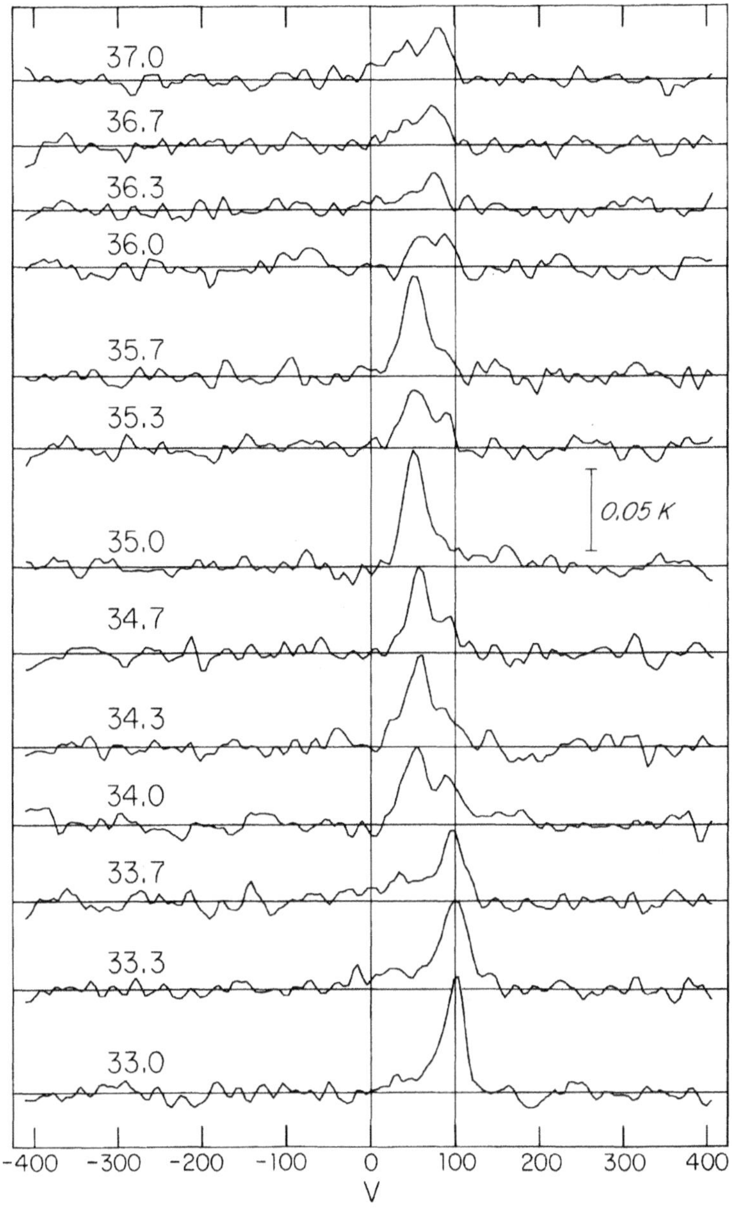

Figure 1: Examples of H166α spectra. The vertical scale is antenna temperature, the horizontal scale is LSR velocity. The longitude of observation is written beside each spectrum.

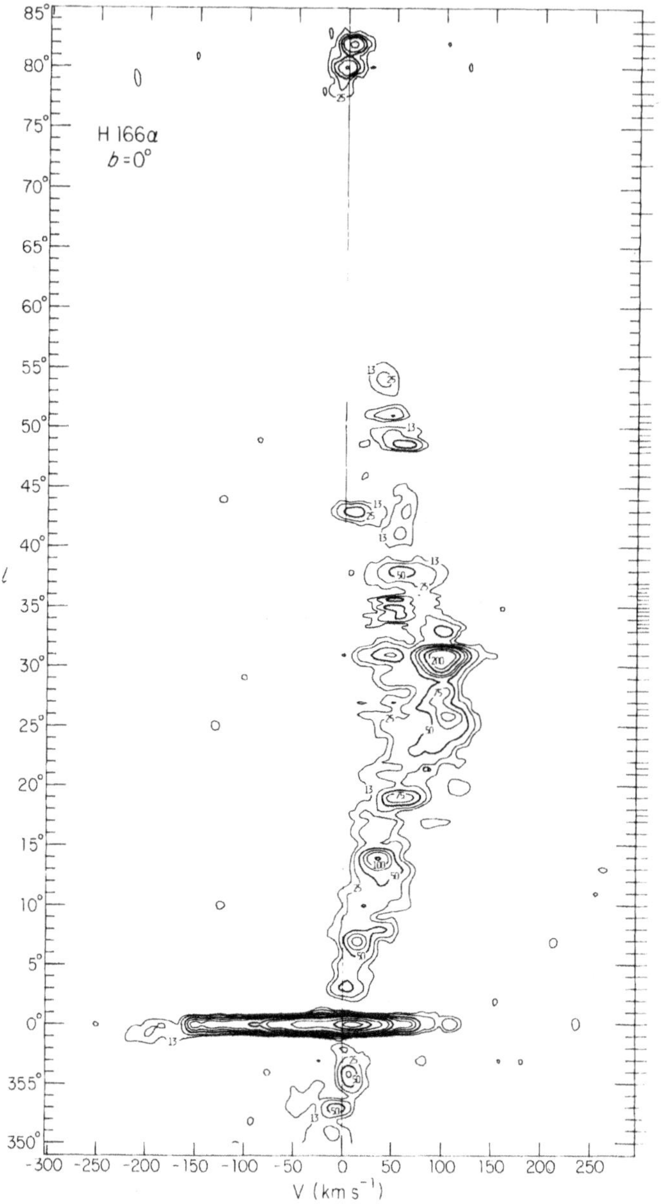

Figure 2: H166α emission at b=0° displayed in velocity-longitude coordinates. Contours are labeled with the antenna temperature of the line in milli-Kelvins. The lowest countour is ∿3 times the rms noise at most positions. Marks along the right border of the figure indicate the longitudes where measurements were made.

This figure gives the general impression of a broadly distributed H166α line component together with a number of bright isolated features. However, artificial smoothness is one result of undersampled observations, and over most of the figure spectra were taken only every degree in longitude.

There is little emission at longitudes >55°, at negative velocities for longitudes >3°, or at high positive velocities for longitudes <25°. Neutral hydrogen covers a much larger range in longitude and velocity and reaches its peak brightness near $\ell=70°$. Thus the ionized gas is not distributed like, or related in any simple way to the HI density. Ionized gas is almost entirely confined to galactocentric radii $4 \lesssim R \lesssim 8$ kpc.

There are significant differences between the distribution of the H166α line and the distribution of molecular lines that trace dense clouds. The most striking difference is the existence of substantial amounts of molecular gas between $2 \lesssim R \lesssim 4$ kpc (Burton and Gordon, 1978), in an area almost totally lacking in recombination line emission. In addition, the "3-kpc expanding arm" is not prominent in H166α even though it contains a lot of molecular and neutral gas.

The Cygnus region near longitude 80° is, as usual, exceptional. It may be a nearby example of the conditions that prevail over much of the inner Galaxy.

Figure 3 shows the average power in H166α vs. longitude. Vertical bars cover the entire range of measurements in a given longitude interval.

The lowest value of the integral in each bin marks the maximum possible contribution of a distributed or diffuse medium. At least one-third of the average power in the line varies on an angular scale $\sim 1°$, and it seems safe to attribute this component to emission from discrete HII regions in the antenna beam.

The origin of the remaining two-thirds of the H166α is more difficult to determine. Even the most recent surveys of discrete HII regions are not complete in that they do not give the position, velocity and size of all dense HII regions that could be detected in H166α. However, the discrete source surveys do give enough information to answer an interesting question: if an HII region lies θ arc-min from an H166α position, what is the probability that the H166α spectrum will have a line at the HII region velocity?

The set of known HII regions was taken from the recent Bonn H110α recombination line survey (Downes et al. 1980); objects in this survey have a continuum size that is typically only a few arc-min. H166α lines are often so broad that individual

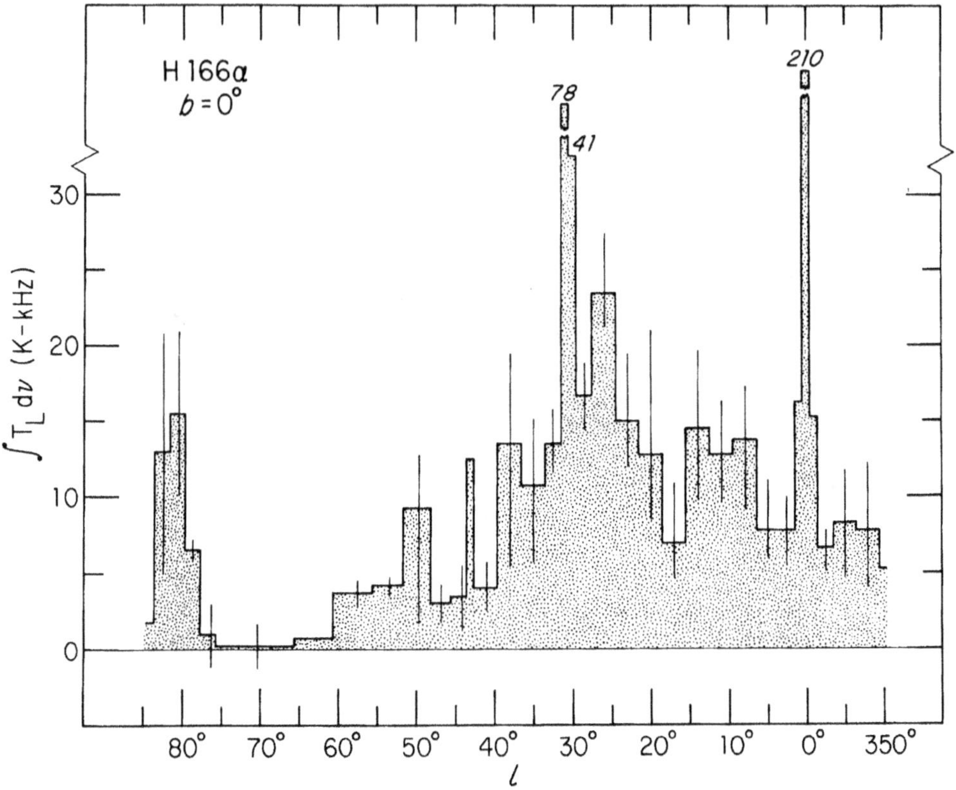

Figure 3: Total power in the H166α line (in Kelvins-kHz of antenna temperature) averaged over intervals of longitude. Vertical bars cover the entire range of measured values in each interval.

components cannot be distinguished, so if there was a reasonable line intensity at the velocity of the H110α line, the HII region was assumed to have been detected.

About 95% of the H110α sources had counterparts in H166α when $\theta \leq 30$ arc-min, and about 85% had counterparts when $\theta \leq 60$ arc-min. These are surprisingly large percentages. A displacement of 30 arc-min from the H166α beam position places the continuum peak of the HII region at the $\sim$1% response level of the beam. HII regions that are 60 arc-min from the beam center are entirely out of the main beam. Most of the H110α nebulae that were not detected in H166α are at longitudes and velocities corresponding to galactocentric radii $\geq$7 kpc.

Another comparison can be made by adding all H110α profiles within a given area to form a "composite" H110α profile. Composite profiles for HII regions within 60 arc-min of the H166α positions generally resemble the H166α profiles. In particular, whenever the composite H110α profile shows significant structure (line wings or multiple peaks), H166α spectra show similar features, although the ratio of line component strengths in the two spectra often differs by a great deal. H166α spectra are usually broader and more complex than the composite H110α spectra, indicating that the H166α observations are seeing gas that was not detected in the higher frequency Bonn survey. Some of the extra emission in H166α is probably due to discrete HII regions whose H110α line was too weak to be detected.

These correlations can be given several interpretations. First, the source of the H166α lines may be so widely distributed that it fills all available velocity space in the inner Galaxy. In this case correlation with H110α (and with any other species that is located in the same general area of the Galaxy) is expected. However, this explanation would not account either for the decrease in correlation as θ increases or for the similarity between the shape of H110α composite profiles and the observed H166α.

Second, discrete HII regions could be more extensive than the radio source size indicates. This seems to be the more likely explanation. Recombination lines from many nebulae can be seen at some distance from the continuum peak, and the lines often appear to trail off rather than show a sharp boundary. The extragalactic HII regions discussed earlier typically have optically determined diameters much larger than the radio sizes of galactic HII regions. Some of this is no doubt selection, but the optical results are probably more accurate because radio measurements will tend to measure the size of the densest fluctuations rather than the overall size of the nebula. In the radio, faint continuum associated with the outer parts of nebulae could easily be confused with the galactic background and with other sources, while any bright knots of emission would be observed and catalogued as distinct HII regions. Incidentally, the notion that many HII regions have large low-density components is expected from models of star formation that place the new stars at the edge of molecular clouds (Zuckerman, 1973).

The association of H166α emission with discrete HII regions can also explain the surprisingly good correlation between the power in H166α and the 5 GHz continuum intensity (see figure 8 in Burton and Gordon, 1978). This correlation implies that much of the H166α comes from hot, relatively dense HII regions.

Hart and Pedlar (1976a) have mapped a large, faint HII region associated with W3, and have suggested that it could be typical of the objects seen in the inner Galaxy. Their measurements are not sufficient to determine if the extended gas is part of the dense W3 nebula, or if it is the remnant of an earlier phase of star formation that occured in the vicinity. But their observations do indicate that the emission measure of the extended component is too large for the object to belong in any of Courtes's categories except that occupied by normal HII regions.

I have not yet considered the possibility that most H166α comes from the distributed ionized gas that connects some large HII regions in nearby galaxies. If this gas is in the Milky Way it could certainly mimic an extended component around dense HII regions. However, on the basis of the extragalactic results, this gas appears too faint and too smoothly distributed to account for the properties of the H166α.

Even if the correlations between discrete HII regions and H166α lines are coincidence, discrete HII regions occupy so much velocity-longitude space in the inner Galaxy that it is questionable if one could ever be certain that an HII region is not lying somewhere along the line of sight. Thus questions of their association with H166α emission must be decided by relatively subtle arguments of HII region size and evolution, arguments that cannot be settled by existing data.

All of the H166α emission does not come from a single type of object; the observations sample a rich environment. For this reason it is impossible to derive a set of physical parameters that describe conditions in all directions, for these conditions are certainly as variable as the ionized gas in the Galaxy is. In the next section I will turn the discussion to one particular feature in the H166α observations and examine it in depth. This gas, whatever its connection with discrete HII regions, is almost certainly not connected with current star formation.

IV. PROPERTIES OF THE GAS NEAR ℓ=36°.

Figure 4 shows a velocity-longitude map of H166α between 33° and 37°. Individual spectra over this range are given in figure 1. Vertical lines are drawn at zero velocity and at the terminal velocity derived from the rotation curve of Burton and Gordon (1978). Contours are labeled in antenna temperature. The dots mark the location of Type I OH masers (Turner, 1979). Type I OH masers are associated with the earliest phases of star formation and are almost always found near very dense, small HII regions.

There are no Type I OH masers, no dense HII regions, and no other signs of recent star formation, such as H_2O masers, at velocities $\gtrsim$60 km/s between longitudes $\sim 33\overset{\circ}{.}7$ and $\sim 36\overset{\circ}{.}9$. The lower velocity line peaks in this area come from HII regions listed in the Bonn survey; they are typical nebulae. But near ℓ=36° the high velocity gas is

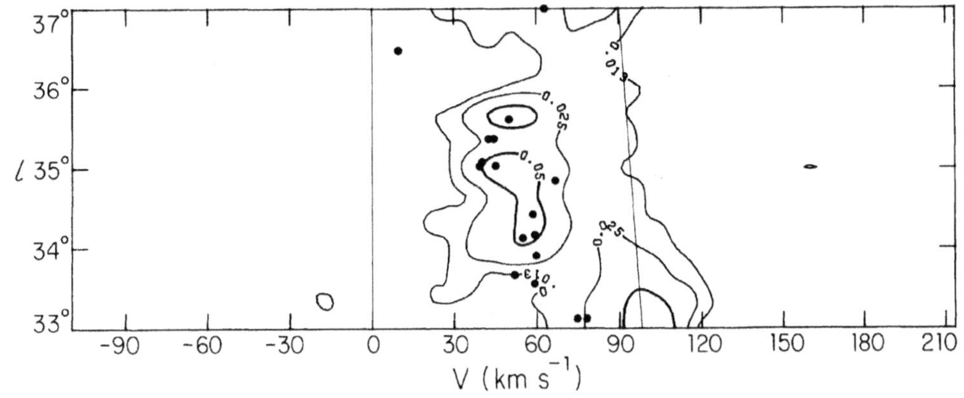

Figure 4. Velocity-longitude map of H166α emission between longitudes 33° and 37° at b=0°. Spectra were taken every one-third degree in longitude. Contours are labeled with the antenna temperature in Kelvins. Dots mark the location of Type I OH masers.

separated both in longitude and velocity from active star-forming regions.

There are theoretical reasons for expecting star formation to be scarce at the high velocities. This part of velocity-longitude space has been identified with the edge of an "interarm" region by everyone who has tried to model the spiral structure of our Galaxy, whether they based their models on low frequency nonthermal continuum emission, HI streaming motions, the kinematics of HII regions or the intensity of γ-rays (see the discussion in Lockman, 1979, and references therein).

The high velocity lines near ℓ=36° are important because they are relatively isolated in velocity-longitude space, and because they may not be coming from traditional HII regions. They could be a model for some of the H166α emission seen in more confused areas of the plane. This material has not been detected except in H166α.

The part of velocity-longitude space that is of interest corresponds to an area with an extent of about 350 pc across the line of sight and up to 8 kpc along the line of sight. (The actual linear extent of the gas could be much less than 8 kpc; velocity crowding compresses a great deal of space into a few km/s.) The constancy of the recombination line emission over this interval suggests that the gas may actually be distributed rather than confined to relatively small angular structures.

Beyond what is seen in figure 4, there is little direct information on this medium. In order to establish its properties I assume that gas like this could exist in other parts of the plane, and therefore other observations can be used to limit, but not to define, its properties. I do not require that this gas account for anything but the emission near 36°, i.e. its absence from other parts of the plane is acceptable. But at the same time, its presence elsewhere should not produce more recombination line emission (or more of anything else, like low frequency continuum opacity) than is observed.

(a) Temperature and Emission Measure

Figure 5 shows some constraints that can be placed on the distributed emission. The lower curve shows the combination of emission measure E and electron temperature T_e required to produce the observed lines if the gas fills the beam, the departure coefficient b=1, and stimulated emission makes a negligible contribution to the line intensity. The last assumption is almost certainly correct in directions of low continuum intensity, such as near longitude 36°. The assumption that b=1 is certainly incorrect, and thus the curve defines a lower limit to $E(T_e)$. If the source of the lines does not fill the antenna beam the curve is likewise a lower limit to $E(T_e)$. Thus the emission measure at a given temperature must lie above the H166α, b=1 curve.

A second constraint is that the source of the H166α must not have too large a continuum optical depth at 80 MHz. I use $\tau_c(0.080\ \mathrm{GHz})=0.5$ as an upper limit to the opacity from this medium. The observed opacity at 80 MHz varies from source to source (Dulk and Slee, 1972), and the value of one-half is taken as representative; it allows for the presence of interstellar sources of opacity that might not be sources of detectable recombination line emission.

A lower limit to the value of the departure coefficient b, that gives an upper limit to the emission measure at a given temperature, is known from purely radiative solutions for the level populations (the zero density case). But for the present observations the "zero density" curve lies so close to the $\tau_c(0.08)$ curve that it gives no extra information, and has not been drawn in figure 5.

The curve that runs at almost constant E is derived from the 5 GHz continuum measurements of Altenhoff et al. (1979). Around longitude 36° the 5 GHz continuum is rather uniform with an average brightness temperature ∿0.2 K. If all of the continuum is thermal and comes from the gas that produces the H166α line, then the emission measure of the gas lies along the curve marked $T_c(5)=0.2$. If some of the continuum is nonthermal, or comes from thermal gas that does not produce substantial H166α emission, then the source of the H166α must have properties that place it below this line.

The resulting range of possible parameters for the H166α source is rather small. Neither high electron temperatures, nor high emission measures are permitted: $T_e \lesssim 3000$ K and E $\lesssim 1400$ $\mathrm{cm^{-6}}$ pc.

If the medium is not distributed uniformly across the antenna

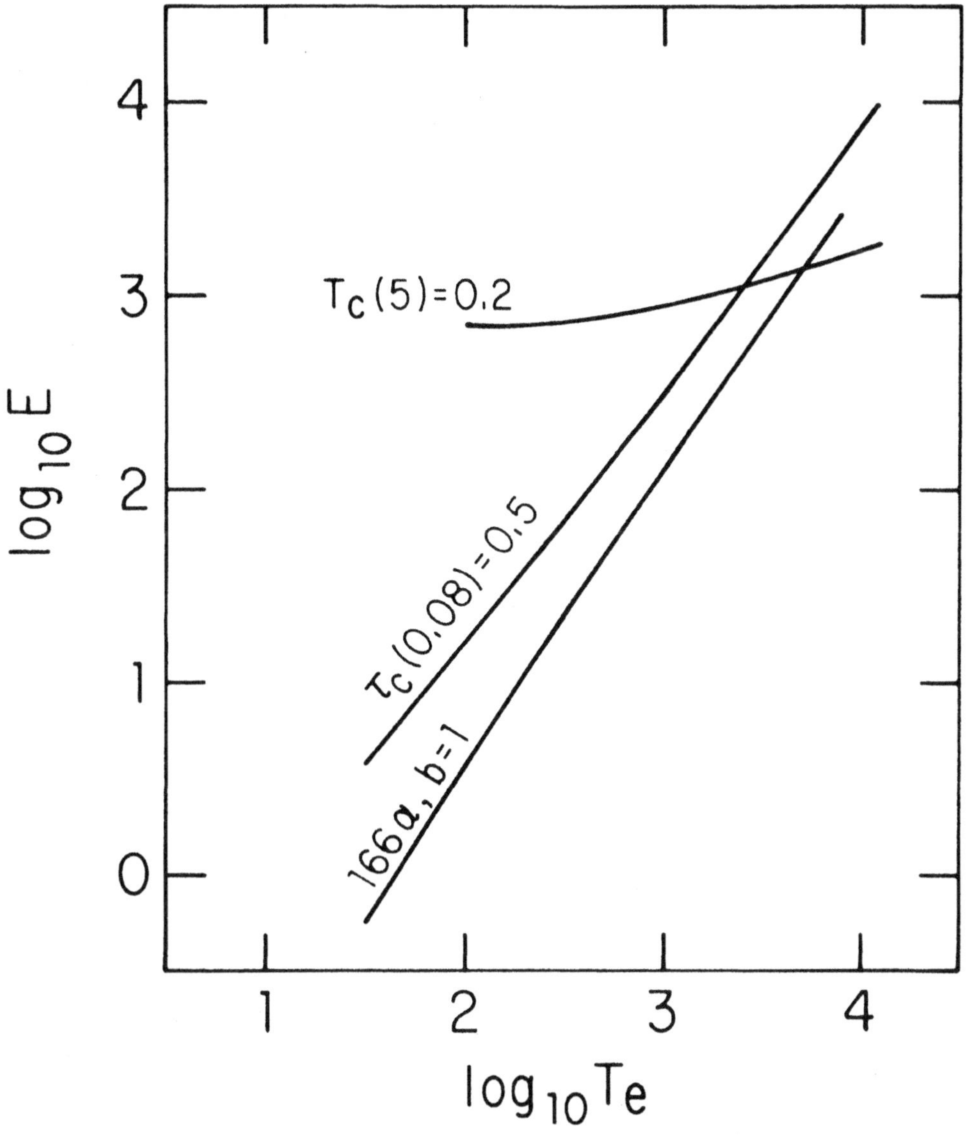

Figure 5. Constraints on the emission measure, E, and the electron temperature, T_e, of ionized gas near longitude 36°. The curves were derived from the 5 GHz continuum brightness temperature at longitude 36°, from the power in the H166α line, and from 0.08 GHz observations of the continuum opacity of the interstellar medium. The source of the H166α lines must have an $E(T_e)$ that lies in the area bound by the three curves.

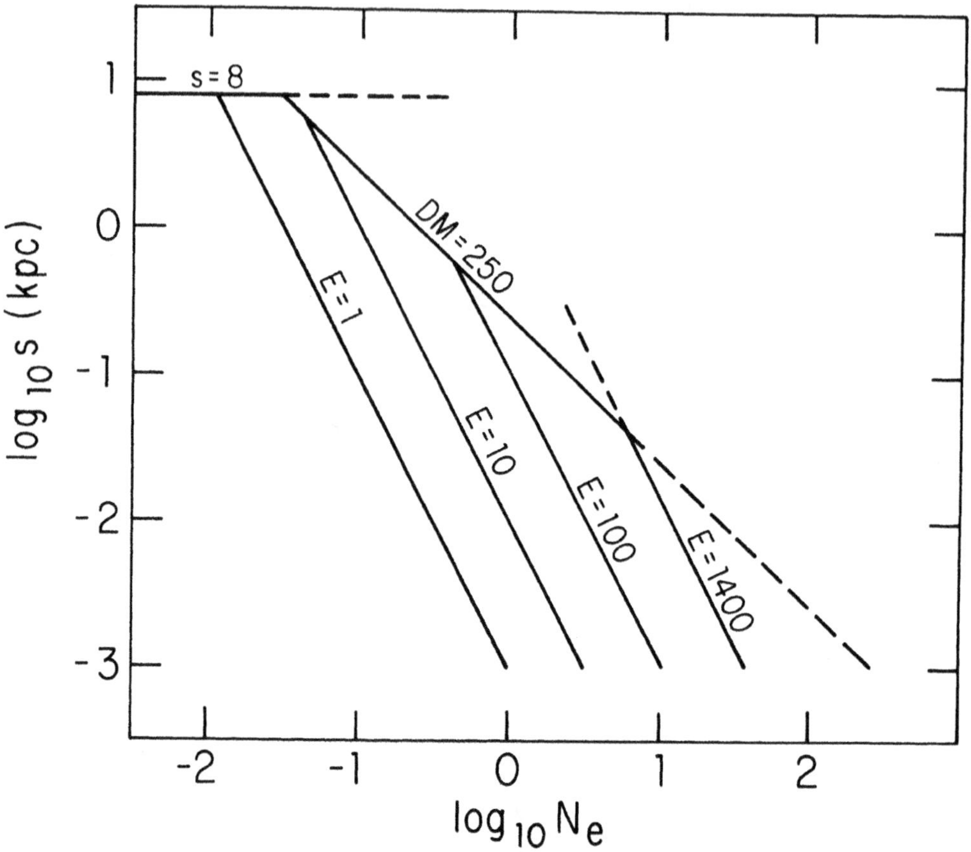

Figure 6. Constraints on the path length, s, and electron density, N_e, of the gas near longitude 36°.

beam, the H166α, b=1 curve must be raised in inverse proportion to the angular dilution factor. Moderate dilution factors (≳0.1), however, will not affect the T_c=0.2 K limit, for this was derived from observations taken with a beam that had a factor of fifty smaller area than the H166α beam. Thus if the medium is clumped across the H166α beam, the maximum permitted electron temperature will be lower that the limits shown in figure 5.

(b) Electron Density and Path Length.

Figure 6 shows constraints that can be placed on the electron density N_e and on the path length s through the gas. The line at s=8 kpc is the maximum path length derived from kinematic considerations. The line E=1400 cm^{-6} pc is the largest emission measure permitted by figure 5. The distributed medium must have some combination of N_e and s that lies to the left of the E=1400 line.

The total electron density integrated through the galactic plane is known from pulsar dispersion measures. Hulse and Taylor (1975) find a maximum dispersion measure ($DM=\int N_e\, ds$) of about 250 cm^{-3} pc for a set of pulsars at $\ell \gtrsim 35°$. [This statement is not strictly true. Two pulsars in the Hulse-Taylor survey have twice the "maximum" dispersion measure, indicating that there are clumps of ionized gas (HII regions) in their directions. Thus ionized gas that has a patchy distribution across this part of the plane must have $DM \sim 250$ cm^{-3} pc, gas that is more smoothly distributed must have $DM \lesssim 250$ cm^{-3} pc. In either case the limit in figure 6 is appropriate for H166α.] The line marked DM=250 shows how this information limits the size and density of the source of H166α; the gas must lie to the left of the line.

Note that I am not requiring that the source of the H166α line provide all of the observed dispersion measure (in contrast to the position taken by Manchester and Mebold, 1977, when they analyzed the recombination lines toward pulsar 1641-45), merely that it not produce more dispersion measure than is observed.

Figure 6 shows that if the distributed gas has a very low density, it must have a small emission measure. Figure 5 shows that if the gas has a small emission measure it must have a low electron temperature to produce the recombination lines. Thus the source of H166α lines near $\ell=36°$ cannot be low density, diffuse, hot gas like that expected from models of the intercloud medium. If the gas is at several thousand Kelvins, it must reside in clouds no larger than a few tens of parsecs along the line of sight.

Up to this point constraints on properties of the distributed emission have been based on rather elementary considerations. However, Shaver (1976) pointed out that very low density gas can have a population inversion so large as to provide embarassingly strong recombination lines, through maser action, at low frequencies where the galactic plane is a source of strong continuum radiation. Pankonin's (1975) measurements suggest that $|\tau_L| \leq 0.01$ for the H248α line at 428 MHz (note that most combinations of T_e and N_e give negative optical depths in the line). Calculations using the departure coefficients of Salem and Brocklehurst (1979) show that this limit requires that the gas near $\ell=36°$ have an electron density >0.5 cm^{-3}. This is in basic agreement with Shaver's conclusion, even though he used different departure coefficients and placed somewhat different restrictions on the source of the emission. A very low electron density medium cannot be the source of these H166α lines, no matter what its electron temperature is.

This result has no bearing on the existence of a low density electron component of the ISM, it just indicates that such gas has not been seen in H166α. It must have $E(T_e)$ below the H166α, b=1 line in figure 5.

Shaver's criterion, that limited the electron density of the gas, is meaningful only if gas identical to that producing the high velocity lines near longitude 36° is widespread throughout the galactic plane. While this is most likely true, it is clear that this type of analysis should be used carefully. It will be discussed further in the next section.

We are left with an area in $s(N_e)$ bound by $N_e \sim 1$ cm^{-3} to the left, by the DM =250 cm^{-3} pc line to the top and by the E=1400 cm^{-6} pc line to the right. This encompasses a broad range of parameters on which we have no other quantitative limits, except that the emission appears smooth over transverse dimension ∿350 pc. If the gas has a similar extent along the line of sight then figure 6 shows that it must have a density ∿1 cm^{-3}, an emission measure of a few hundred cm^{-6} pc, and, from figure 5, a temperature ∿1000 K.

Alternative solutions, though less palatable, are not firmly excluded. Higher emission measure gas (corresponding to $T_e \sim 3000$ K) lies in the domain where s is tens of parsecs or less. To have the observed angular smoothness, these clouds must be large sheets of ionized gas arranged nearly face-on toward us. This might be the case if the gas were in a large shell rather than a uniform sphere. Gas with a density lower than 1 cm^{-3} may be barely permissible considering the uncertainties in arguments about low frequency maser action. Still, the low frequency observations do not allow gas much cooler than 500 K to be the source of these H166α lines.

The relatively low temperature implied by these results is not too different from temperatures derived by other groups, looking in other directions, at other frequencies. Mebold <u>et al.</u> (1976) find $T_e \sim 3000$ K for 2.6 GHz lines from a local minimum in the galactic continuum radiation at $\ell=24°$. If a slightly different baseline were removed from their data, the temperature estimate would be even lower. Manchester and Mebold (1977) find from observations of a recombination line near 5 GHz, that the gas in the direction of pulsar 1641-45 can be no hotter than 3300 K. Again, this is a conservative limit, and any relaxation of their assumptions will decrease it. They also estimate a minimum T_e, but this was done by assuming that all of the dispersion measure toward the pulsar came from the source of the recombination lines. This assumption may not be correct. In sum, there is quite a bit of evidence for the existence of moderate density ionized clouds that are, at the hottest, about 3000 K, and at the coolest, about 500 K.

In this analysis I have not used the correlation found by

Jackson and Kerr (1975) between the 5 GHz line and continuum intensities near longitude 26°.5. This result may be important, but I do not think that it tells us much about the gas near $\ell=36°$. The H166α emission near 36° is so weak that its (LTE) intensity at 5 GHz is a factor of five below the Jackson and Kerr detection limit. Thus the results of Jackson and Kerr refer to a medium with an emission measure at least five times greater than the emission measure of the gas near 36°. They are probably seeing small HII regions.

V. OTHER OPINIONS

It is a reviewer's duty to acknowledge opinions other than his own. Therefore this section is devoted to a discussion of a few works in this field that I have not yet given the notice they deserve.

The first of these was mentioned earlier. Hart and Pedlar (1976a) measured the H166α line in the W3/W4/W5 region with "the intention of providing observations of a well-understood HII complex which could be compared with H166α observations of the inner regions of the Galaxy." The idea that W3 is "well-understood" was probably meant to be taken in relation to our complete ignorance of conditions in the inner Galaxy; we actually have little understanding of this complicated HII region. Hart and Pedlar simply made an analogy between the bright W3 region (with its attendant lower density gas) and the galactic plane. The strenuous discussion of §III ends with just their conclusion: much of the H166α in the inner Galaxy comes from low-density gas associated with bright HII regions (see also Hart, 1977).

Shaver (1976) took a more quantitative approach. He was interested in determing whether the observed recombination lines came from hot or cold gas, and used observations of low frequency lines toward continuum sources to put a lower limit of a few cm^{-3} on the density of the gas. But he could find no quantitative way to distinguish between hot and cold gas of this density, and therefore used Jackson and Kerr's (1975) result to argue for a hot medium. Earlier in this article I suggested that the Jackson and Kerr result should be applied cautiously, for their observations were not sensitive enough to detect the high-velocity gas near longitude 36°. Nonetheless, I think that Shaver's general conclusion is valid: much of the H166α comes from hot (or warm; at least it is not cold) gas that could properly bear the label of an HII region.

Mezger (1978) made a more radical suggestion. He claimed to find evidence in continuum and H166α surveys for an "extended low-density fully ionized HII region" that fills "most of the interstellar space in our Galaxy". Mezger's model has a temperature of 7000 K, an rms (though not necessarily local) density ~ 0.3 cm^{-3}, and a vertical extent of 100 pc above and below the plane. This medium, rather than HII regions, is supposed to be the dominant source of H166α emission.

I think that Mezger has overestimated the scale height of the H166α emission by a factor of about two, and has grossly underestimated the contribution of normal HII regions to the observations. But a more quantitative problem with this model is that its rms density is an order of magnitude higher than that inferred from pulsar dispersion measures (Taylor and Manchester, 1977). Mezger's model gives an integral electron density through the Galaxy that differs from the measured value (Hulse and Taylor, 1975) by more than a factor of ten. The Galaxy cannot be full of gas that has an rms density of a few tenths cm^{-3}. The only way to make this model consistent both with the observed line and continuum emission, and with the pulsar dispersion measures, is to decrease the extent of the ionized gas by about two orders of magnitude, and increase its (local) density by about an order of magnitude. These changes do not alter the emission measure of the gas, but give it a more correct rms density and total dispersion measure. The resultant ionized structures though, have sizes of a few hundred parsecs, rather than the proposed tens of kiloparsecs. They are large, low-density HII regions and not an all-pervasive medium; this model has been reduced to those proposed by Hart and Pedlar, and by Shaver. There is certainly an extended low-density electron gas in the Galaxy, but it seems that its properties are such that it has not yet been observed in recombination lines.

VI. CONCLUDING REMARKS

Throughout this article I have tried to avoid presenting a single, all-encompassing explanation for the low-frequency recombination lines. I think that we must accept that the observations sample a diverse medium and attempts at unified explanations will thus most likely be futile. Still, a few general facts seem clear: much if not most of the H166α emission comes from hot rather than cold gas. Some of the lines may come from clouds that are hundreds of parsecs in extent and that are not connected with sites of current star formation. The low-frequency observations have detected ionized gas that has not been seen in higher frequency surveys.

Large, low-density HII regions, like the one near $\ell=36°$, might be quite common in the inner parts of the Galaxy, but it is not clear that optical observations in the solar neighborhood can provide us with a local example of the medium. We can draw analogies with gas seen in nearby (and hopefully similar) spiral galaxies, but we must be careful to make quantitative comparisons. For example, the (often unstated) assumption that dense galactic HII regions seen in high-frequency radio surveys are equivalent to HII regions seen in spiral arms of nearby galaxies has caused much misunderstanding; a simple comparison of the size and emission measure of these nebulae shows that they are quite different.

In view of the considerable amount of telescope time required to measure a low-frequency spectrum, it is appropriate to ask if the results justify the effort. Perhaps the best answer comes from comparisons of the sensitivity of the various methods available for detection of lines from ionized gas. In the absence of obscuration, optical measurements are plainly superior (see Courtes, 1977), but optical observations are out of the question for study of the inner Galaxy.

In the radio the most comprehensive surveys have been made at 5 GHz (in the H109α or H110α lines) and at 1.4 GHz in the H166α line. The most sensitive 5 GHz survey (Downes et al., 1980) has an rms noise of about 30 mK. The most sensitive H166α survey has an rms noise of about 5 mK (Lockman, 1980). Because the intensity of recombination lines from optically thin gas varies inversely with frequency, the H166α observations are twenty times more sensitive to large, optically thin nebulae than the 5 GHz observations. For very small nebulae, or for those whose optical depth is significant at 1.4 GHz, the smaller beamwidth and higher frequency of the H110α surveys gives them the advantage in sensitivity. There are also HII regions, like W3(OH), that are so small and dense that they must be observed at frequencies higher than 5 GHz. Thus any investigation of the ionized gas in the Galaxy requires that many different observations be made. Because they are so sensitive to certain nebulae, I think that low-frequency recombination line observations are just as important as the higher frequency work for locating ionized gas in the Galaxy and for studying the overall characteristics of star formation.

Despite the efforts of the last decade, only a small fraction of the galactic plane has been observed in any recombination line. Improvements in telescope systems should allow a factor of five reduction in the time it takes to obtain an adequate spectrum, and this should allow us to make reasonably complete studies of specific areas without the enormous effort that such programs have required in the past. I suspect that low-frequency recombination line observations will be pursued vigorously in the coming decade.

REFERENCES

Altenhoff, W.J., Downes, D., Pauls, T. and Schraml, J. 1978, Astron. Astrophys. Suppl., 35, pp. 23-54.
Burton, W.B. and Gordon, M.A. 1978, Astron. Astrophys., 63, pp. 7-27.
Cesarsky, D.A. and Cesarsky, C.J. 1973, Ap.J., 184, pp. 83-91.
Courtes, G. 1977, in Topics in Interstellar Matter, ed. H. van Woerden, Dordrecht:Reidel, pp. 209-242.
Downes, D. and Wilson, T.L. 1974, Astron. Astrophys., 34, pp. 133-139.
Downes, D., Wilson, T.L., Bieging, J. and Wink, J. 1980, Astron. Astrophys. Suppl., in the press.
Dulk, G.A. and Slee, O.B. 1972, Aust. J. Phys., 25, pp. 429-441.

Gordon, M.A. and Cato, T. 1972, Ap.J., 176, pp. 587-596.
Goss, W.M. 1968, Ap.J. Suppl., 15, pp. 131-202.
Gottesman, S.T. and Gordon, M.A. 1970, Ap.J. (Letters), 162, pp. L93-L97.
Hart, L. 1977, in Topics in Interstellar Matter, ed. H. van Woerden, Dordrecht:Reidel, pp. 187-193.
Hart, L. and Pedlar, A. 1976a, M.N.R.A.S., 176, pp. 135-143.
Hart, L. and Pedlar, A. 1976b, M.N.R.A.S., 176, pp. 547-559.
Hulse, R.A. and Taylor, J.H. 1975, Ap.J. (Letters), 201, pp. L55-L59.
Jackson, P.D. and Kerr, F.J. 1971, Ap.J., 168, pp. 29-35.
Jackson, P.D. and Kerr, F.J. 1975, Ap.J., 196, pp. 723-731.
Lockman, F.J. 1976, Ap.J., 209, pp. 429-444.
Lockman, F.J. 1979, Ap.J., 232, pp. 761-781.
Lockman, F.J. 1980, in preparation.
Manchester, R.N. and Mebold, U. 1977, Astron. Astrophys., 59, pp. 401-404.
Matthews, H.E., Pedlar, A. and Davies, R.D. 1973, M.N.R.A.S., 165, pp. 149-172.
Mebold, U., Altenhoff, W.J., Churchwell, E. and Walmsley, C.M. 1976, Astron. Astrophys., 53, pp. 175-177.
Mezger, P.G. 1978, Astron. Astrophys., 70, pp. 565-574.
Pankonin, V. 1975, Astron. Astrophys., 38, pp. 445-450.
Salem, M. and Brocklehurst, M. 1979, Ap.J. Suppl., 39, pp. 633-651.
Shaver, P.A. 1976, Astron. Astrophys., 49, pp. 1-16.
Taylor, J.H. and Manchester, R.N. 1977, Pulsars, San Francisco: Freeman.
Turner, B.E. 1979, Astron. Astrophys. Suppl., 37, pp. 1-332.
Zuckerman, B. 1973, Ap.J., 183, pp. 863-869.

SURVEYS OF GALACTIC HYDROGEN RADIO RECOMBINATION LINE EMISSION

T.L. Wilson
Max-Planck-Institut für Radioastronomie, Bonn

Large scale surveys of recombination line emission from the galaxy are made in order to:

(1) delineate the distribution of ionized gas in our galaxy,
(2) to study the properties of recombination line emission in ionized gas far from the sun,
(3) study the physical conditions in distant HII regions,
(4) enlarge the sample of distant HII regions, and
(5) catalog newly detected, flat spectrum non-thermal sources which are recognized by the absence of recombination line emission.

In the following discussion, large scale surveys of the radio recombination lines of hydrogen will be presented. The aim of this review is to:

(1) examine the already existing data,
(2) consider the interpretation of the results,
(3) discuss the relation between these data and other astronomical surveys, and
(4) determine what we learn from such catalogs. A discussion of some of the results of the new H110α/H_2CO survey (Downes et al 1979 hereafter the H110α survey) will be used as an illustration of points (2) to (4).

Previous Data

In Table 1 are listed all of the surveys published to date. The entries in the Table are self-expanatory. These surveys may be divided into two parts: those directed at observing discrete HII regions, and those aimed at measuring emission from extended, ionized gas. The first type of survey (e.g. that of Reifenstein et al 1970) is usually done at frequencies higher than 5 GHz, and is biased toward compact ($\lesssim$1 - 2 pc) high density ($n_e \gtrsim 10^3$ cm^{-3}) HII regions. The surveys of the extended gas are usually carried out at lower frequencies (∿1.4 - 1.6 GHz) where the compact, high density sources are optically thick and hence will not contribute very much to the line emission. However, the compact HII regions are usually associated with lower density ionized gas; that is,

P.A. Shaver (ed.), Radio Recombination Lines, 205–223.

Published Hydrogen Recombination Line Surveys of Extensive Regions of the Galactic Plane Outside the Galactic Center

Coverage in galactic longitude	Line transition	Line frequency (MHz)	Telescope beam size (arc min)	Type of survey	Number of sources in survey	Sampling interval or source finding list	Reference
351^{0} through 0^{0} to 209^{0}	159 → 158	1651.542	31.4' x 35.6'	discrete source	39	Westerhout (1958) Downes & Rinehart (1966)	Dieter (1967)
348^{0} through 0^{0} to 84^{0}	110 → 109	5008.923	6.5'	discrete source	120	Altenhoff 11 cm survey (Altenhoff et al (1970)) Downes & Rinehart (1966)	Reifenstein et al (1970)
261^{0} through 0^{0} to 353^{0}	110 → 109	5008.923	4.0'	discrete source	149	Thomas and Day 1969 a, b Day et al (1969) Beard (1966) Hill (1968)	Wilson et al (1970)
9.4^{0} to 130^{0}	158 → 157	1683.2	33'	source-free areas in plane	---	11 areas free from discrete sources ---	Gordon and Cato (1972)
358^{0} to 50.5^{0}	167 → 166	1424.734	21'	uniformly spaced data	---	data taken every 1^{0} at $b = 0^{0}$	Lockman (1976)
5^{0} to 60^{0}	167 → 166	1424.734	31' x 33'	uniformly spaced data	---	every 1^{0}, $5^{0}<\ell<22^{0}$ every 0.5^{0}, $22^{0}<\ell<32^{0}$ every 1.0^{0}, $32^{0}<\ell<60^{0}$	Hart and Pedlar (1976)
6^{0} to 111.5^{0}	86 → 85	10522.04	2.8'	discrete source	21	Reifeinstein et al (1970)	Viner et al (1976)
1.1^{0} to 209^{0}	110 → 109 92 → 91	5008.923 8584.82	2.6' 1.5'	discrete source	36 6	Altenhoff et al. (1978)	Churchwell et al. (1978)
0.7^{0} to 212^{0}	67 → 66 66 → 65	22364.2 23404.3	0.7'	discrete source	19	Various interferometer maps	Wilson et al. (1979 a,b)

the transition from a high density ionized gas to a neutral gas usually goes via a low density ionized gas, and not a sharp high density ionized-gas-neutral gas boundary. Hence, to some extent, the surveys of the low density emission are affected by the presence of high density HII regions.

Ideally, in any study, one wishes to use a sample of sources which have the same intrinsic properties. However, selection effects cannot be avoided. The influence of these selection effects depends on just what problem one wishes to investigate with the data. For example, one should be wary of source comparisons. If the nearby, well-studied HII region Orion A were located 7 kpc from the sun, it would be one of the weaker sources in the H110α survey. In the H109α surveys of Reifenstein et al (1970) and Wilson et al (1970) (in this review the "H109α survey" refers to that of Reifenstein et al (1970), and the "H109α surveys" to both papers), a source having the intrinsic properties of Orion A would not be detected if it were placed more than 5 kpc from the sun. On the other hand, sources having the intrinsic properties of M 17 would be included in the H109α surveys if located within 20 kpc of the sun. Hence comparisons between nearby HII regions and the typical sources cataloged in these surveys should be treated with caution. Lockman and Brown (1978) have analyzed the data in the existing H109α surveys, and also point out the possible presence of such selection effects. In regard to the gradient of T_e^* with distance from the galactic center, proposed by Churchwell and Walmsley (1975), two systematic effects mentioned by Lockman and Brown (1978) are the masering in the recombination emission at 5 GHz (this will be discussed in the reviews of Churchwell and also Walmsley in this volume), and the use of a non-uniform sample of sources.

In another context, namely, the spatial distribution of ionized gas, one must first decide whether a given source is a member of a complex of sources or whether it is an independent HII region, and second, one must differentiate between intrinsically weak HII regions (which are included in surveys only because they are near the sun) and distant, intrinsically powerful sources, which truly represent the structure. We will next consider the individual recombination line surveys in detail.

The Extended Gas

The two definitive surveys were carried out by Hart and Pedlar (1976) and Lockman (1976). The generally accepted interpretation is that the gas is a low density, fully ionized medium with an LTE electron temperature in the range 4000-6000 K, which is located mostly between 3.5 and 6 kpc from the galactic center. In early surveys of this extended emission (Gottesman and Gordon 1970, Gordon and Gottesman 1971), it was thought to arise from emission in a partially ionized, cold (kinetic temperature $\sim$50 - 1000 K) gas. The demise of this theory was caused by the realization that:

(1) sufficient thermal continuum radiation is present in the galactic plane to allow for explanation in terms of many weak HII regions, and
(2) the recombination line intensity is correlated with the thermal continuum intensity (Matthews et al 1973, Jackson and Kerr 1975, Hart and Pedlar 1976).

Finally, the characteristics expected from line formation which involves large non-LTE effects in a cold, partially ionized gas were never observed, leading one to accept the more mundane solution. Because of the weakness of the line emission, and the $\gtrsim$30' telescope beams, only the most general description of the structure of the emission region can be given, and this structure agrees with that obtained by Westerhout (1958) from continuum measurements alone.

Lockman (1976) has pointed out that the distribution of ionized gas obtained from his survey of the extended H166α line emission differs from the distribution obtained from the H109α line survey of Reifenstein et al (1970). A comparison of the H109α survey with the more complete H110α survey shows that the source sample of the H109α survey was not complete. This will be discussed in more detail later.

Discrete Source Surveys

The surveys of discrete sources are best carried out at high frequencies, where the telescope beam is small. This condition is important because the galactic plane is crowded with sources. The two "standard" surveys are the H109α surveys (i.e. Reifenstein et al (1970) and Wilson et al (1970)). The northern survey of Reifenstein et al (1970) missed detecting the H109α line in some sources, and the corrections to this survey are given by Wynn-Williams et al (1971), Wilson and Altenhoff (1972) and Dickel and Milne (1972).

In Fig. 1 (a) we plot the number of sources in the 11 cm continuum survey (this is contained in Altenhoff et al (1970), which was used as the initial finding list for the survey of Reifenstein et al (1970). Below this, we show the number of sources which were detected in the H109α line, and those for which no line was found. In Fig. 1 (b), we plot the same quantities for the new 4.875 GHz continuum (Altenhoff et al 1979) and the H110α surveys made with the 100-m telescope. The most striking feature of the comparison is the increase in the number of sources in the new compared to the old surveys: there are ∿4.5 times more continuum sources, ∿3.3 times more sources where a recombination line was also detected, and ∿2.5 times more sources which do not show a recombination line. The increase in the ratio of the number of thermal sources (i.e. those exhibiting an H110α line) to those without a line is probably the result of resolving complexes of HII regions into individual components. Fig. 2 and 3 show the galactic longitude range $\ell = 22.5^{o}$ to $\ell = 25^{o}$ for the two surveys, and this comparison emphasizes the increase in the number of sources in the H110α survey, compared to the H109α survey. However, even in the H110α survey, only the strongest ∿23% of the total number of discrete continuum sources from the Altenhoff et al (1979) survey are included (see Fig. 1). Hence any differences between the distribution of H110α data and the H166α results of Lockman (1976 and this volume) may be caused by the incompleteness of the H110α sample for weak sources.

Applications of the Results

Two applications of the survey data are:
(1) an investigation of the gradient in the LTE electron temperature, T_e^*, with distance from the galactic center, and
(2) an attempt to obtain the spiral structure of the Galaxy.

LTE Electron Temperature, T_e^*

This section will deal with the LTE electron temperature, T_e^*, because corrections for non-LTE effects depend on the true electron density, which cannot be determined directly. In Fig. 4 we plot the T_e^* values from the H109α survey against the H110α T_e^* data for the HII regions common to both surveys (Any difference in T_e^* from H109α or H110α data should be insignificant). Churchwell et al (1978) had noted that the T_e^* values in the H109α survey are systematically low by ∿2000 K. Because

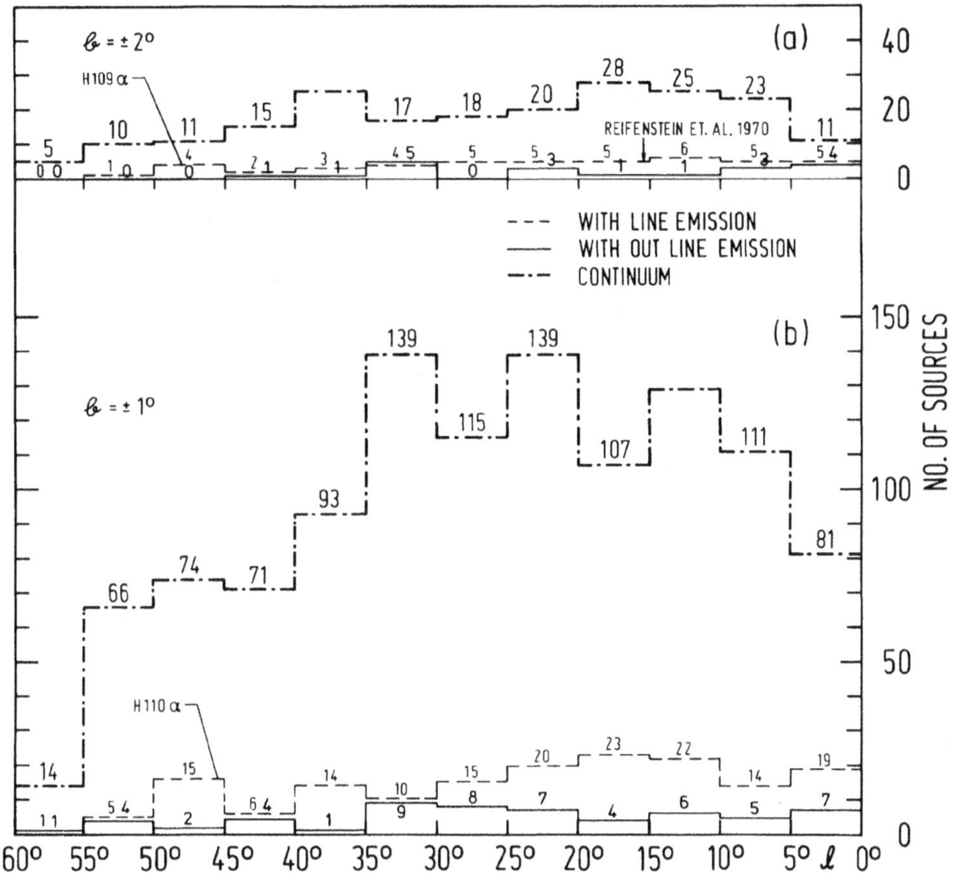

Fig. 1(a). The histograms of continuum sources, sources with, and sources without recombination line emission for 5° intervals of galactic longitudes, for $b = \pm 2^{\circ}$. The continuum data are taken from the 2.7 GHz survey of Altenhoff et al (1970), and the H109α line data from Reifenstein et al (1970).

Fig. 1(b). Corresponding histograms, in which the 5 GHz continuum data is taken from Altenhoff et al (1979), and the H110α data from Downes et al (1979). The range in b is $\pm 1^{\circ}$.

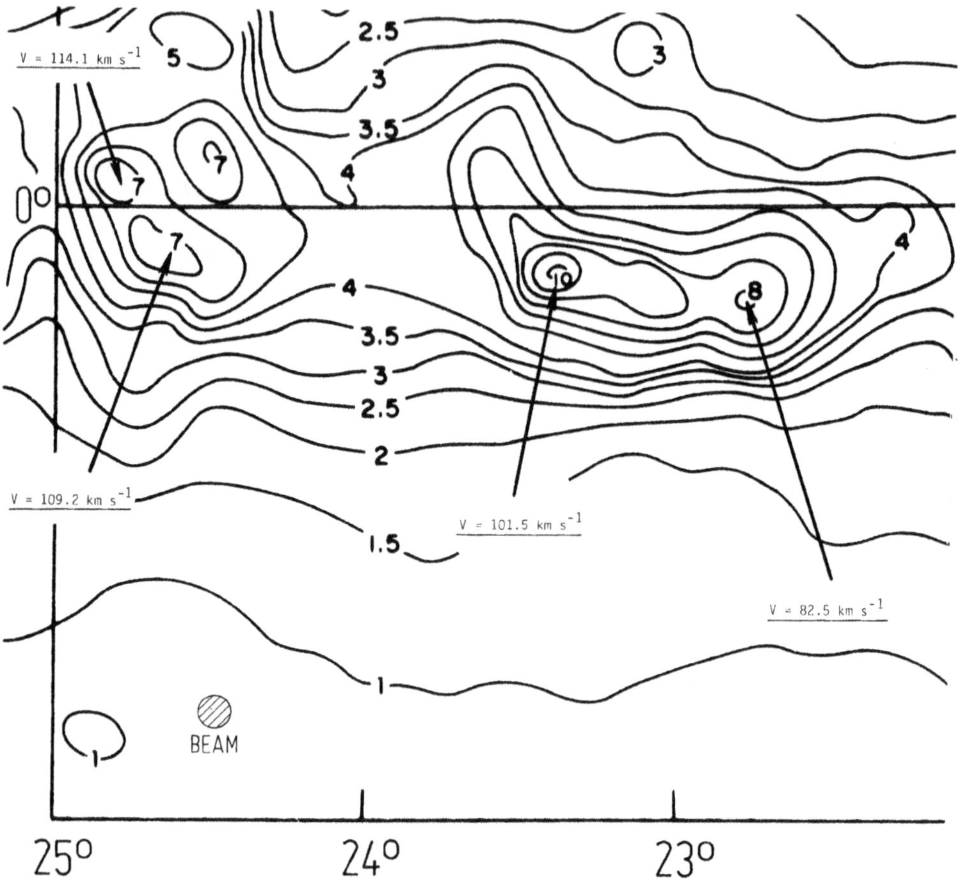

Fig. 2. A map of the galactic plane, between ℓ = 22° and 25°, at 2.7 GHz (Altenhoff et al 1970). The arrows indicate the positions where H109α (Reifenstein et al 1970) were taken; the radial velocities are shown above and below the arrows.

of the better signal-to-noise ratios, the H110α results are to be prefered. For the H110α sources in common with the H109α/He109α survey of Churchwell et al (1978), which are the strongest sources, the T_e^* values in the H110α survey are 950 K (= 12%) larger. The H110α survey and the survey of Churchwell et al (1978) were carried out at about the same time, with the same telescope. However the recombination line and continuum measurements for a given source were not made as near in time as possible. (Such an observing procedure was used for the H65α-H66α surveys by Wilson et al 1979 a,b). This ∿12% difference between the T_e^* results of the H110α survey and those of Churchwell et al (1978) is not understood, and is probably the best indication of the size of the errors in T_e^* for

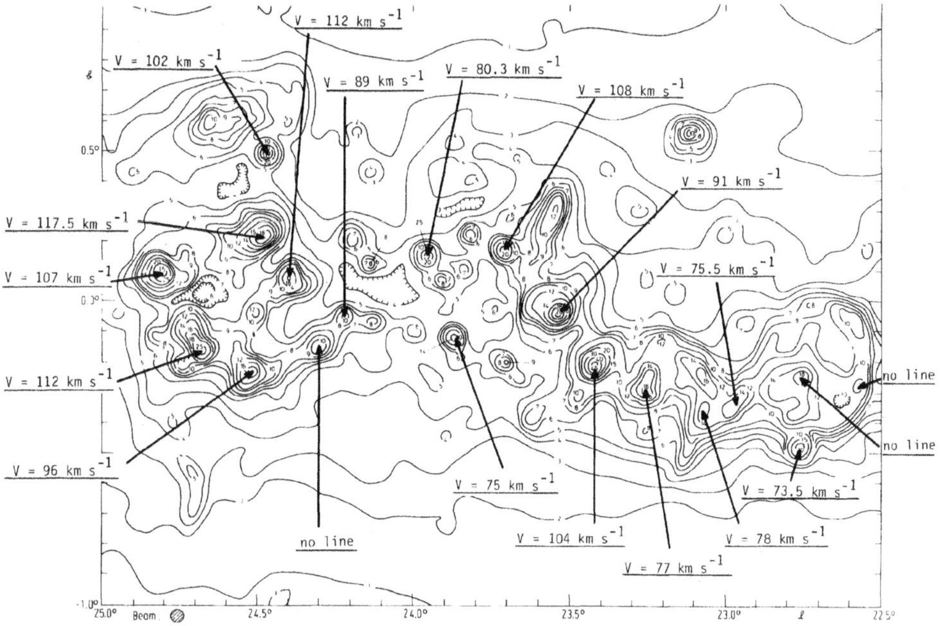

Fig. 3. A 5 GHz map (Altenhoff et al 1979) of the galactic plane over the same longitude interval as in Fig. 2. The arrows indicate the positions where H110α data were taken (Downes et al 1979). A comparison with Fig. 2 shows the effect of the increased resolution and sensitivity in the new surveys.

the stronger HII regions. The errors for the weaker sources are certainly larger, because:
(1) the signal-to-noise ratio of the data is lower, and
(2) there exist uncertainties in how much of the H110α line emission is formed in the ∿0.2 to 0.5 K galactic background. In the T_e^* values obtained from the H110α data, all of the underlying galactic background has been included in the continuum temperature, T_c, which is required to calculate T_e . Since T_e^* is proportional to T_c, this effect will increase T_e^* by ∿10 - 25% for the weaker sources (those with $T_c \leq 2$ K). In spite of the uncertainties, the determination of T_e^* in the H110α survey is still much more accurate than that of the H109α survey.

We have also compared the Emission Measures of HII regions in the H109α and H110α surveys. In only two cases, for the sources at ℓ = 12.807°, b = - 0.204° and at ℓ = 34.254°, b = 0.144°, do the new measurements exceed the older values by more than a factor of three. This shows, that at least as observed with a 2.6' resolution, the sources contained in H109α survey are not composed of many compact, high Emission Measure clumps.

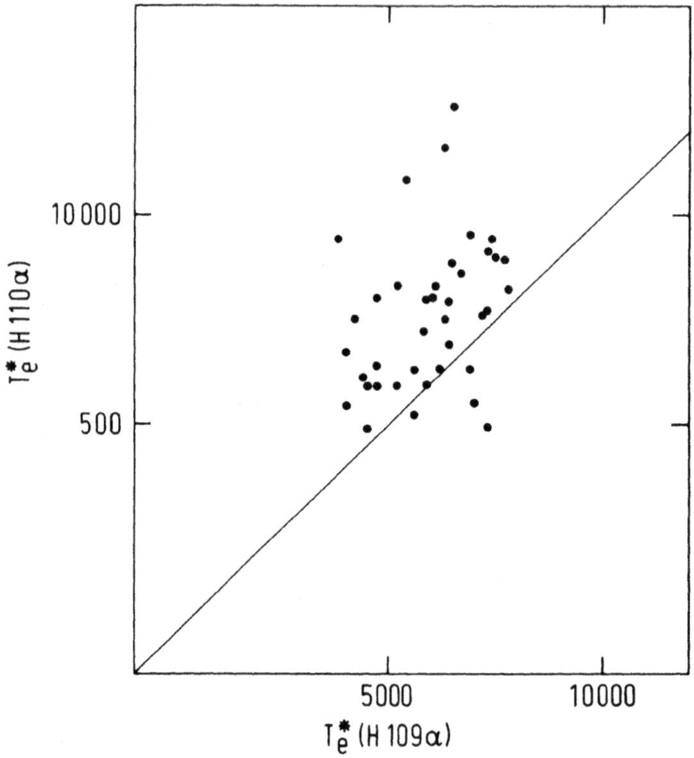

Fig. 4. A plot of T_e^* from the H109α survey (Reifenstein *et al* (1970) versus T_e^* from the H110α survey (Downes *et al* 1979).

Churchwell and Walmsley (1975) have proposed, on the basis of the H109α surveys, that a gradient in the electron temperature of HII regions with distance from the galactic center is present in our galaxy. The search for such an effect is motivated by the presence of gradients in the abundance of metals such as C, N, O, S in HII regions in some S_{cd} galaxies. (see e.g. Burbidge and Burbidge 1975).(However such gradients are not present in all galaxies (see e.g. Pagel *et al* 1978)). The relative abundance of these elements affects the cooling and hence the electron temperature. The absence or presence of an electron temperature gradient in our galaxy will give another parameter with which one can compare our galaxy with other galaxies.

Lockman and Brown (1978) have objected to the claims of Churchwell and Walmsley (1975) for several reasons, the first being that of maser enhancement of the recombination line intensity, causing T_e^* to differ significantly from the true electron temperature, T_e. Such an effect might give a dependence of T_e^* on the distance of an HII region from the sun, D_{Sun}, or, in many cases on D_{GC} (This will be discussed later).

Table 2. H110α sources for which D_{Sun} has been determined

Galactic Coordinates ℓ	b	D_{Sun} Distance from sun (kpc)	D_{GC} Distance from Galactic Center (kpc)	Near or Far kinematic Distance	u Excitation parameter (cm^{-2} pc)	T_e^* LTE Electron Temperature (K)
5.956	-1.265	1	10.0	Near	44	6500
5.973	-1.178	1	9.1	Near	38	7700
6.165	-1.168	1	9.4	Near	17	4700
6.553	-0.095	17 (?)	6.8	Far	135	7300
6.979	-0.250	3.3	6.7	Near	60	7500
7.472	+0.060	27	17.2	Unique	156	3400
8.362	-0.303	6	4.4	Near	60	7300
8.666	-0.351	6	4.2	Near	52	11000
10.073	-0.412	6	4	Adopted a)	65	6000
10.159	-0.349	6	4	Adopted a)	162	5700
10.190	-0.426	6	4	Adopted a)	76.6	5600
10.315	-0.150	6	4	Adopted a)	104	5400
10.617	-0.384	6	4	Adopted a)	84	10500
10.664	-0.467	6	4	Adopted a)	57	5300
10.458	+0.024	6	4	Near	53	6000
12.204	-0.116	16 (?)	6.6	Far	120	7000
12.745	-0.153	5	5	Adopted b)	86.6	4500
12.807	-0.204	5	5	Adopted b)	115	7900
12.865	-0.230	5	5	Adopted b)	63	6100
12.909	-0.277	5	5	Adopted b)	40	4700
13.186	+0.045	5.8	4.5	Adopted b)	71	5700
13.381	+0.071	6 (?)	7.6	Far	53	4700
13.998	-0.128	4	6.4	Near	50	5800
14.600	+0.014	4	6.2	Near	98	4700
14.626	+0.087	4	5.9	Near	83	5300
14.445	-0.651	1.5	8.5	Near	25	10400
15.032	-0.687	2	8	Near	161	9100
15.085	-0.680	2	8	Near	182	7600
15.181	-0.625	2	8	Near	40	9500
15.198	-0.768	2	8	Near	42	10000
15.143	-0.940	2	8	Near	37	8700
16.936	+0.758	2.8	7.2	Near	64	6100
16.984	+0.934	2.8	7.4	Near	59	5000
16.995	+0.868	2.8	7.6	Near	52	8100
18.143	-0.289	4.5	5.6	Near	65	5800
18.185	-0.397	4.5	6.0	Near	53	6200
18.231	-0.254	4.5	5.9	Near	76	7600
18.258	-0.297	4.5	5.9	Near	64	5300
18.305	-0.391	4.5	7.0	Near	42	7300
18.252	+1.892	3.2	7.1	Near	59	5800
18.881	-0.493	5.4	5.0	Near	110	5900
18.936	-0.440	5.4	4.9	Near	90	5500
18.954	-0.019	5.4	5.9	Near	54	5200
19.066	-0.281	5.4	5.3	Near	75	5200
19.608	-0.235	4.5	6.6	Near	68	7900
19.614	-0.132	4.5	5.4	Near	57	7300
19.671	-0.137	4.5	5.7	Near	50	7900
20.074	-0.141	4.5	6.3	Near	42	13000
20.0733	-0.087	13.8	5.7	Far	159	5900
20.988	+0.092	17.0 (?)	8.4	Far	96	8200
22.760	-0.485	12.5	5.2	Far	103	6100
22.947	-0.315	12.5 (?)	5.2	Far	115	5100
22.982	-0.356	12.5 (?)	5.1	Far	115	7800
23.254	-0.268	12.5 (?)	5.1	Far	112	8600
23.421	-0.214	9	4.2	Tangent	124	6000
23.538	-0.041	11.6	4.7	Far	147	4300
23.706	+0.171	9	4.2	Tangent	77	5800

a) Wilson (1974), b) Gardner *et al* (1975)

Table 2 cont.

Galactic Coordinates		D_{Sun}	D_{GC}		u	T_e*
ℓ	b	Distance from sun (kpc)	Distance from Galactic Center (kpc)	Near or Far kinematic Distance	Excitation parameter (cm^{-2} pc)	LTE Electron Temperature (K)
23.869	-0.119	12.6 (?)	5.3	Far	78	6900
23.956	+0.152	12.2 (?)	5.1	Far	96	5600
24.217	-0.053	12	4.8	Far	68	7300
24.392	+0.072	9	4.1	Tangent	79	5500
24.467	0.489	9	4.4	Tangent	100	6400
24.484	+0.211	9	4.1	Tangent	130	4700
24.677	-0.161	9	4.2	Tangent	87	5700
24.805	+0.098	9	4.3	Tangent	113	5800
25.382	-0.177	13.5 (?)	6.2	Far	205	7100
25.396	+0.034	18.5 (?)	10.4	Far	100	8800
25.405	-0.254	13.2 (?)	6.0	Far	95	8100
25.700	+0.031	14.0 (?)	6.6	Far	102	12500
25.766	+0.211	9	4.3	Tangent	118	6200
26.103	-0.069	15.3 (?)	7.7	Far	144	11900
27.276	+0.148	15.2 (?)	7.8	Far	88.7	7400
27.491	+0.189	15.0 (?)	7.7	Far	118	7200
28.600	+0.015	10.7	5.1	Far	116	8300
28.658	+0.030	9.2	4.9	Far	109	8500
28.801	+0.174	9.2	4.8	Tangent	72	6500
29.944	-0.042	9.2	5.1	Far	150	6100
30.227	-0.145	9.2	5.1	Far	81	9100
30.539	+0.024	13.6	7.1	Far	92	6800
30.602	-0.106	8.6	5.1	Tangent	77	6800
30.694	-0.261	8.6	5.2	Tangent	84	6800
30.776	-0.029	7	5.5	Near	197	6000
31.401	-0.259	7	5.5	Near	61	6000
31.411	+0.309	8.5	5.2	Tangent	60	4900
32.797	+0.192	15.6 (?)	9.0	Far	131	9500
33.914	+0.111	8.3	5.6	Tangent	58	5300
34.254	+0.144	3.7	7.2	Near	77	8700
35.063	-1.506	3.2	7.7	Near	52	6000
35.194	-1.750	3.2	7.6	Near	76	8500
35.346	-1.827	3.2	7.6	Near	50	8800
35.603	-0.033	3.5	7.4	Near	46	8600
35.663	-0.030	3.5	7.3	Near	58	9400
37.361	-0.228	13.2 (?)	8.0	Far	78	7600
37.439	-0.040	11.5 (?)	7.1	Far	112	10200
37.538	-0.113	12 (?)	7.2	Far	91	5000
37.671	+0.132	8	6.1	Tangent	77	6600
37.763	-0.216	11.5 (?)	7.1	Far	109	8400
37.871	-0.399	11.5 (?)	7.1	Far	122	11300
43.169	+0.002	13.9	9.5	Far	293	8700
45.125	+0.136	9.7	7.6	Far	116	10100
45.451	+0.060	9.7	7.7	Far	126	7400
45.475	+0.130	9.7	7.7	Far	83	9500
48.596	+0.042	12.0	9.2	Far	142	7800
48.930	-0.286	6.6	7.5	Tangent	111	8200
49.060	-0.260	6.6	7.6	Tangent	74	9100
49.076	-0.377	6.6	7.6	Tangent	80	8900
49.204	-0.345	6.6	7.6	Tangent	132	7700
49.384	-0.298	7	7.9	Tangent	165	7400
49.407	-0.193	7	8.1	Tangent	114	5900
49.437	-0.465	7	7.7	Far	88	8000
49.486	-0.381	7	7.7	Far	252	7000
49.582	-0.382	6.5	7.6	Tangent	89	7800
49.588	-0.456	6.5	7.6	Tangent	60	8400
50.125	-0.670	6.5	7.7	Tangent	61	7700
53.184	+0.155	11.4 (?)	9.7	Far	95.5	6500

A second objection is concerned with the source sample: the sources far from the sun may be intrinsically different from those near the sun, or these small samples do not represent the whole galaxy.

In the simplest situation, the effect of recombination line masering for a given HII region would cause changes in T_e^* with observing frequency. Such changes would lead to variations in the gradients obtained for T_e^* vs. D_{GC}, if the source sample is the same for surveys carried out at different frequencies. However this hypothesis is not supported by the measurements. For example, the gradient of T_e^* with D_{GC} is 312 K kpc^{-1} at 5 GHz (from the H109α survey of Churchwell et al 1978), 380 K kpc^{-1} at 10.6 GHz (from the H86α survey of Lichten et al 1979) and 400 kpc^{-1} at 22-23 GHz (from the H65/66α survey of Wilson et al 1979 a,b). The correlation coefficients for all of these data are larger than 0.7.

In order to better test for the presence of a gradient in T_e^*, the sample should be enlarged, and as a check of systematic effects, one should also examine in greater depth the correlation of T_e^* with D_{Sun}. One can attempt to place the sources in the galaxy, using the radial velocities of the HII regions together with a velocity-distance law (such as the Schmidt (1965) Model, which is based on motions in circular orbits about the galactic center). One of the problems with such kinematic distances is that the distance of the HII region is determined uniquely with respect to the center of dynamics, the galactic center, but not with respect to the sun. For sources within 10 kpc of the galactic center, there are usually 2 possible distances from the sun, and absorption line measurements are required to place an HII region at either the nearer or farther of the 2 distances. Radhakrishnan et al (1972, and the following papers) give a description of the technique of HI absorption line measurements. For a discussion of the relative merits of HI, OH and H_2CO absorption results, see Wilson et al (1978). Optical identification of the HII regions will also resolve the ambiguity. In some cases the kinematic distances obtained from the radial velocity of the HII region and absorption lines are contradictory and an adopted distance is used. There are 116 HII regions in the H110α survey for which estimates of D_{Sun} are available. The galactic coordinates and source parameters are given in Table 2. In the cases where the assignment of D_{Sun} is not certain, we have placed a question mark after this value.

In Fig. 5 (a) T_e^* is plotted versus D_{GC} and in Fig. 5 (b) versus D_{Sun}. The correlation coefficient for T_e^* versus D_{GC} is 0.23, and for T_e^* versus D_{Sun} is 0.01. Regardless of the cause, if there were a dependence of T_e^* on D_{GC} and if all of the sources were located at the near kinematic distance, there would be a dependence of T_e^* on D_{Sun}. This is not the case because 39 of the 116 sources with available D_{Sun} values have been placed at the far kinematic distance. A least-squares linear fit to these data gives

$$T_e^* = 240\ D_{GC} + 5900.$$

We have taken D_{GC}, the distance from the galactic center, to be the error-free independent variable. Because of the low correlation coeffi-

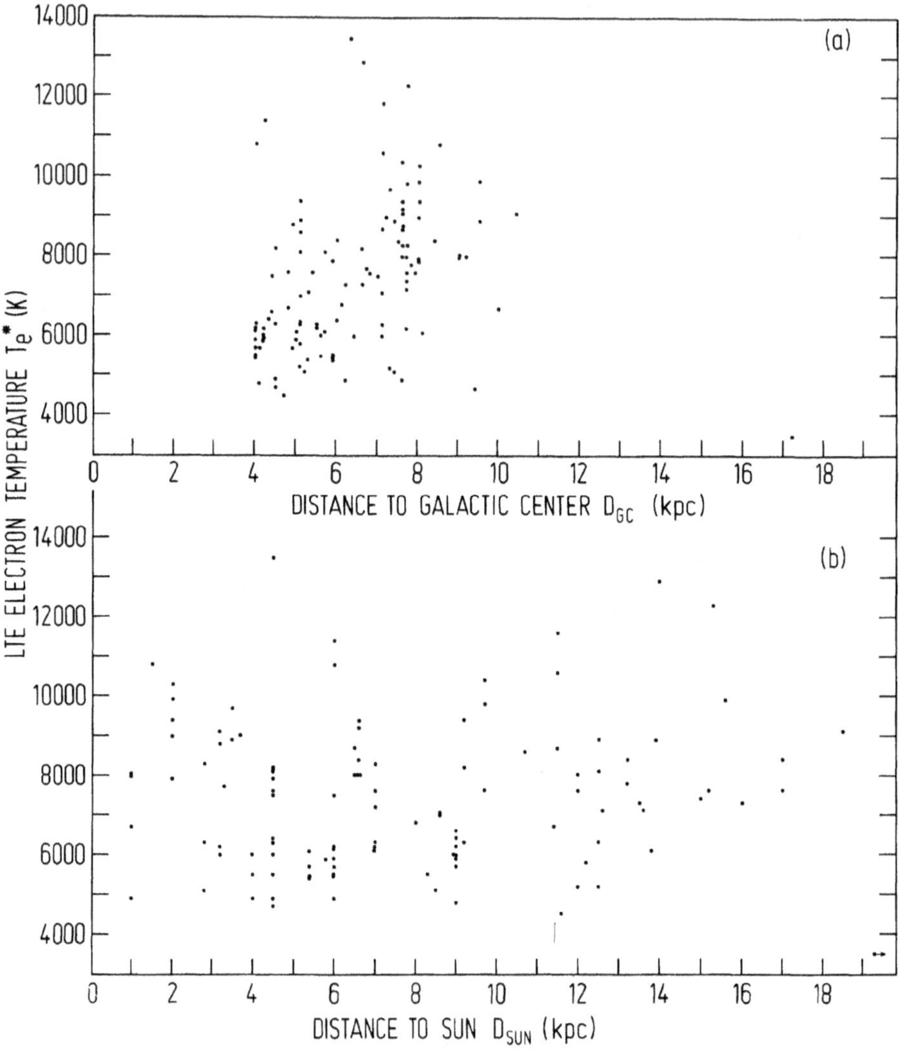

Fig. 5(a). A plot of T_e^*, obtained from the H110α survey, versus the distance to the galactic center.
Fig. 5(b). T_e^* plotted against distance from the sun. The HII region at the extreme right in both plots is at $\ell = 6.553^o$, $b = -0.095^o$.

cient, a fit to the T_e^*- D_{Sun} results is meaningless. For another subset of the sources in the H110α survey, namely the 100 HII regions between 4 and 9 kpc from the galactic center, the gradient is 340 K kpc^{-1}. The large scatter in the data causes the 100 K kpc^{-1} difference between the gradients.

As discussed by Wilson <u>et al</u> (1979), some scatter in T_e^* from source to source is to be expected because of variations in the election density and surface temperatures of the exciting stars. It is probable that the errors in the measured T_e^* values are larger than 1000 K and this error can only decrease the correlation coefficient for the results. However the dependence of T_e^* on D_{GC} and the lack of such a dependence on D_{Sun} implies that the T_e^*- D_{GC} gradient is not caused by systematic effects which depend on D_{Sun}. Furthermore the agreement of the T_e^* gradients from H109α, H110α, H86α and H65α/H66α surveys would imply that the gradient in the LTE electron temperature T_e^* reflects a gradient in the true electron temperature, T_e.

Although the electron temperature gradient does not seem to be caused by any selection effects, such effects <u>are</u> present in the source sample. In Fig. 6 (a) and (b) are plotted the Excitation Parameter, u ($\simeq$ 13.58 $S^{1/3}$ $D_{Sun}^{2/3}$, where S is the continuum flux density of the HII region), versus D_{GC} and D_{Sun}. There is no trend in u with D_{GC}, but there is most certainly a trend with D_{Sun}. This is because the H110α survey is flux density limited, and that is, no H110α measurements were made for sources which had flux densities less than about 1.5 Jy. Hence the lower envelope of the curve in Fig. 6 (b) is u = 15 $D_{Sun}^{2/3}$.

In spite of such selection effects, which must affect all recombination line surveys, the gradient in T_e^* with D_{GC} seems to be real. The contribution of the H110α survey is that a gradient in T_e^* with D_{GC} is present in a sample of sources larger than that used by Churchwell <u>et al</u> (1978).

The Distribution of HII Regions in the Galaxy

We will use the H110α results to study the distribution of the HII regions in the northern part of the galaxy. We assume that the sources move in circular orbits; hence within 6° of galactic longitude ℓ = 0, most of the source velocity is at right angles to the line of sight, and hence in Fig. 7 we have plotted the radial velocities for the longitude range from 6° to 50°. There is no obvious spiral pattern to be seen in this diagram. This is in agreement with the conclusions of Lockman (1979), who analysed the H109α surveys of Reifenstein <u>et al</u> (1970) and Wilson <u>et al</u> (1970). It might be argued that Fig. 7 contains both intrinsically weak and intrinsically strong HII regions, and that a spiral pattern will be traced out only by the intrinsically strong HII regions, which are presumably formed in spiral arms. Such objects may have large peculiar motions. However, the lifetime of these sources is short, and hence they will be observed close to their sites of formation. To differentiate between weak and strong thermal sources, we use the criterion given by

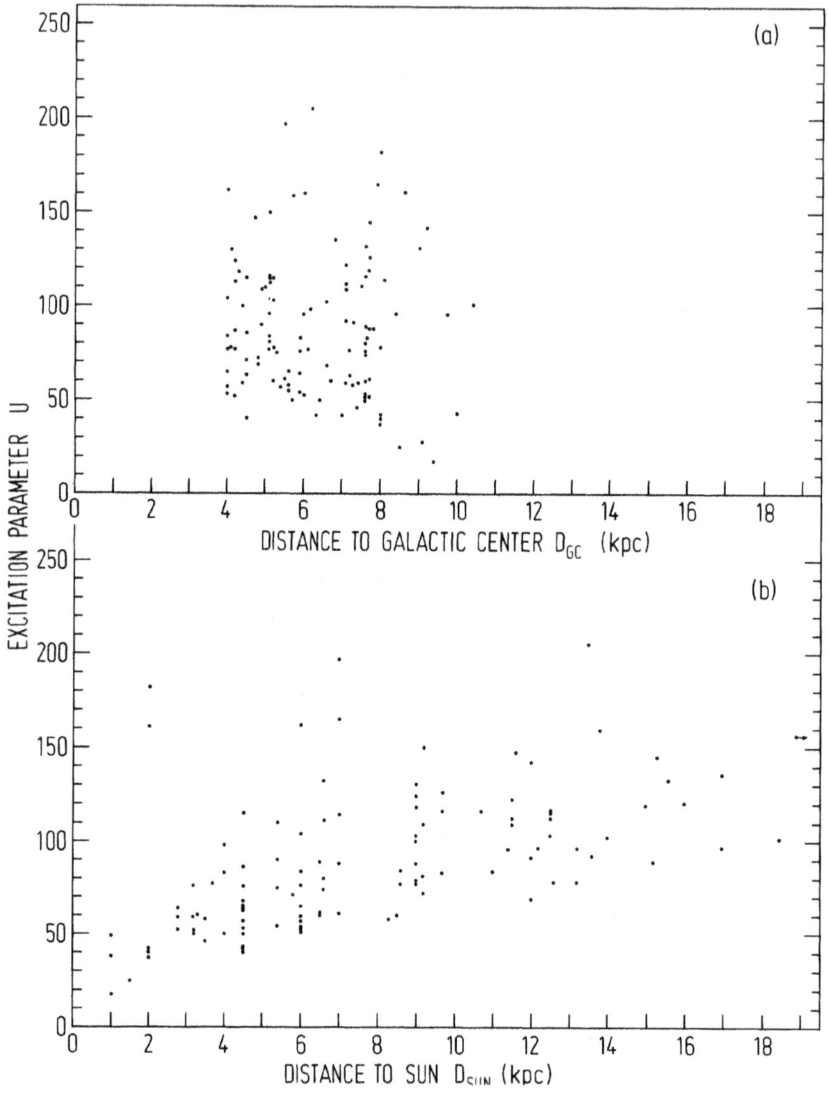

Fig. 6(a). A plot of the Excitation Parameter, u, (= $13.58\ S^{1/3}\ D_{Sun}{}^{2/3}$) versus distance to the galactic center.
Fig. 6(b). u plotted against distance from the sun.

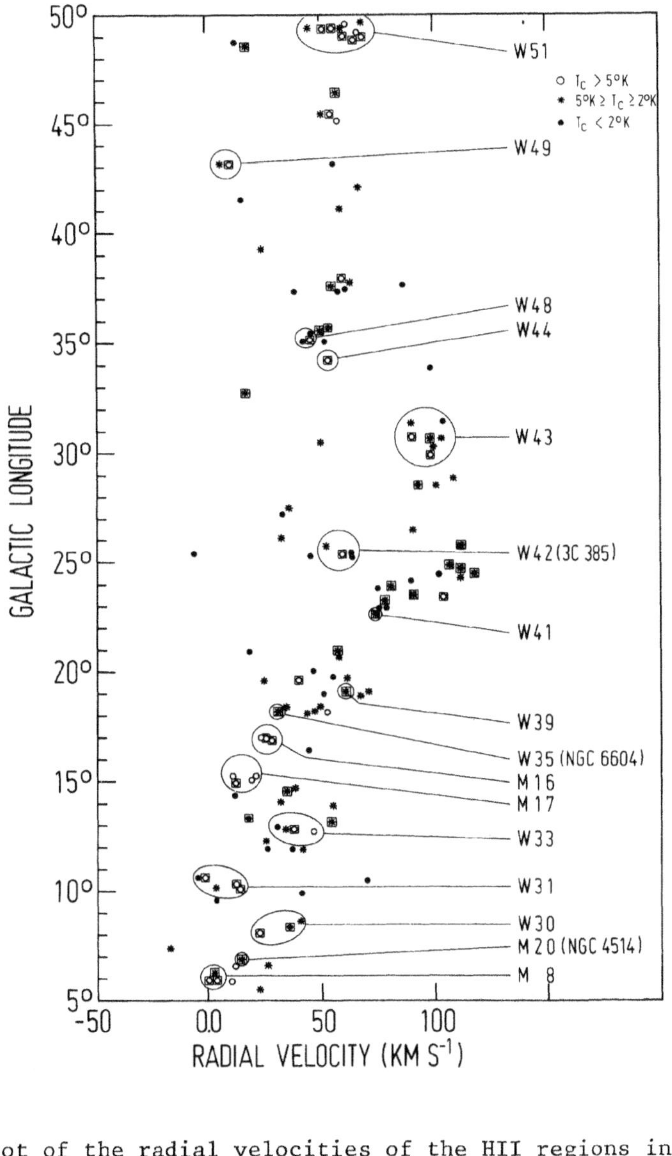

Fig. 7. A plot of the radial velocities of the HII regions in the H110α survey of galactic longitudes 6° to 50°. The source symbols are explained in the figure. A square around the symbols indicates that the source was cataloged in the H109α survey. The commonly used names of some of the sources are shown.

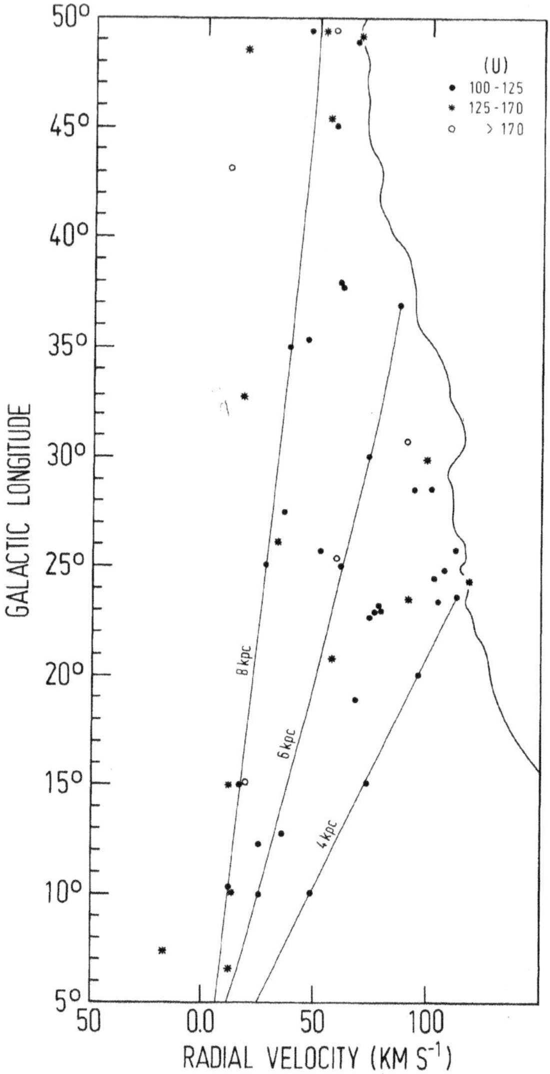

Fig. 8. The longitude-velocity plot of those HII regions for which: (1) estimates of the distance from the sun are available, and (2) the excitation parameter is larger than 100. The lines marked "4 kpc", "6 kpc" and "8 kpc" were calculated using the Schmidt (1965) model for those distances from the galactic center. The irregular curve in the right-hand part of the diagram is the maximum velocity of the HI (Burton 1971).

Mezger (1970)

$$S_\nu \, D^2_{Sun} \geq 400 \text{ Jy (kpc)}^2 \qquad (1)$$

D_{Sun} is known with some certainty only for the sources listed in Table 2 and we have plotted only those sources from Table 2 which satisfy relation (1), which is equivalent to $u \geq 100$. The number of objects is now considerably reduced, and the ability to link up these points into a continuous arm feature is again difficult. We have drawn on these figure curves which represent distances of 4,6 and 8 kpc from the galactic center (under the assumption that the Schmidt Model (1965) is applicable). The radial velocities of most of the sources are consistent with a distance of 4-6 kpc from the galactic center in the Norma region. This is a region where the integrated intensity of CO (and by interference H_2) reaches a maximum (see, e.g., Scoville and Solomon 1975). There is another concentration of sources at about 8 kpc from the galactic; this corresponds to the Sagittarius arm. In the outer parts of the galaxy, that is between 8 and 10 kpc from the galactic center, there are only a few sources. However among these is the W49 complex, one of the most powerful in the galaxy.

The maximum radial velocity obtained from HI studies (Burton (1971)) is also plotted in Fig. 8. As can be seen, all of the H110α radial velocities are lower than this maximum showing that there are no large non-circular motions of HII relative to that of HI, at least near the so-called subcentral points, (where the radial velocity reaches its maximum). The agreement of kinematic and stellar-type distance determinations (see e.g. Georgelin and Georgelin 1976) implies that the magnitude of non-circular motions in the galaxy is about the nominally used value of $\sim$10 km s^{-1}. Much larger deviations, in the order of 20-30 km s^{-1}, have been reported for the W31 (see e.g. the discussion of Wilson 1974, Greissen and Lockman 1979) and W33 regions (Gardner et al 1975, Goss et al 1978). Such large non-circular motions are not present throughout the entire galaxy, however. Non-circular motions of $\sim$10 km s^{-1} correspond to a distance uncertainty of 1 kpc, in the Schmidt Model for longitudes between 20° and 50°. With such an uncertainty, the spiral arms can be drawn in only a very rough fashion. From these considerations, the picture of our galaxy in the ionized gas proposed by Georgelin and Georgelin (1976) is probably an overinterpretation. In the northern half of the galaxy most of the dense HII regions are located in the Norma arm, between 4-6 kpc from the galactic center, with a second, somewhat smaller collection of sources in the Sagittarius arm, at $\sim$7-8 kpc. A few other sources, such as W49 and W3 are probably in a third spiral arm, located about 10-12 kpc from the galactic center. However the results of the H110α survey do not define a spiral pattern.

ACKNOWLEDGEMENTS

The author greatfully thanks Miss L.M. Ziurys who spent a great deal of time, helping with the text and carefully checking the tables and figures. He also thanks Frl. K. Salm for typing the text and tables. A part of this review was prepared at the University of California, Berkeley. Radio Astronomy research at Berkeley is supported in part by NSF grant number AST 78-210337.

REFERENCES

Altenhoff, W.J., Downes, D., Goad, L., Maxwell, A., Rinehart, R.: 1970, Astron. Astrophys. Suppl. 1, 319.
Altenhoff, W.J., Downes, D., Pauls, T., Schraml, J.: 1979, Astron. Astrophys. Suppl. 35, 23.
Burbidge, E.M., Burbidge, G.R.: 1975, in HII Regions and Related Topics, ed. T.L. Wilson, D. Downes, Springer-Verlag , p. 304.
Burton, W.B.: 1971, Astron. Astrophys. 10, 76.
Churchwell, E., Walmsley, C.M.: 1975, Astron. Astrophys. 38, 451.
Churchwell, E., Smith, L.F., Mathis, J., Mezger, P.G., Huchtmeier, W.: 1978, Astron. Astrophys. 70, 719.
Dickel, J.R., Milne, D.K.: 1972, Austral. J. Phys. 25, 539.
Dieter, N.H.: 1967, Astrophys. J. 150, 435.
Downes, D., Wilson, T.L., Bieging, J., Wink, J.: 1979, Astron. Astrophys. Suppl. (in prep.).
Gardner, F.F., Wilson, T.L., Thomasson, P.: 1975, Astrophys. Letters 16, 29.
Georgelin, Y.M., Georgelin, Y.P.: 1976, Astron. Astrophys. 49, 57.
Gordon, M.A., Gottesman, S.T.: 1971, Astrophys. J. 168, 361.
Gordon, M.A., Cato, T.: 1972, Astrophys. J. 176, 587.
Goss, W.M., Matthews, H.E., Winnberg, A.: 1978, Astron. Astrophys. 65, 307.
Gottesman, S.T., Gordon, M.A.: 1970, Astrophys. J. 162, L93.
Greissen, E., Lockman, F.J.: 1979, Astrophys. J. 228, 740.
Hart, L., Pedlar, A.: 1976, Mon. Not. Roy. Astron. Soc. 176, 547.
Jackson, P.D., Kerr, F.J.: 1975, Astrophys. J. 196, 723.
Lichten, S.M., Rodriguez, L.M., Chaisson, E.J.: 1979, Astrophys. J. 229, 524.
Lockman, F.J.: 1976, Astrophys. J. 209, 429.
Lockman, F.J., Brown, R.L.: 1978, Astrophys. J. 222, 153.
Lockman, F.J.: 1979, Astrophys. J. (in press).
Matthews, H.E., Pedlar, A., Davies, R.D.: 1973, Mon. Not. Roy. Astron. Soc., 165, 149.
Mezger, P.G., 1970, in IAU Symposium 38, The Spiral Structure of Our Galaxy, ed. W. Becker, J. Contopoulos, D. Reidel Publ. Co. , p. 107.
Pagel, B.E.J., Edmunds, M.G., Fosbury, R.A.E., Webster, B.L.: 1978, Mon. Not. Roy. Astron. Soc. 184, 569.
Radhakrishnan, V., Brooks, J.W., Goss, W.M., Murray, J.D., Schwarz, U.J.: 1972, Astrophys. J. Suppl. 24, 1.

Reifenstein, E.C. III, Wilson, T.L., Burke, B.F., Mezger, P.G., Altenhoff, W.J.: 1970, Astron. Astrophys. 4, 357.
Schmidt, M.: 1965, in Galactic Structure, ed. A. Blaauw, M. Schmidt, Univ. of Chicago Press, p. 513.
Scoville, N.Z., Solomon, P.M.: 1975, Astrophys. J. 199, L105.
Westerhout, G.: 1958, Bull. Astron. Inst. Netherlands 14, 215.
Wilson, T.L., Mezger, P.G., Gardner, F.F., Milne, D.K.: 1970, Astron. Astrophys. 6, 364.
Wilson, T.L., Altenhoff, W.J.: 1972, Astron. Astrophys. 16, 489.
Wilson, T.L.: 1974, Astron. Astrophys. 31, 83.
Wilson, T.L., Pankonin, V., Dickey, J.: 1978, Astron. Astrophys. 68, 303.
Wilson, T.L.; Bieging, J., Wilson, W.E.: 1979a, Astron. Astrophys. 71, 205.
Wilson, T.L., Pauls, T.A., Ziurys, L.M.: 1979b, Astron. Astrophys. 77, L3.
Wynn-Williams, C.G., Downes, D., Wilson, T.L.: 1971, Astrophys. Letters 9, 113.
Viner, M.R., Clarke, J.N., Hughes, V.A.: 1976, Astronom. J. 81, 512.

THE EVIDENCE FOR A GRADIENT OF H II REGION ELECTRON TEMPERATURES WITH GALACTIC RADIUS

Ed Churchwell
Washburn Observatory
University of Wisconsin-Madison
Madison, Wisconsin U.S.A.

INTRODUCTION

In the past several years a substantial effort has been devoted to the determination of radial abundance and temperature gradients in spiral galaxies. Searle (1971), Shields (1974), Rubin, et al. (1973), Smith (1975), Pagel, et al. (1979) and others have established beyond a reasonable doubt that the relative metal abundances (e.g., O, N, Ne and S) systematically increase from the edge to the nucleus of several spiral galaxies (e.g., M31, M33, M51, M101, NGC 300), see Figure 1. The electron temperatures deduced from the optical [O III] or [N II] lines show an accompanying decrease as the abundances increase (Smith 1975, Pagel et al. 1979). Similar abundance gradients have been derived for our galaxy based on optical studies of planetary nebulae, (D'Odorico, Peimbert and Sabbadin, 1976; Torres-Peimbert and Peimbert, 1977; see Figure 2); of H II regions (Hawley, 1978; Peimbert, Torres-Peimbert, and Rayo, 1978; see Figure 3), and of the kinematic and photometric properties of F, G and K stars (Mayor, 1976). In external spirals and in our own galaxy O, Ne and S appear to have about the same gradient but N is a factor of two or more steeper.

Motivated by the apparent abundance gradients in several spiral galaxies, Churchwell and Walmsley (1975) (Paper I) analyzed the northern (Reifenstein, et al. 1970) and southern (Wilson et al. 1970) H109α surveys for evidence of a galactic radial gradient in H II region electron temperatures. With the simplest assumptions possible (optically thin nebulae with line emission in thermodynamic equilibrium), marginal evidence was found for a gradient of T_e^* (the electron temperature averaged over the telescope main beam when $\tau < 1$ and TE prevails) with galactic radius R_G in both the northern and southern surveys. Encouraged by this result, Churchwell et al. (1978) (Paper II) used high quality H109α data taken with the 100 meter antenna (resolution ≈ 2.6' arc) to further investigate possible T_e^* and H_e^+/H^+ gradients with R_G. (Since Dr. Mezger already discussed the helium problem, I will confine my discussion to the results related to temperature only.) T_e^* was found to have a gradient (with much smaller dispersion than the earlier data) of ∿320K/kpc in both

P.A. Shaver (ed.), Radio Recombination Lines, 225–237.

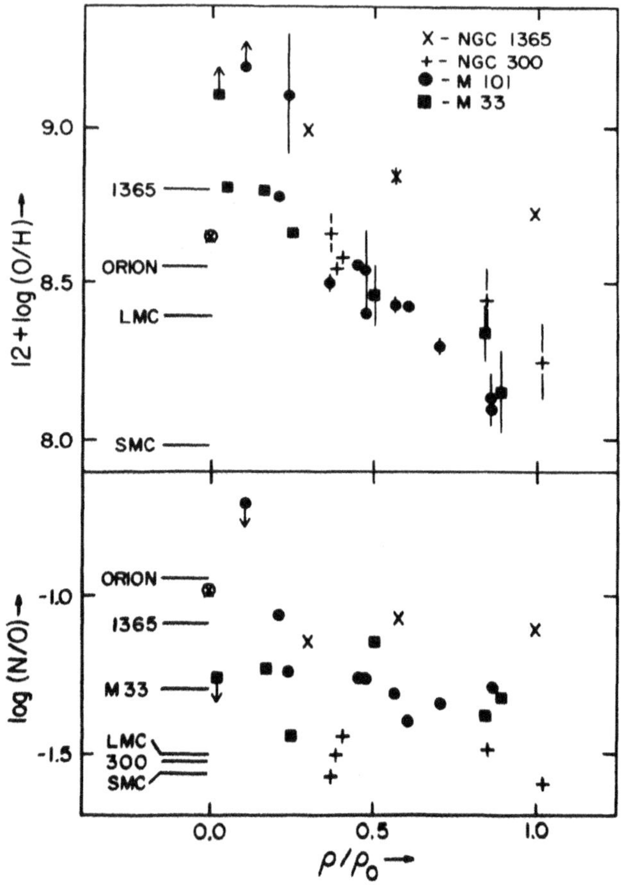

Figure 1: (Top) A plot of the oxygen-to-hydrogen abundance ratio as a function of the normalized Holmberg radius for the spiral galaxies M33, M101, NGC300 and NGC1365. The symbols for each galaxy are designated in the figure. Both figures are taken from Pagel et al. (1979). The arrows indicate limits. (Bottom) The nitrogen-to-oxygen abundance ratio for the same galaxies as above. In both illustrations the appropriate mean abundance ratio is indicated on the left side for Orion, the LMC, the SMC, NGC1365, etc.

-- -- --

the H109α and H137β lines; the mean 137β line temperatures however, were 500-800K higher than those of the 109α line. This was interpreted as an indication of a small but noticable departure from thermodynamic equilibrium in the H109α line. Several possible systematic, instrumental, and physical effects were considered in Paper II which might cause such a gradient, and it was concluded that: (1) the apparent T_e^* gradient is probably indicative of a true electron temperature gradient with R_G; and (2) the most likely effect which alone could reasonably produce the T_e^* gradient is a gradient in the relative metal abundances (especially oxygen which is the primary coolant in H II regions).

Determinations of electron temperatures, He/H abundances, and other physical properties of H II regions from radio recombination lines have been criticised by Lockman and Brown (1976, 1978), Brown and Gómez-González (1975) and Brown, Lockman and Knapp (1978) on grounds that the detailed geometry, temperature, density, and brightness distributions must be known in an H II region before recombination lines can be properly interpreted. Predictably, considerable controversy over the interpretation of radio

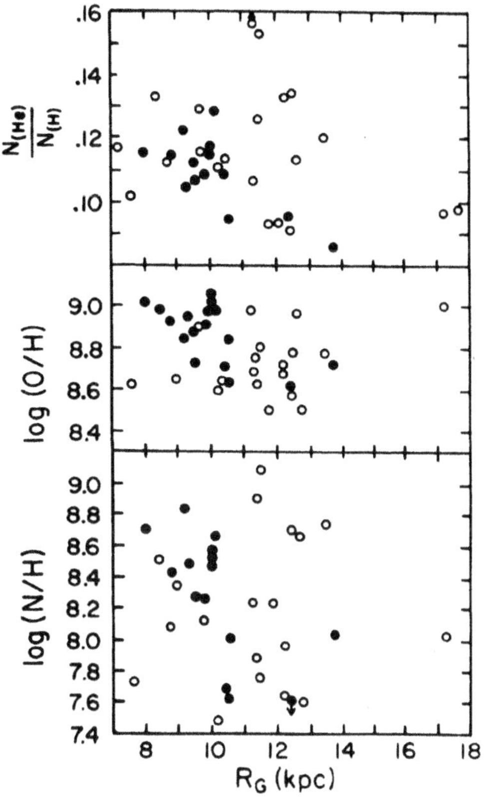

Figure 2: Abundances determined from planetary nebulae by Torres-Peimbert and Peimbert (1977) as a function of galactic radius (R_G). (The open circles indicate low accuracy measures and the filled circles indicate high accuracy determinations). (Top) The helium-to-hydrogen abundance ratio by number versus R_G. (Middle) The oxygen abundance relative to hydrogen (on a scale where log H=12) versus R_G. (Bottom) The nitrogen abundance on the same scale as in (middle) versus R_G.

recombination lines (RRL's) has developed with the result that galactic gradients of any kind derived from simple analyses of RRL's are largely viewed with suspicion if not complete disbelief. Counter arguments to those of Brown and co-workers have been presented by several investigators, most notably Shaver (1979) and Mezger and co-workers in a series of papers dealing with the He^+/H^+ ratio. A certain degree of entrenchment appears to have set in; however, it is hoped that the new data now available will help to resolve these issues.

This review is divided into two major parts: 1) A re-evaluation of the evidence for a galactic H II region temperature gradient based on several new high frequency RRL surveys; and 2) the interpretation of this gradient in terms of an abundance and ionization (i.e., T*) gradient in the galaxy.

NEW EVIDENCE FOR A T_e-R_G GRADIENT

It was pointed out in Paper II and by other observers, that at λ6 cm RRL's can be affected in some H II regions (the most compact ones) by collision broadening, by optical depths $\tau \geq 1$ (both of which would elevate T_e^* above the actual volume averaged $\langle T_e \rangle$ in the nebula) and by stimulated emission (which would depress T_e^* below $\langle T_e \rangle$). At λ's ≲ 3 cm it is generally agreed that the above problems are negligibly small in all but the most compact H II regions. Further, the question of temperature gradients within Orion A has been investigated by Pauls and Wilson (1977), Chaisson and Dopita (1977) (both in the H66α line),

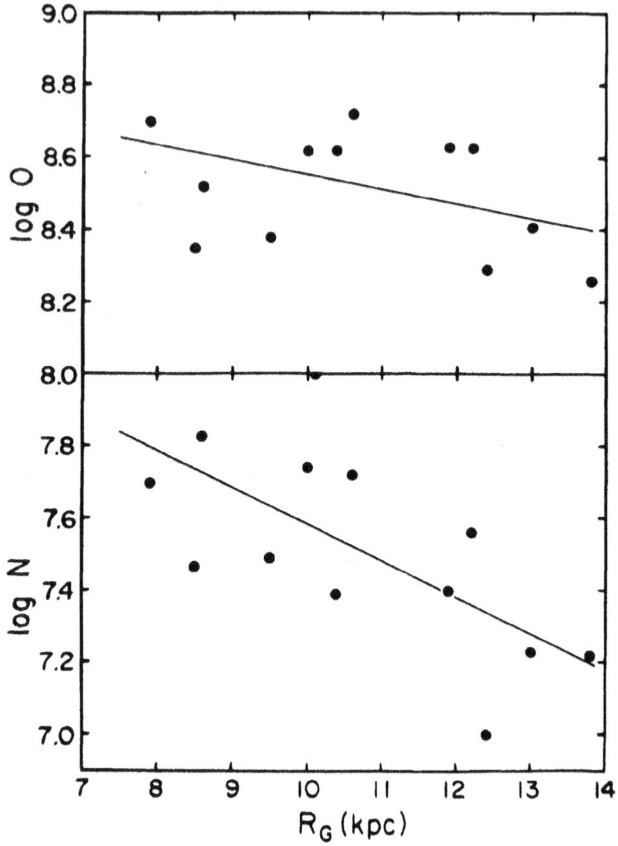

Figure 3: The relative oxygen (top) and nitrogen (bottom) abundances versus R_G as determined from galactic H II regions by Hawley (1978). The scale is log H = 12.

Pankonin, Walmsley, and Harwit (1979) in the H76α line, and by myself using unpublished H109α data taken by Churchwell, Mezger and Smith. In all cases T_e^* was found to be surprisingly constant across the nebula. Thus, even in a nebula known to have substantial density fluctuations, there is no evidence whatever for a radial temperature gradient. Therefore, with the exceptions of geometrical uncertainties (which cannot be unambiguously resolved) and possible selection effects, none of the major criticisms of the 6 cm data are important at $\lambda \lesssim 3$ cm.

The Data

The results of recent RRL surveys at $\lambda \leq 6$ cm are shown in Figure 4. They are as follows: Top left - H109α (open circles) - resolution = 2.6' arc by Churchwell et al. (1978, Paper II) and H110α (solid dots with error bars - the number of sources averaged together is given below each error bar) by Downes et al. (1980); Middle left - H85α - resolution ≈ 2.8' arc by Viner et al. (1976); Lower left - H86α - resolution ≈ 3.5' arc by Lichten et al. (1979); Upper right - H76α - resolution ≈ 1' arc by Thum et al. (1979); Middle right - H66α resolution ≈ 40" arc by T.L. Wilson, Bieging, and W.E. Wilson (1979) -Δ; T.L. Wilson, Pauls and Ziurys (1979) -Δ; and Thum et al. (1979) -□ ; Middle right - H66α - resolution ≈ 2.3' arc by Waltman and Johnston (1973) -Ⓞ; and Lower right = H56α - resolution ≈ 1.9' arc Berulis et al. (1975). In all cases T_e^* is plotted against galactic radius, R_G. The adopted distances are basically the same in each survey and in general the sources were either the same or a subset of those observed in Paper II. The H110α survey of Downes et al. (1980) is a large scale survey similar to the northern and southern H109α surveys but with better spatial resolution.

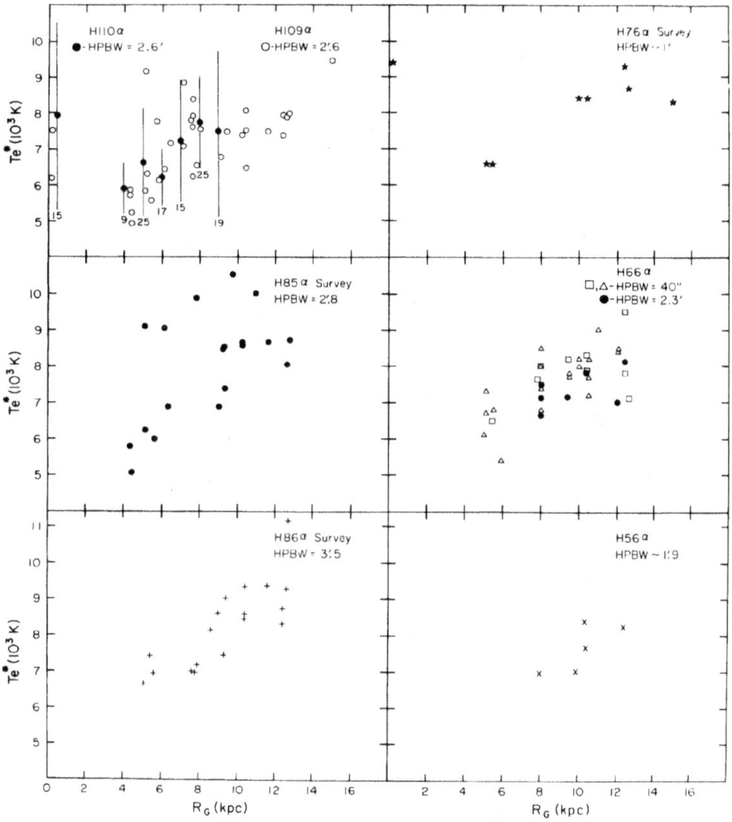

Figure 4: The LTE electron temperature, T_e^*, versus R_G: upper left, the H109α survey of Churchwell et al. (1978) designated by open circles and the H110α survey of Downes et al. (1980) designated by filled circles with error bars, the number of sources averaged together are given below each error bar; middle left, the H85α survey of Viner et al. (1976); lower left, the H86α survey of Lichten et al. (1979); upper right, the H76α survey of Thum et al. (1979); middle right, the H66α surveys of Thum et al. (1979); Wilson, Bieging, and Wilson (1979), Wilson, Pauls, and Ziurys (1979), and Waltman and Johnston (1973, lower right, the H56α data of Berulis et al. (1975).

The signal-to-noise ratios of the individual sources are consequently lower than those of Paper II and the errors larger.

All these surveys show a T_e^* gradient quite consistent both in intercept and in slope with that of the 6 cm survey of Paper II independent of transition and resolution. When the data are all plotted on the same graph there is a peak-to-peak dispersion of ∿2000K. A linear least squares fit to the combined data of all surveys (excluding DR21, Sgr B2, and W3(OH) due to possible optical depth effects) results in a gradient $T_e^* = 5570K+240R_G$ where R_G is in kpc and T_e^* is in K . The standard deviations of the intercept and slope are, respecitively, 280K and 30K. The correlation coefficient is 0.59. None of the surveys show a significant correlation of T_e^* with distance from the sun ($D_\odot$).

Due to collisional de-excitation of coolants, inclusion of all sources, independent of density, to determine T_e^* could introduce some scatter in the T_e^*-R_G relation or possibly could impose a systematic dependence of T_e^* on R_G. To avoid these possibilities, Mezger, et al. (1979) used only nebulae with $n_e(rms) \leq 10^3 cm^{-3}$ and RRL surveys with $\lambda \leq 3$ cm to derive the dependence of T_e^* on R_G. They found that from about 7 kpc to 4 kpc the T_e^* gradient rapidly steepens so that on average nebulae near 4 kpc are cooler than those in the galactic center. The detailed shape of the T_e^*-R_G relation is at the moment somewhat uncertain, but there can be no doubt that on average T_e^* decreases from $R_G \sim 15$ kpc to $R_G \sim 4$ kpc.

Selection Effects

Spatial Resolution. These surveys have resolutions ranging from 3.5' arc to 40" arc, and yet not only is the T_e^* vs R_G gradient essentially the same for all surveys, but also the T_e^* derived for a given nebula is, within the errors, generally the same at H109α and H66α (see Wilson, Bieging, and Wilson 1979). This demonstrates rather clearly that, contrary to the arguments given by Brown et al. (1978), resolution effects are not an important determinant of the T_e^* vs R_G gradient for the sample of H II regions observed. Nor does the transition observed appear to be important when $n \leq 109$. The more distant H II regions observed have on average a lower electron density as seen with a 2.6' arc beam than nearby ones, probably because most H II regions have extended low-density halos which dominate the RRL emission when the whole nebula is in the beam. Obviously different volumes of gas in a given H II region are selected both as a function of frequency and resolution but this seems to have little effect on the derived value of T_e^* for the nebula. Perhaps temperatures are basically constant throughout most H II regions as appears to be the case in Orion.

The Malmquist Effect. Another selection effect pointed out by Brown et al. (1978) as a primary determinant of T_e^* with distance is the "Malmquist effect". This effect is the following: at progressively greater distances, intrinsically brighter nebulae (i.e., greater excitation parameters) are selectively observed; or, in other words the weakest sources observable at 10 kpc are intrinsically brighter than the weakest ones observable at 1 kpc. The result of this, they argue is that nearby H II regions will have on average a higher T_e than more distant ones. It is certainly true that the 6 cm survey (Paper II) shows a well defined correlati of excitation parameter (U) with distance from the sun ($D_\odot$), see figure 5. The solid line is a linear least squares fit with a slope of 13.55 ± 1.81, an intercept of 30.9 ± 12.1, a correlation coefficient r=0.78 and a standard deviation σ_U=36.5. A plot of U vs R_G, however, shows much less correlation ($r \approx 0.34$). Of more interest is the fact that neither T_e^*(H109α) nor T_e^{model} (Paper II) are correlated with U ($r < 0.13$). It therefore is quite improbable that the Malmquist effect is responsible for the T_e^*-R_G correlation.

In conclusion of this section, the new short wavelength data pres-

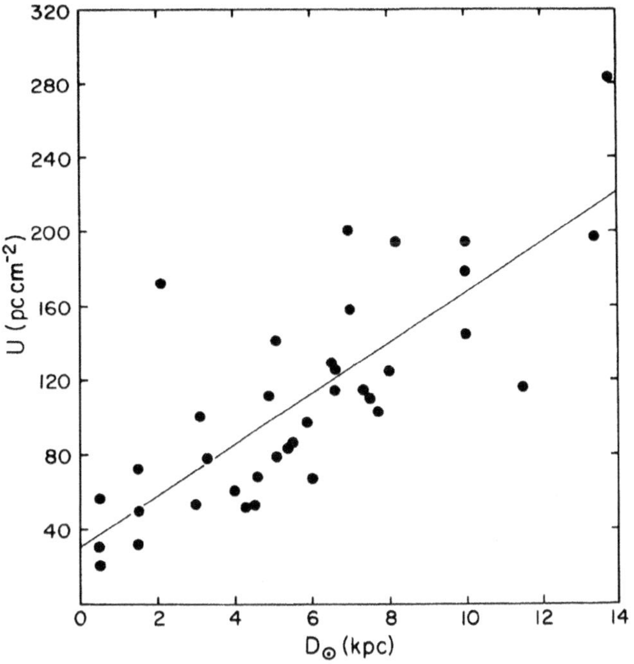

Figure 5: Excitation parameter (U in pc cm^{-2}) versus distance from the sun ($D_{\odot}$ in kpc) for H II regions in the 6 cm survey (Paper II). The solid line is a linear least squares fit to the points.

ently available seems to indicate that at $\lambda \lesssim 6$ cm a simple LTE analysis of RRL's gives a reasonably accurate measure of the mean electron temperature in H II regions. Both Walmsley and Shaver (this volume) have shown that stimulated emission and collision broadening effects begin to occur at about the same frequency. Thus, although there is evidence that both effects occur in some H II regions at 6 cm, they apparently roughly compensate for each other so that T_e^* (H109α) ≈ T_e^* (H66α). At n $\lesssim$ 85α neither of these effects are likely to be important. Neither the "Malmquist effect", resolution effects, nor different transitions (n $\lesssim$ 109) appear to have any systematic effect on T_e^*. I therefore conclude that the present best interpretation of the T_e^*-R_G relation is that it indicates a gradient of the mean electron temperatures of H II regions with galactic radius.

IMPLICATIONS OF A GALACTIC RADIAL TEMPERATURE GRADIENT

In the remainder of this review I will assume that a galactic temperature gradient exists and will consider some possible implications of such a gradient. In Paper II it was argued that neither a gradient in the effective temperature (T_*) of the ionizing stars nor a gradient in the absorption properties of dust in H II regions could account for the observed T_e gradient. It was however, shown on the basis of models by Balick and Sneden (1976) and Mathis (unpublished) that an increase in the relative abundance of heavy elements Z by about a factor of two from 13 kpc to 5 kpc could produce the observed T_e gradient. This is illustrated in Figure 6 where the T_e-R_G gradient is superimposed on models of Balick and Sneden (1976) which predict nebular equilibrium temperatures as a function of stellar effective temperature for abundances varying from Z = 0.32 $Z_{\odot}$ to Z = 10 $Z_{\odot}$. For Z = $Z_{\odot}$, a change in T_* from 32,000K to 50,000K can increase the temperature of the surrounding nebula by less than 1000 K, but a change in Z by a factor of 3 or so can easily repro-

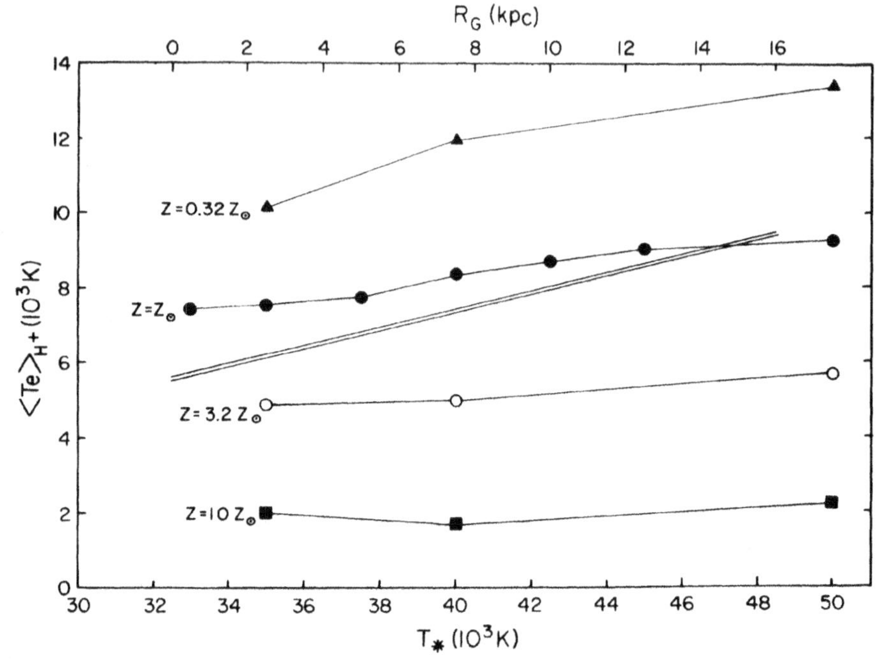

Figure 6: The mean nebular temperature $\langle T_e \rangle$ versus stellar effective temperature T_* for metal abundances ranging from $0.32\ Z_\odot \leq Z \leq 10\ Z_\odot$ in H II regions with a uniform density of 10^3 cm^{-3}. The data were taken from the models of Balick and Sneden (1976). The double line curve is the observed T_e^* versus R_G (scale at the top) relation superimposed on the models. This figure illustrates that over the range of relevant spectral types, the observed galactic temperature gradient cannot be produced by a change in T_* alone, but a small gradient in Z can easily account for the T_e gradient.

duce the T_e-R_G gradient. The Balick and Sneden (1976) models do not include line blanketing in the UV nor geometry appropriate for extended atmospheres, so one might expect the results to change somewhat when these effects are included. However, these are likely to be changes of degree rather than changes of trend. Thus, the general trend implied by the models are probably correct, but reliance on number predictions based on them would be unwise. Let us therefore grant that a gradient in Z is the basic reason why T_e varies with R_G and consider some implications of this.

Due to the T_e gradient, the [O III] line intensity ratio (λ5007 + λ4959)/λ4363 should increase by about a factor of 6 in nebulae with a uniform density $N_e = 10^3$ cm^{-3} near the galactic center relative to its value at R_G = 10 kpc and the [N II] (λ6548 + λ6583)/λ5755 ratio should

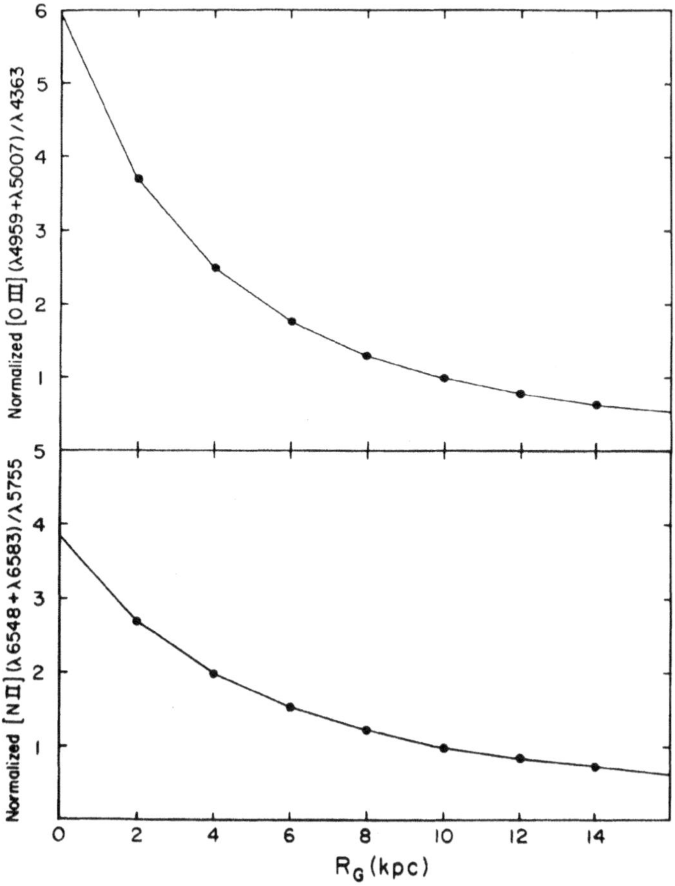

Figure 7: (Top) the expected intensity ratio [O III] (λ4959 + λ5007)/λ4363 (relative to its value at 10 kpc) versus R_G for idealized H II regions with a uniform electron density of 10^3 cm^{-3} and a galactocentric electron temperature gradient of T_e = 5570 + 240 R_G. The circled point is the point to which all others have been normalized. (Bottom) the expected variation with galactic radius of the [N II] (λ6548 + λ6583)/λ5755 intensity ratio in H II regions with the same conditions as in the top.

- - - - - -

increase by a factor of $\sim$ 4 in the same nebulae over the same R_G range (see Figures 7). Unfortunately, it is impossible to test this observationally in the Galaxy because of interstellar extinction at visible wavelengths. Increases are observed in several spiral galaxies but there are ionization problems which are discussed below.

The 88μ and 52μ fine structure lines of [O III] are not significantly attenuated by interstellar dust and should be observable to distances as great as RRL's in nebulae with $N_e < 10^4$ cm^{-3}. The 88μ line intensity is essentially independent of nebular temperature for $N_e \gtrsim 10$ cm^{-3} but it is a rapidly decreasing function with increasing density for $N_e > 10^{2.5}$ cm^{-3} because of collisional de-excitation, (see for example, Simpson, 1975). In H II regions, one would expect oxygen to be dominantely in the form of O^{++} even for nebulae ionized by stars as cool as 37,000K. Therefore in nebulae with $N_e \lesssim 10^3$ cm^{-3} the 88μ and 52μ [O III] lines should increase in approximately direct proportion to the increase in O abundance until optical depth effects become important. Thus, for

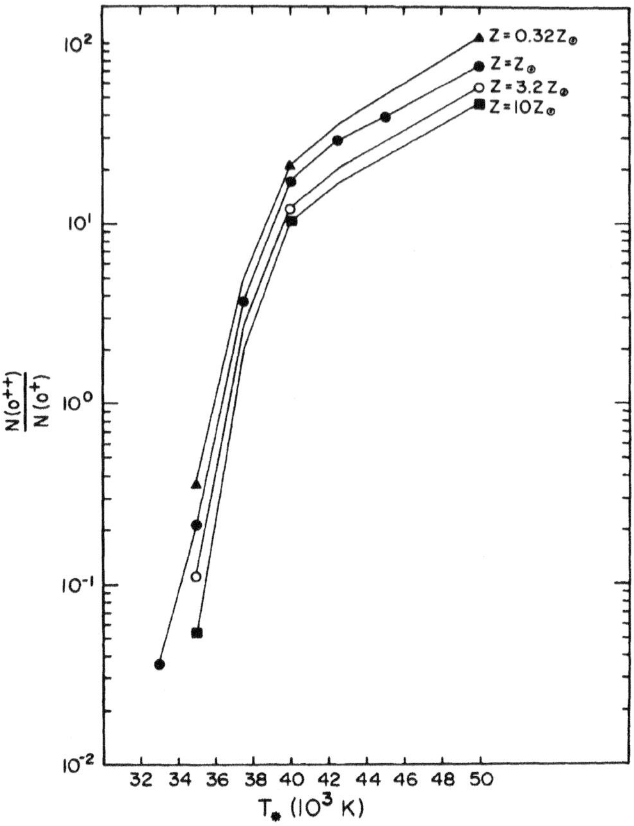

Figure 8: The relative O^{++} to O^{+} abundance ratio versus stellar effective temperature for 0.32 $Z_{\odot} \leq Z \leq 10$ $Z_{\odot}$ and $n_e = 10^3 cm^{-3}$ from the Balick and Sneden (1976) models. This shows that unless T* < 40,000K $n(O^{++}) > n(O^{+})$ for all relevant Z values.

\- - - - - -

an oxygen abundance gradient d log (O/H)/ $dR_G \simeq -0.04$, one would expect the 88μ line intensity to increase by about a factor of 4 from 15 kpc to 0 kpc in typical H II regions with $N_e = 10^3$ cm^{-3}, $\tau < 1$, and EM = 2 x 10^5 pc cm^{-6}. In fact, observations (Dain, et al. 1978) show relatively intense 88μ line emission from the bright optically visible H II regions in the solar neighborhood, but in the galactic center region only upper limits are observed which are more than a factor of 10 less than the 88μ line fluxes of nearby H II regions. Searches by Townes and coworkers have also failed to detect [O III] and [N III] infrared fine structure lines in the galactic center region. Although both Sgr A and Sgr B2 are known to contain high density compact components in which the 88μ lines would be quenched, from RRL and radio continuum observations it is known that both sources have a large low density component where the [O III] λ88μ line should be quite strong. The question is, why not? The answer appears to be low ionization (i.e., oxygen is mostly O^{+}, and helium is mostly He°). The Balick and Sneden (1976) models show that for T* $\geq$ 40,000K essentially all oxygen will be in the form of O^{++} regardless of the metal abundances, however as T* decreases below ∿ 40,000K the far UV metal absorption edges in stellar atmospheres become progressively deeper. Increased metal abundances enhances the absorption edges even more, so that when T* $\lesssim$ 37,000K $N(O^{++})/N(O^{+}) < 1$ in the surrounding nebula. This is illustrated in Figure 8. Here the ratio of O^{++}/O^{+}, as calculated by Balick and Sneden, is plotted as a function of T* for 0.32 $Z_{\odot} \leq Z \leq 10$ $Z_{\odot}$.

Shields and Tinsley (1976) and Shields and Searle (1978) have noted that in M101 the observed increase in H_β equivalent width with galactic radius and the increase in [O II] and decrease in [O III] strengths in the central regions requires both an increase in Z and a decrease in T_* toward the nucleus. They achieve a gradient of T_* with radius by invoking an upper mass limit (M_u) on star formation which decreases with decreasing R_G. Kahn (1974) has argued that Mu is basically determined when the radiation pressure on grains in an opaque cocoon surrounding a protostar prevents further accretion onto the star. By assuming that the grain opacity is proportional to Z, Shields and Tinsley (1976) showed that Mu $\propto Z^{-\frac{1}{2}}$ and therefore that a gradient in Mu may be a direct result of the gradient in Z.

From an observational point of view, a quite analogous situation appears to hold in our galaxy. a T_e gradient derived from RRL observations and direct abundance measurements at optical wavelengths in nearby H II regions and PN's strongly indicate a systematic increase in Z toward the galactic center. Oort (1977) has argued that the CO/H_2 abundance ratio in the galactic center region is significantly higher than in the solar neighborhood which probably indicates higher O/H and C/H abundances in the galactic center than at $R_G \sim 10$ kpc. Further, the low abundances of He^+ and O^{++} require low ionization in the galactic center which cannot be explained by the increase in Z alone. A decrease in T_* is required toward the galactic center. This might be an indirect consequence of the Z gradient via its effect on the absorption properties of dust which probably plays a dominate role in determining Mu. Panagia (1979) has proposed that interstellar dust could perhaps control the upper mass range of the IMF (initial mass function) via its influence on the temperature of dark clouds. Since the Jeans mass is a strong function of temperature ($\alpha\ T^{2/3}$), this is a plausible idea but needs to be pursued in more detail.

Mezger, Smith and Churchwell (1974), Paper II, and Sarazin (1976) argued that the low ionization in the galactic center was probably due to selective absorption by dust primarily because of the correlation of He^+/H^+ with far infrared flux. This explanation, however, appears less likely in the light of recent experimental and theoretical results and will not be considered further here (see the reviews by Mezger and Panagia for further discussion of this).

SUMMARY

New short wavelength RRL surveys (H109α, H110α, H86α, H76α, H66α, and H56α) all show roughly the same T_e^* gradient with R_G. Selection effects such as the "Malmquist effect", different transitions, and different angular resolutions do not appear to have any detectable influence on T_e^*. Comparisons of T_e^* determined from different transitions for a given nebula appear to be about the same within the errors. T_e^* is constant across the face of Orion at H109α, H76α and H66α. On the basis of the above, I conclude that T_e^* determined at $\lambda \leq 6$ cm is probably a reasonably accurate approximation to a beam-averaged electron temperature in H II regions and that the T_e^*-R_G gradient is indicative of a similar gradient in the actual mean temperatures of H II regions.

The low ionization in the galactic center (as evidenced by low abundances of He^+ and O^{++}) in conjunction with a radial temperature gradient is quite analogous to that in M101 and appears to be best explained by a combination of an increasing metal abundance and decreasing stellar effective temperature with decreasing galactic radius. It seems likely that all the gradients so far discussed (T_e, T_*, and Mu) are probably directly or indirectly results of a Z gradient.

Further work in this area should give high priority to more sensitive and extensive observations of the [O III] 88μ and 52μ lines and other forbidden infrared fine structure lines with better velocity resolution. As indicators both of abundance and ionization equilibrium, measurements would be very useful of [N II] (121.7μ), [N III] (57.31μ), [S III] (33.65μ, 18.68μ), [S IV] (10.52μ), [Ne II] (12.79μ) and [Ne III] (15.38μ, 36.10μ) in H II regions with a wide range of R_G values.

Broadband IR observations are still the best direct probe of interstellar dust and should be given high weight especially in dark dust clouds where star formation is thought to be occurring. High quality optical spectra, radio continuum and RRL data, should be combined to study the detailed properties of H II regions within ∿ 3 kpc of the sun. Special consideration should be given to H II regions thought to have low T_e values.

From the theoretical side one of the things needed most are realistic model atmosphere calculations which include non-LTE effects, extended atmosphere effects, line blanketing and the metal edges in the UV. Such models should be calculated for a range of abundances and the emergent spectrum used to estimate nebular parameters such as T_e and ionization equilibrium.

REFERENCES

Balick, B., Sneden, C. 1976, Ap.J., 208, 336.
Berulis, I.I., Smirnov, G.T., Sorochenko, R.L. 1975, Sov. Ast. Letters, 1, 187.
Brown, R.L., Gómez—González, J. 1975, Ap.J., 200, 598.
Brown, R.L., Lockman, F.J., Knapp, G.R. 1978, Ann. Rev. Astron. Astrophys., 16, 445.
Chaisson, E.J., Dopita, M.A. 1977, A&A, 56, 385.
Churchwell, E., Smith, L.F., Mathis, J., Mezger, P.G., Huchtmeier, W. 1978, A&A, 70, 719.
Churchwell, E., Walmsley, C.M. 1975, A&A, 38, 451.
Dain, F.W., Gull, G.E., Melnick, G., Harwit, M., Ward, D.B. 1978, Ap.J., 221, L17.
D'Odorico, S., Peimbert, M., Sabbadin, F. 1976, A&A, 47, 341.
Downes, D., Wilson , T.L., Bieging, J.H., Wink, J. 1980, in preparation.
Hawley, S.A. 1978, Ap.J., 224, 417.
Kahn, F.D. 1974, A&A, 37, 149.
Lichten, S.M., Rodriguez, L.F., Chaisson, E.J. 1979, Ap.J., 229, 524.

Lockman, F.J., Brown, R.L. 1976, Ap.J., 207, 436.
Lockman, F.J., Brown, R.L. 1978, Ap.J., 222, 153.
Mayor, M. 1976, A&A, 48, 301.
Mezger, P.G., Pankonin, V., Schmidt-Burgk, J., Thum, C., Wink, J. 1979, A&A, in press.
Mezger, P.G., Smith, L.F., Churchwell, E. 1974, A&A, 32, 269.
Oort, J.H. 1977, Ann. Rev. Astron. & Astrophys., 15, 295.
Pagel, B.E.J., Edmunds, M.G., Blackwell, D.E., Chun, M.S., Smith, G. 1979, M.N.R.A.S., submitted.
Panagia, N. 1979, Mem. Soc. Astron. Ital., 50, 79.
Pankonin, V., Walmsley, C.M., Harwit, M. 1979, A&A, submitted.
Pauls, T., Wilson, T.L. 1977, A&A, 60, L31.
Peimbert, M., Torres-Peimbert, S., Rayo, J.F. 1978, Ap.J., 220, 516.
Reifenstein, E.C. III, Wilson, T.L., Burke, B.F., Mezger, P.G., Altenhoff, W.J. 1970, A&A, 4, 357.
Rubin, V.C., Kumar, C.K., Ford, W.K. Jr., 1972, Ap.J., 177, 31.
Sarazin, C.L. 1976, Ap.J., 208, 323.
Searle, L. 1971, Ap.J., 168, 327.
Shaver, P.A. 1979, A&A, in press.
Sheilds, G.A. 1974, Ap.J., 193, 335.
Shields, G.A., Searle, L. 1978, Ap.J., 222, 821.
Shields, G.A., Tinsley, B.M. 1976, Ap.J., 203, 66.
Simpson, J.P. 1975, A&A, 39, 43.
Smith, H.E. 1975, Ap.J., 199, 591.
Thum, C., Mezger, P.G., Pankonin, V. 1979, A&A, submitted.
Torres-Peimbert, S., Peimbert, M. 1977, Rev. Mex. Astron. Astrofis., 2, 181.
Townes, C.H. 1979, IAU Symposium No. 87, Mont. Tremblant, Quebec, Canada.
Viner, M.R., Clarke, J.N., Hughes, V.A. 1976, A.J., 81, 512.
Waltman, E.B., Johnston, K.J. 1973, Ap.J., 182, 489.
Wilson, T.L., Bieging, J., Wilson, W.E. 1979, A&A, 71, 205.
Wilson, T.L., Mezger, P.G., Gardner, F.F., Milne, D.K. 1970, A&A, 6, 364.
Wilson, T.L., Pauls, T.A., Ziurys, L.M. 1979, A&A, submitted.

ON THE ELECTRON TEMPERATURE OF EXTENDED, LOW DENSITY HII REGIONS

A. Pedlar and R. D. Davies
Nuffield Radio Astronomy Laboratories
Jodrell Bank, Macclesfield, Cheshire

1. INTRODUCTION

Radio recombination lines are probably the most reliable method of determining the electron temperatures of low density HII regions. In high density objects with high emission measures, considerable error can be introduced by the enhancement of the line intensity by stimulated emission (Seaton 1974). Such effects are small in low emission measure objects ($E<10^4$ pc cm^{-6}). This has been demonstrated by observations of low frequency lines in the cases of W1, W16, W35 and W80, where no strong evidence for enhancement was found in a comparison of 220α and 166α lines (Pedlar & Davies 1972, Pedlar 1980). This implies that the 166α measurements lead to an electron temperature which is not seriously in error.

In this paper we shall be mainly concerned with the 166α results of Pedlar (1980) which can be combined with previous measurements (Pedlar & Matthews 1973, Hart and Pedlar 1976a) to provide a sample of 15 objects all in the local or Perseus spiral arms. In Table I we give the average 166α LTE electron temperatures deduced for these regions, together with the approximate emission measures deduced from the brightness temperature of the free-free continuum.

2. THE VARIATION OF ELECTRON TEMPERATURE AS A FUNCTION OF EMISSION MEASURE

Most of the electron temperatures lie in the range 4000 to 7000 K. Their mean value is 5700 ± 300 K, which is significantly lower than the value of 7500 to 8100 K which Churchwell et al. (1978) deduce to be the mean electron temperatures of HII regions in the local and Perseus spiral arms respectively. Most of the sources which Churchwell et al. observed were small diameter, high emission measure ($E > 10^5$ pc cm^{-6}) objects which are presumably younger than those studied in the present survey.

P. A. Shaver (ed.), Radio Recombination Lines, 239–244.

Table I Properties of extended, low density HII regions

SHARPLESS NO.	MEAN LTE ELECTRON TEMPERATURE (K)	APPROX. EMISSION MEASURE (PC CM^{-6})
S115	4250	8.5×10^3
S117	6900	3.0×10^3
S119	4750	1.6×10^3
S131	5600	1.8×10^3
S132	5600	6.4×10^3
S140	5100	2.1×10^3
S142	4900	5.8×10^3
S155	6100	-
S157	6900	5.5×10^3
S171	6400	1.3×10^4
S190, S199	} 6900	2.1×10^3
S220	5300	1.1×10^3
S261	4000	2.5×10^3
S275	7900	2.6×10^3
S280	5100	2.3×10^3

In Fig. 1 we have used the data from Table I, together with that of Churchwell et al. (1978) and Kazes et al. (1977), to show the variation of electron temperature with emission measure for HII regions situated between 9 and 13 kpc from the Galactic centre. Although there is a scatter of electron temperatures of approximately ± 1000 K, there is no doubt that the electron temperature is strongly correlated with the emission measure. A similar effect was noted by Gardner et al. (1970) in high emission measure objects. The effect is unlikely to be due to non-LTE effects, as we would expect the highest emission measure objects to give the largest line enhancements and hence the lowest apparent electron temperatures. Although there are large uncertainties in obtaining electron densities from emission measures, we have used the distances obtained by Georgelin & Georgelin (1970), to estimate the rms electron densities in the low brightness HII regions, and in Fig. 2 we have plotted these against electron temperature. Again a strong correlation is observed, which suggests that the average electron temperature increases from approximately 6000 K for rms electron densities less than 100 cm^{-3} to approximately 10,000 K for rms densities of 5×10^3 cm^{-3}. This increase in electron temperature is most likely due to the collisional de-excitation of metastable states of coolant ions such as OII etc., which reduces the cooling function in the higher density HII regions, resulting in a higher electron temperature. The HII region models of Stasinka (1978), for an O/H ratio of 7×10^{-4}, predict an increase from 6500 K to 7500 K if the density increases from 100 to 10,000 cm^{-3}. This can be reconciled with the above measurements if the degree of clumping is higher in high emission measure HII regions.

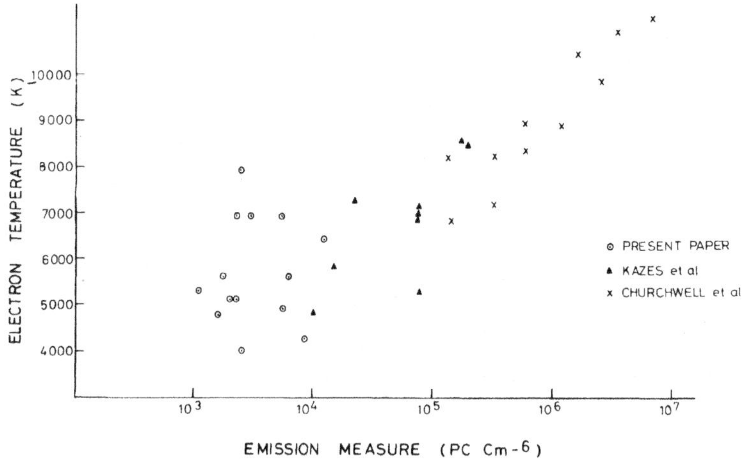

Fig. 1 The variation of electron temperature with emission measure for a sample of HII regions between 9 and 13 kpc from the Galactic centre.

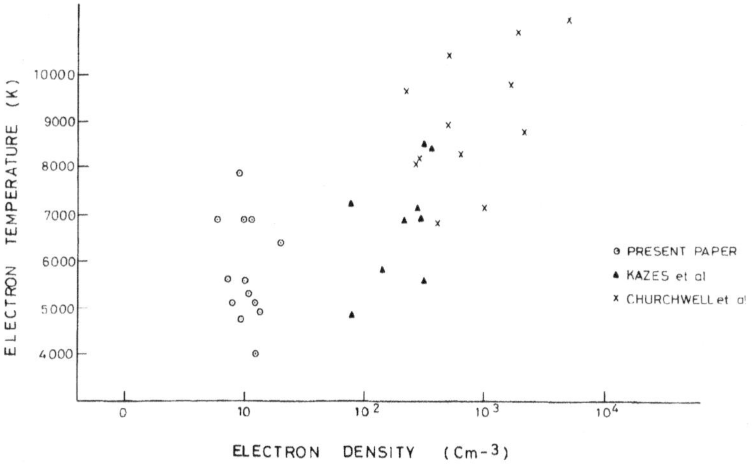

Fig. 2 The variation of electron temperature with rms electron density for a sample of HII regions between 9 and 13 kpc from the Galactic centre.

3. A POSSIBLE GRADIENT IN THE ELECTRON TEMPERATURE AS A FUNCTION OF GALACTOCENTRIC DISTANCE

A number of workers (Churchwell et al. 1978, Viner, Clark & Hughes 1976, Lichten, Rodriguez & Chaisson 1979) have reported a gradient in the electron temperature of high emission measure ($E > 10^5$ pc cm^{-6}) HII regions as a function of Galactocentric distance, R. Their data indicate that the mean electron temperature of such regions increases from 6000 K at R = 5 kpc to $\sim$ 8500 K at R = 13 kpc. It is interesting to speculate whether a similar gradient exists in the low density HII regions.

The present H166α data are taken from HII regions either in the local arm (R $\approx$ 10 kpc) or the Perseus arm (R $\approx$ 11.5 kpc). The average electron temperature for extended, low brightness HII regions in the local arm is 5500 $\pm$ 330 K and for the Perseus arm 5900 $\pm$ 560 K. The error is the standard deviation of the mean, and hence although the trend is consistent with the gradient of 310 K/kpc obtained by Churchwell et al. (1978), the statistical significance of the present data is small.

It is difficult to obtain reliable electron temperatures of low brightness HII regions in the inner (R < 8 kpc) parts of the Galaxy. The main problem is the unknown amount of nonthermal emission which makes the estimation of the thermal continuum temperature uncertain. However, most observations of "diffuse" recombination lines from the inner part of the Galaxy have resulted in electron temperatures significantly less than 5000 K. Hart & Pedlar (1976b), using a statistical argument, deduced the average electron temperature of diffuse ionized gas between 4 and 6 kpc to be 3100 $\pm$ 1700 K. Measurements of the H110α line by Jackson & Kerr (1975) eliminated the nonthermal background contribution and obtained an average electron temperature of 4400 $\pm$ 600 K Mebold et al. (1976) have investigated a region near $\ell = 24^o$ and deduced that the electron temperature is <6500 K and possibly 3000 K. Hirabayashi (1974) has used high frequency continuum data over the region $\ell < 35^o$ to deduce an average electron temperature of 3000 K. Observations of low frequency lines from the directions of the SNR 3C391 and W49B (Pankonin 1975) suggest that the electron temperature of intervening low density ionized gas is $\lesssim$4000 K.

S54 (NGC 6604, W35) is one of the few extended, low emission measure objects in the inner part of the Galaxy (R = 7.8 kpc) which can be studied without too much confusion from nonthermal emission. Radio recombination lines (Pedlar et al. 1978) and low frequency continuum (Shaver 1970) indicate its electron temperature to be 4000 K. Recently Shaver, McGee and Pottasch (1979) have reported electron temperatures approximately equal to 4000 K for two southern nebulae in the inner part of the Galaxy.

Thus there appears to be considerable evidence for a decrease in the electron temperature of diffuse low density HII regions from 6000 K at 12 kpc to 4000 K at 5 kpc. This is shown in Fig. 3 together with the corresponding curve for the higher density HII regions.

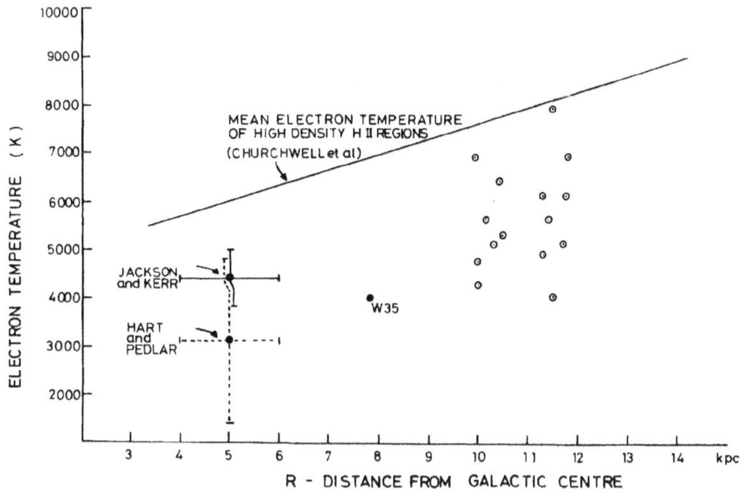

Fig. 3 The variation of the electron temperature of extended low density HII regions as a function of galactocentric distance. The observations of Jackson & Kerr and Hart & Pedlar have each been averaged over the region R = 4 to 6 kpc.

The increase in electron temperature for the high density regions has been interpreted as evidence for a metallicity gradient across the Galaxy (Churchwell et al. 1978), resulting in more coolant ions in HII regions in the inner region of the Galaxy. This results in an increase in the cooling function and a decrease in the electron temperature. Thus if the low density HII regions have the same abundances as their high density counterparts, then it is not too surprising to find that their electron temperatures also decrease in the inner parts of the Galaxy. The surprisingly low temperature (≈4000 K) would appear to be due to a combination of the increase in the abundance of coolants and a lack of collisional de-excitation of the coolant ions. The theoretical calculations of Stasinka (1978) would suggest that such low temperatures are only possible for O/H ratios somewhat in excess of 10^{-3}.

REFERENCES

Churchwell, E., Smith, L.F., Mathis, J., Mezger, P.G. and Hutchmeier, W., (1978). Astron.Astrophys. 70, 719.
Gardner, F.F., Milne, D.K., Mezger, P.G. and Wilson, T.L.,(1970), Astrophys.Lett. 6, 87.
Georgelin, Y.P. and Georgelin, Y.M., (1970). Astron.Astrophys. 6, 349.
Hart, L. and Pedlar, A., (1976a). Mon.Not.R.astr.Soc. 176, 135.
Hart, L. and Pedlar, A., (1976b). Mon.Not.R.astr.Soc. 176, 547.
Hirabayashi, H., (1974). Pub.Astron.Soc.Japan, 26, 263.
Jackson, P.D. and Kerr, F.J., (1975). Astrophys.J. 196, 723.
Kazes, I., Walmsley, C.M. and Churchwell, E., (1977). Astron.Astrophys. 60, 233.
Lichten, S.M., Rodriguez, L.F. and Chaisson, E.J., (1979). Astrophys.J., 229, 524.
Mebold, U., Altenhoff, W.J., Churchwell, E. and Walmsley, C.M., (1976). Astron.Astrophys. 53, 129.
Pankonin, V., (1975). Astron.Astrophys., 38, 445.
Pedlar, A., (1980). (Submitted to Mon.Not.R.astr.Soc.)
Pedlar, A., and Davies, R.D.(1972). Mon.Not.R.astr.Soc., 159, 129.
Pedlar, A., Davies, R.D., Hart, L. and Shaver, P.A., (1978). Mon.Not. R.astr.Soc. 182, 473.
Pedlar, A. and Matthews, H.E., (1973). Mon.Not.R.astr.Soc., 165, 381.
Seaton, M.J., (1974). Q.Jl.R.astr.Soc., 15, 370.
Shaver, P.A., (1970). Astrophys.Lett., 5, 167.
Shaver, P.A., McGee, R.X. and Pottasch, S.R. (1979), Nature, 280, 476.
Stasinka, G., (1978). Astron.Astrophys., 66, 257.
Viner, M.R., Clarke, J.N. and Hughes, V.A., (1976), Astron.J., 81, 512.

PART V

EXTRAGALACTIC RADIO RECOMBINATION LINES

PROSPECTS FOR OBSERVATIONS OF EXTRAGALACTIC RADIO RECOMBINATION LINES

P.A. Shaver
European Southern Observatory c/o CERN CH-1211 Geneva 23
Switzerland

ABSTRACT. The possibility of detectable extragalactic radio recombination lines is explored. Stimulated recombination lines may be detectable from the nuclei of active galaxies and quasars at the mJy level. The major uncertainty concerns the geometry of the nuclear emission-line region - its size and homogeneity. Low-frequency recombination lines may also be detectable from optical absorption-line regions in the directions of distant radio sources.

INTRODUCTION

Radio recombination lines may provide a powerful new tool in extragalactic research. At present they appear to be just at the limit of detectability, but in the future they may become more accessible. They would be important in a variety of problems: in the study of the ionized gas in galaxies and quasars, in the study of the gas associated with optical absorption-line systems, and in redshift determination for unidentified sources. They may provide a test of the cosmological hypothesis for quasars, a direct measurement of q_o, and evidence regarding the time-variability of fundamental physical quantities.

Searches for extragalactic radio recombination lines will be facilitated by the fact that there are many of them, and most of the work can be done within protected radio astronomy bands. By contrast searches for the HI line must cover a wide range of unprotected frequencies. The difficulty lies in the fact that the radio recombination lines are expected to be very weak, and those which originate within galaxies and quasars will be very broad, $\gtrsim 10^2 - 10^3$ km s^{-1}.

SPONTANEOUS AND STIMULATED EMISSION

In the optically-thin case, the flux in the line is given by

P.A. Shaver (ed.), Radio Recombination Lines, 247–258.

$$\Delta S_L \sim b_m \Delta S_L^* (1-\beta_n \tau_c/2) - \tau_L S_o \qquad (1)$$

where ΔS_L^* is the LTE value and S_o is the flux of a background continuum source. The other symbols have their usual meanings. The first term on the right-hand side represents the intrinsic line emission (i.e. associated only with the ionized gas), and the second terms represents line emission stimulated by the background continuum radiation. The intrinsic line emission is generally dominated by spontaneous emission (i.e. $b_m (1-\beta_n \tau_c/2) \sim 1$), so when there is no background source the line flux from a galaxy or quasar is given simply by

$$\Delta S_L^* \sim 2.10^{-36} \nu L(H\beta)/\Delta V_L' T_e^{0.62} D^2 \text{ mJy} \qquad (2)$$

with frequency ν in GHz, Hβ luminosity L(Hβ) in ergs s^{-1}, electron temperature T_e in K, distance D in Mpc (cosmological subtleties are ignored here), and $\Delta V_L' = \Delta V_D(1+1.48\gamma)$ where ΔV_D is the Doppler line width (FWHM) in km s^{-1} and γ is the ratio of pressure (or radiation) broadening to Doppler broadening. Thus if we require $\Delta S_L > 1$ mJy (very optimistic at present even for an Arecibo-type telescope), then for $\Delta V_D = 500$ km s^{-1}, $\gamma = 0$, $T_e = 10^4$ K, and $\nu = 5$ GHz,

$$D < [7.10^{-41} L(H\beta)]^{\frac{1}{2}} \text{ Mpc} \qquad (3)$$

Typically L(Hβ) $\sim 3.10^{40}$ erg s^{-1} for normal galaxies, giving D < 1.5 Mpc; for active galaxies and quasars L(Hβ) $\sim 10^{42}$ erg s^{-1}, and D < 8 Mpc. Thus, spontaneous radio recombination line emission will only be detected from relatively nearby systems.

By contrast, the detectability of stimulated recombination line emission is only limited by the intensity of the background source. For pure stimulated emission,

$$\Delta S_L \sim - \tau_L S_o \qquad (4)$$

and the line-to-continuum ratio is the optical depth in the line. Again requiring $\Delta S_L > 1$ mJy, and assuming a continuum flux S_o of $\sim 1 - 10$ Jy, it is only necessary that $|\tau_L| > 10^{-4}$-10^{-3}, and it will be shown below that such optical depths are possible. Fig. 1 shows line intensities for hypothetical galaxies or quasars, and the great enhancement provided by stimulated emission is obvious.

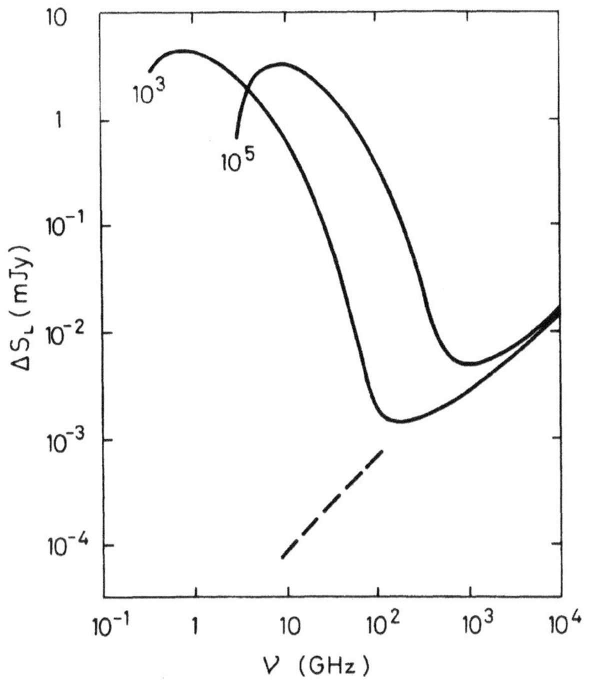

Figure 1. Possible Hnα line intensities from a quasar at z ∿ 1 , for the two model emission-line regions from Table 1 with $N_e = 10^3$ and 10^5 cm^{-3}. The dashed line represents pure spontaneous emission. The quasar has a 5 GHz flux density of 10 Jy and spectral index -0.75.

NUCLEI OF ACTIVE GALAXIES AND QUASARS

There are two distinct types of emission-line region within the nuclei of active galaxies and quasars (e.g. Baldwin, 1975; Osterbrock, 1978,a,b). The "broad-line region" is dense ($\gtrsim 10^8$ cm^{-3}) and small ($\lesssim$ 1 pc), and emits permitted optical lines of widths $\gtrsim$ 3000 km s^{-1} (FWHM); this region becomes opaque in the infrared (Fig. 2), and is of little interest for radio recombination line observations. The other region (the "narrow-line region") is typified by densities of ∿ 10^3 - 10^5 cm^{-3} and sizes of ∿ 10^2 - 10^3 pc, and emits relatively narrow optical lines (∿ 500 km s^{-1}). The fact that it is expected to become opaque at ∿ 0.1 - 10 GHz (Fig. 2), where considerable structure is often seen in the spectra of extragalactic radio sources, suggests that it may influence the radio spectra by free-free absorption. In that case it may also give rise to detectable radio recombination lines by virtue of stimulated emission.

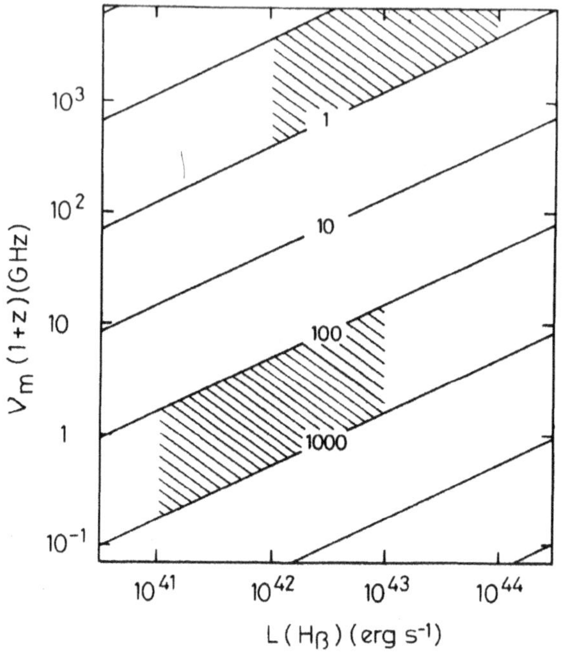

Figure 2. Turnover frequency ν_m of a nonthermal source with spectral index -0.8 due to free-free absorption in a surrounding sphere of thermal plasma with Hβ luminosity L(Hβ) and diameter 2R. The straight lines signify constant diameter (pc), as indicated; the top and bottom cross-hatched areas represent the broad- and narrow-line regions respectively.

Table 1. Models of Emission-Line Regions

N_e(cm^{-3})	10^2	10^3	10^4	10^5
T_e(K)	15000	15000	15000	15000
ΔV_D(km s^{-1})	500	500	500	500
L(Hβ) (erg s^{-1})	10^{42}	10^{42}	10^{42}	10^{42}
2R (pc)	$10^{3.5}$	10^3	$10^{2.5}$	10^2
f	3.10^{-3}	8.10^{-4}	3.10^{-4}	8.10^{-5}
ν_1(GHz)	0.1	0.3	1	3

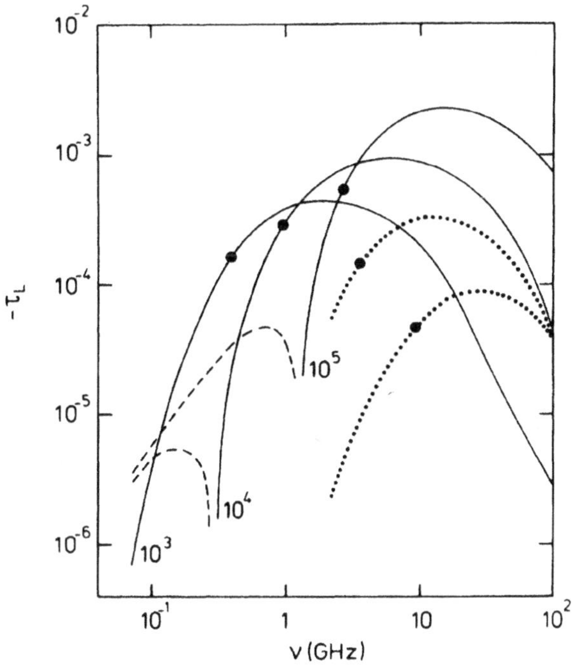

Figure 3. Hydrogen α-line optical depth as a function of frequency for the three model emission-line regions from Table 1 with $N_e = 10^3$, 10^4, and 10^5 cm^{-3} (Shaver, 1978). Dashed curves signify positive τ_L, and the dots show where $\gamma = 1$. The dotted lines show the effect of the continuum radiation field in the $N_e = 10^4$ cm^{-3} case, for $S_o = 1$ and 10 Jy at z = 1.

The optical depth at line centre can be written

$$\tau_L \sim 10^{-32}\ L(H\beta) A_{X^+}\ b_n \beta_n / \nu \Delta V_L'\ T_e^{1.62} R^2 \qquad (5)$$

where A_{X^+} is the abundance of the ion X^+. For a typical narrow-line region, $L(H\beta) \sim 10^{42}$ erg s^{-1}, $\Delta V_D \sim 500$ km s^{-1}, $T_e \sim 15000$ K and $2R \sim 10^2 - 10^3$ pc; eq. 5 has been evaluated in the case of hydrogen for a range of electron densities and frequencies, and the results are shown in Fig. 3. The optical depth reaches 10^{-4}-10^{-3} over a considerable frequency range, indicating that the lines are in principle detectable.

Heavy elements may also generate detectable radio recombination lines. The low abundances are compensated by the huge overpopulation of high energy levels by dielectronic recombination, which is very efficient at these comparatively high temperatures ($\gtrsim$ 15000 K). Fig. 4 gives some idea of the temperature dependence of the lines for various elements.

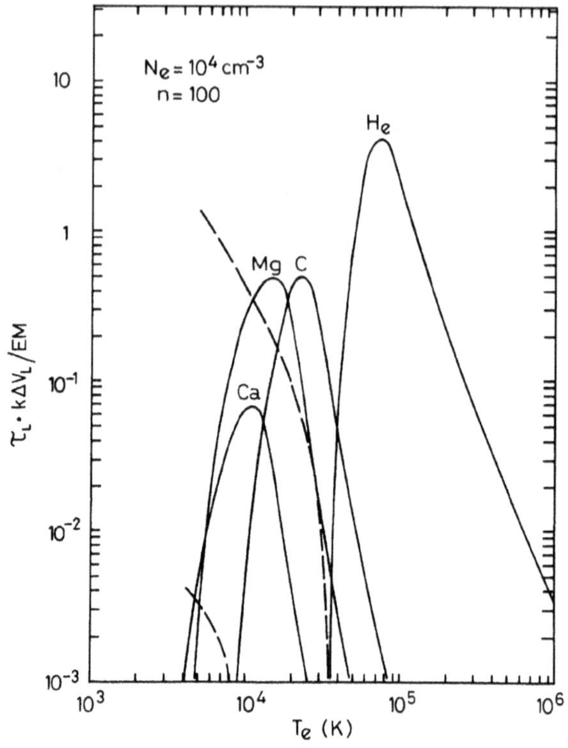

Figure 4. Temperature dependence of the line optical depth for various atoms. Dashed lines signify negative values. Solar abundances have been used for the ions, allowing for collisional ionization. The b_n-factors were computed using the program of Shaver (1976). Dielectronic recombination coefficients are from Shore (1969); somewhat lower values are suggested by more recent work (Brooks et al., 1978; Jacobs et al., 1979).

At 15000 K it appears that the Mg recombination lines would be dominant, and Fig. 5 shows their optical depth as a function of density and frequency. Again values of 10^{-4}–10^{-3} are common, although in this case the lines at their strongest would appear in absorption.

Complications. There are two factors, both associated with the geometry of the emission-line region, which could reduce the optical depth below these values. The first is the intense radiation field of the central continuum source, which can diminish the line optical depth by broadening the line through radiation damping, and by altering the energy level populations.

The linewidth due to radiation damping is

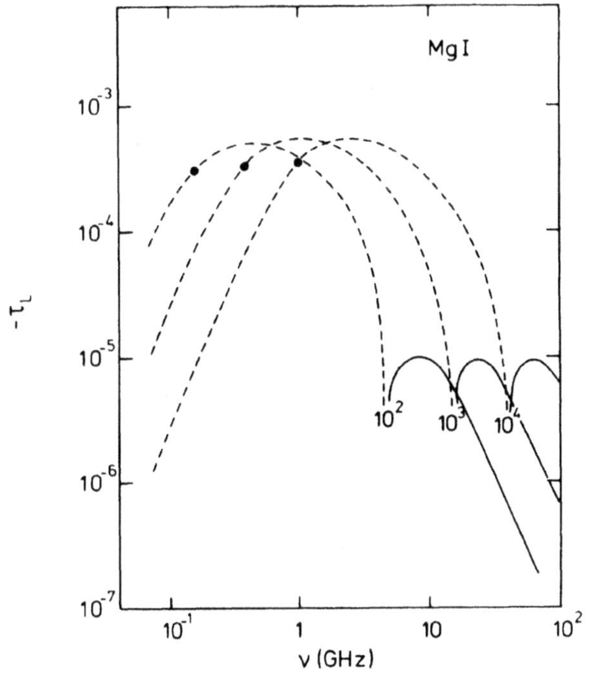

Figure 5. Magnesium α-line optical depth as a function of frequency for three model emission-line regions from Table 1 with $N_e = 10^2$, 10^3, and 10^4 cm^{-3} (Shaver, 1978). Abundances and atomic parameters are as in Fig. 4. Dashed curves signify positive τ_L, and the dots show where γ=1.

$$\Delta V_R \sim \left(\frac{n}{100}\right)^7 \frac{S_o}{\theta^2} \text{ km s}^{-1} \tag{6}$$

and that due to pressure broadening is

$$\Delta V_P \sim \left(\frac{n}{100}\right)^{7.4} \left(\frac{N_e}{1000}\right) T_e^{-0.1} \text{ km s}^{-1} \tag{7}$$

Θ is the angular size of the emission-line region in arcsec, S_o is in Jy and N_e is in cm^{-3}. The n-dependence in these two equations is almost the same, and the T_e-dependence is weak; the relation $\Delta V_R/\Delta V_P \sim 3000$ $S_o/N_e\Theta^2$ is valid within a factor of two for n in the range 50 - 200 and T_e in the range 10^4 - 10^5 K. For $N_e = 10^4$ cm^{-3}, S_o = 1 Jy, and Θ = 0."1

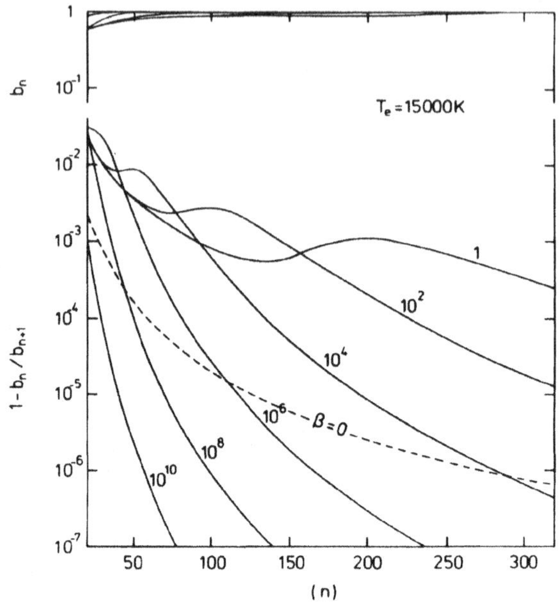

Figure 6. Departure coefficients for hydrogenic atoms plotted against principal quantum number n, for T_e = 15000 K and several electron densities (cm^{-3}) as indicated (Shaver, 1978). The dashed line is the locus of points for which stimulated emission and absorption exactly cancel each other (β= 0), for Δn << n.

(corresponding to 400 pc at z ∿ 1), the ratio $\Delta V_R/\Delta V_P$ ∿ 30 and radiation damping is much more important than pressure broadening.

The radiation field also affects the energy level populations, by increasing the rates of stimulated emission and absorption; ultimately, in the strong field limit, these become equal and there is no net line emission. This condition can be expressed as $N_n/\omega_n = N_m/\omega_m$, or $\beta_n = 0$, whe ω_n is the statistical weight for level n. It is indicated in Fig. 6 by the dashed line; above that line the optical depth is negative, below it positive. As the intensity of the radiation field is increased, the factors $1 - b_n/b_{n+1}$ move towards this line.

The combined effects of the radiation field on the line optical depth are shown in Fig. 3 by the dashed line, for the case $N_e = 10^4$ cm^{-3} with S_o = 1 and 10 Jy at z = 1. The optical depth is substantially reduced. To minimize such effects one would prefer small S_o and large θ (or R), but this would produce weak lines for other reasons (eqs. 4 and 5).

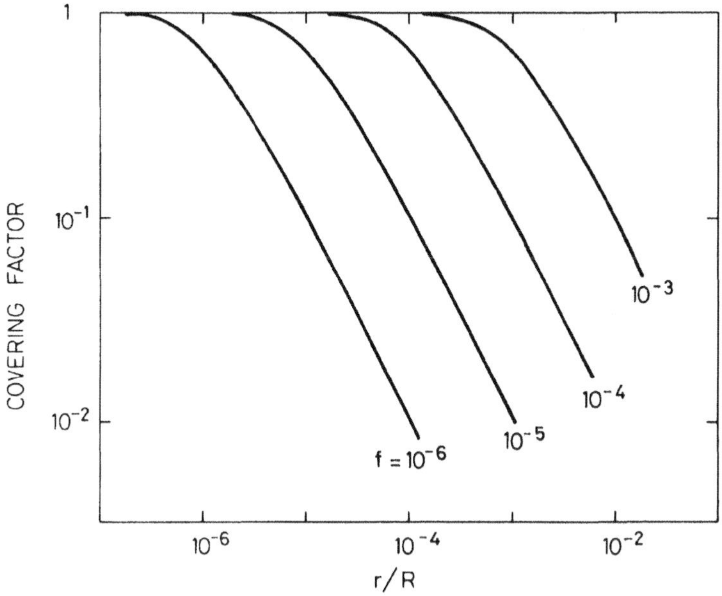

Figure 7. Covering factor plotted against the ratio of cloud radius r to total radius R for various values of the filling factor f.

The second complication concerns the structure of the emission-line region. If the filling factor is small (as appears to be the case), and if the clouds of gas are sufficiently large, only a small fraction of the line of sight to the central radio continuum source will contain ionized gas (the "covering factor" will be small). This is illustrated in Fig. 7, which shows the covering factor as a function of the filling factor f and cloud size r. When r is small (for given f) the continuum source is covered uniformly by ionized gas, and eq. 5 applies. When r is large th coverage becomes incomplete. The critical dividing point is obviously where $r/R = f$. As filling factors for the line-emission regions of active galaxies and quasars are thought to be very small ($\sim 10^{-6} - 10^{-4}$), and as there is no clear evidence for line-of-sight ionized gas in their radio properties (spectra, polarization, etc), the covering factor may indeed be small for most sources. Alternatively, if radio recombination line emission were detected from such a source, it would prove that these other radio properties must also be affected by line-of-sight ionized gas.

Detection of a radio recombination line from a quasar would also support the cosmological interpretation of its redshift. The distance can in principle be obtained directly from

$$D^2 \propto \tau_L^3/f^2 \; F(H\beta) \qquad (8)$$

where F(Hβ) is the measured Hβ flux (Shaver, 1978); an analogous expression has been derived by Gunn (1978) for the hot intracluster gas observed in X-rays. A large optical depth τ_L would be required for detectable radio recombination line emission, and this would in turn imply a large distance, from eq. 8. The uncertainty introduced by the filling factor f could perhaps be reduced by a statistical approach.

ABSORPTION-LINE REGIONS

The gas which causes the absorption lines in the optical spectra of distant quasars may also give rise to detectable radio recombination line emission. The optical lines are narrow (< 100 km s^{-1}) and of low ionization. HI absorption lines have been observed at the absorption-system redshift in a few cases, and separate components with linewidths of 5-10 km s^{-1} are seen (Wolfe and Davis, 1979, and references therein). Physical conditions in these absorption-line regions appear to be similar to those derived from ultra-violet interstellar absorption-line spectra of stars in our own galaxy: normal abundances, and column densities of

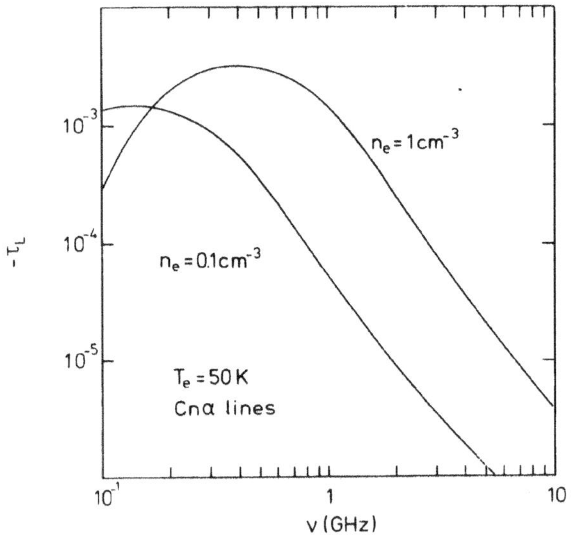

Figure 8. Carbon (CI)α-line optical depths plotted against frequency for two values of electron density as indicated (Shaver, 1978). Parameters are intended to be representative of absorption-line regions: T_e = 50 K, ΔV_D = 5 km s^{-1}, and $N(C^+) = 10^{17}$ cm^{-2}.

$N(Mg^+) \sim 10^{16}$ cm^{-2} and $N(HI) \sim 10^{21}$ cm^{-2} for T = 50 K (Smith et al., 1977, Wolfe and Wills, 1977). Carbon radio recombination lines have been observed in the clouds which produce optical interstellar lines in our galaxy (Crutcher, 1977), and for these clouds $N(C^+) \sim 10^{17}$ cm^{-2}, $N_e \sim 10^{-1}$ cm^{-3}, and $T \sim 50$ K (Morton, 1975; Snow, 1976; Black and Dalgarno, 1977). It thus seems likely that carbon radio recombination lines may also be detectable in the high-redshift absorption spectra of quasars.

The optical depth of such a carbon recombination line is

$$\tau_L \sim 2.10^{-16} N_e N(C^+) b_n \beta_n / \nu \Delta V_L' \, T_e^{2.5} \tag{9}$$

Fig. 8 shows τ_L computed for $N(C^+) = 10^{17}$ cm^{-2}, $\Delta V_D = 5$ km s^{-1}, $\gamma = 0$, $T_e = 50$ K, and $N_e = 0.1$ and 1 cm^{-3} at various frequencies. Values as high as 10^{-3} are attained, indicating that these lines may also be detectable. The difficulty in this case is that the lines are narrow and at their strongest below 1 GHz where the galactic background radiation dominates; both factors limit the sensitivity which can be achieved.

Detection of such a line would give a lower limit for the distance from the quasar to the absorption-line region. This follows from the effect of the continuum radiation field on the linewidth and energy level populations, as discussed above. For n = 200 and $S_o = 10$ Jy at z = 1, the absorption-line region must be at least 50 kpc from the radio nucleus (eq. 6). Comparable limits can be obtained in a similar way using HI absorption lines (Bahcall and Ekers, 1969; Wolfe and Davis, 1979). If a carbon recombination line and an HI line were both detected from the same absorption-line region, it would also be possible to set tighter limits on the constancy of certain fundamental physical quantities over cosmological time scales, by comparing the line frequencies (cf. Wolfe et al., 1976; Wolfe and Davis, 1979). And narrow lines at high redshift such as these might someday be used to measure q_o directly (cf. Saslaw, 1977).

INTERGALACTIC MEDIUM

In principle there should also be stimulated recombination line emission from the general intergalactic medium in the directions of strong radio sources, but the optical depths would be impossibly low due to the very high temperatures of $\sim 10^7$–10^8K (Shaver, 1975). Dielectronic recombination can greatly enhance the intensities of helium recombination lines at temperatures near 10^5K (Fig.4; Burgess and Summers, 1969), but intergalactic temperatures are much too high in this case also.

There is the further possibility of hydrogen and helium recombination lines formed during the recombination era of the primordial plasma (Dubrovich, 1975). Redshifted H10α -H15α lines would appear at radio frequencies, but the line intensities and widths are estimated at $\sim 10^{-4}$K and 2.10^4 km s^{-1} - observationally impossible for the foreseeable future.

CONCLUSIONS

Stimulated radio recombination lines may be detectable from active galaxies and quasars out to very large redshifts. Line intensities as high as 1 - 10 mJy might be anticipated. Those regions causing the optical absorption lines in the spectra of quasars may produce detectable recombination lines, but these would be strongest at frequencies below 1 GHz where detection is difficult. The most likely sources of detectable line emission are the optical emission-line regions in the nuclei of galaxies and quasars; the major uncertainties in estimating possible line intensities concern the structure of the emission-line region - its size and clumpiness. The observational problem is basically that of detecting very weak and very broad lines. This pushes present instrumentation to the limit, but there is no fundamental reason why these technical challenges cannot be met in the not-too-distant future.

REFERENCES

Bahcall, J.N., Ekers, R.D. 1969, Astrophys. J. 157,1055
Baldwin, J.A. 1975, Astrophys.J. 201,26
Black, J.H., Dalgarno, A. 1977, Astrophys. J. Suppl. 34,405
Brooks, R.L., Datla, R.U., Griem, H.R. 1978, Phys.Rev.Letts. 41,107
Burgess, A., Summers, H.P. 1969, Astrophys. J. 157,1007
Crutcher, R.M. 1977, Astrophys. J. 217, L109
Dubrovich, V.K. 1975, Soviet Astron.Letts. 1,196
Gunn, J.E. 1978, in "Observational Cosmology" (ed. J.E. Gunn, M.S. Longair, M.J. Rees; Geneva Observatory), p.3
Jacobs, V.L., Davis, J., Rogerson, J.E., Blaha, M. 1979, Astrophys. J. 230,627
Morton, D.C. 1975, Astrophys. J. 197,85
Osterbrock, D.E. 1978a, Physica Scripta 17,285
Osterbrock, D.E. 1978b, Bull. Amer.astr. Soc. 9,647
Saslaw, W.C. 1977, in "Radio Astronomy and Cosmology" (ed. D.L. Jauncey, D. Reidel), p. 379
Shaver, P.A. 1975, Pramana 5, 1
Shaver, P.A. 1976, Astron.Astrophys. 46,127
Shaver, P.A. 1978, Astron.Astrophys. 68,97
Shore, B.W. 1969, Astrophys. J. 158,1205
Smith, H.E., Burbidge, E.M., Junkkarinen, V.T. 1977, Astrophys. J. 218,611
Snow, T.P. 1976, Astrophys. J. 204,759
Wolfe, A.M., Brown, R.L., Roberts, M.S. 1976, Phys.Rev.Lett. 37,179
Wolfe, A.M., Davis, M.M. 1979, Astron. J. 84,699
Wolfe, A.M., Wills, B.J. 1977, Astrophys. J. 218,39

RADIO RECOMBINATION LINES IN EXTRAGALACTIC SOURCES

M. B. Bell
Herzberg Institute of Astrophysics
National Research Council of Canada, Ottawa

1. INTRODUCTION

Attempts to detect Hydrogen recombination lines from extragalactic sources using narrow 10MHz observing windows began at least as far back as 1971, when E. Seaquist and I searched for emission lines in 3C273. Other unreported searches with negative results were almost certainly made, but meaningful observations began only in late 1975 when wider observing windows (30-90 MHz) became available. Although many improvements in instrumentation and observing techniques have been made in the last four years mere detection of the wide, weak recombination lines found in external galaxies is still a struggle. However detections have now been made in at least two sources with sufficient accuracy to enable us to define the line producing mechanism and, with some improvements in instrumentation many more detections may be made soon, in what is turning out to be an exciting new field of astronomy.

2. DETECTION OF RECOMBINATION LINES IN EDGE-ON GALAXIES

The first detection of a recombination line from an external galaxy (beyond the Magellanic Clouds) was made in April 1976 by Shaver Churchwell and Rots (1977) who detected a weak (15 mJy) H166α line in the edge-on galaxy M82 using the Westerbork Synthesis Radio Telescope. Their spectrum was not an impressive one, as can be seen from figure 1, but their detection was soon followed by more convincing detections of this and other lines both at ARO and at Effelsberg. They interpreted the H166α line results to indicate that the known H II regions M82I and M82II are foreground objects responsible for both the turnover in the continuum spectrum and for the observed emission line, through stimulation by the background nonthermal continuum. They pointed out that their interpretation could be tested by observing higher frequency recombination lines in M82. For spontaneous emission the flux in the line will increase linearly with frequency. For stimulated emission it will decrease with the strength of the background continuum source. In September, 1976, five months after the original detection at Westerbork, the M82 recombination line was confirmed by Bell & Seaquist

P.A. Shaver (ed.), Radio Recombination Lines, 259–268.

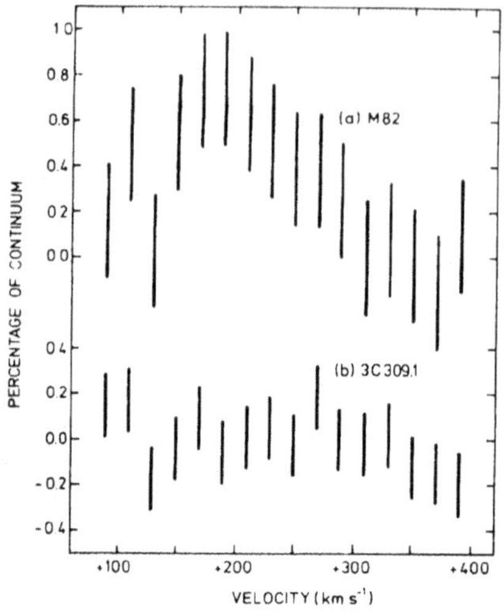

Fig. 1. H166α spectra of (a) M82 and (b) reference source 3C309.1. The lengths of the error bars correspond to 2σ for a single observation. The velocity is heliocentric. (Reprinted from Shaver, Churchwell and Rots, 1977, Astron. Astrophys. 55, *435.)*

(1977) who detected a weak (21 mJy) H102α emission line at ARO. Soon after, Chaisson & Rodriguez (1977) reported detection of a relatively strong (71 mJy) H92α emission line using the Haystack antenna. This was followed by detection of the H110α line at Effelsberg (Shaver Churchwell & Walmsley, 1978). In general the results indicated a decrease in peak line strength with increasing frequency which is opposite to the linear increase expected for spontaneous emission in an unresolved source. Figure 2, from Shaver Churchwell & Walmsley (1978) compares the variation of the measured peak line strength with the predicted variation for the two different line producing mechanisms. It is obvious from the figure that the observed H166α, H110α & H102α lines cannot be due to spontaneous emission, and as originally suggested by Shaver, Churchwell & Rots (1977) must, for the most part, be caused by stimulated emission.

The intensity of the H92α line at 8.3 GHz reported by Chaisson & Rodriguez (1977) appears to be inconsistent with the observations of the lines at 4.9, 6.1 and 14.7 GHz, given the usual theory of recombination line emission, and for that reason the H92α line has not been included.

In June 1977 the H85α line was detected at the Algonquin Radio Observatory, (Bell & Seaquist, 1978). Since the H110α & H85α measurements were made with essentially identical beamwidths the observed intensity ratio for these lines could not be attributed to resolution effects in the presence of extended disk emission in M82. The strength of the H85α line (10 mJy), H102α and H110α lines suggested further that the line reported by Chaisson and Rodriguez had been overestimated.

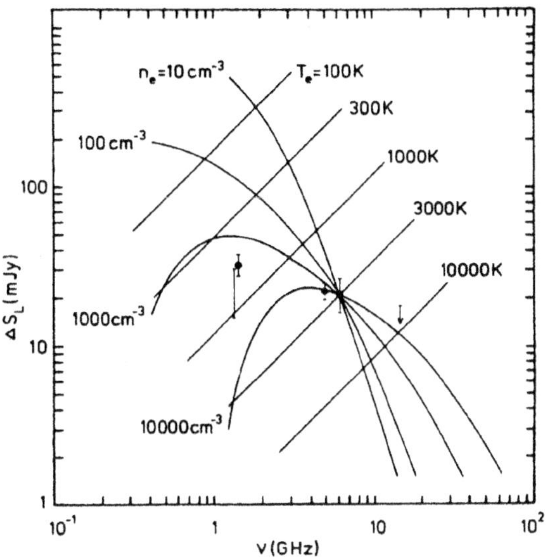

Fig. 2. Peak α-recombination line flux density plotted against frequency for M82. The straight lines represent pure LTE spontaneous emission for a range of electron temperatures, and the curves represent pure stimulated emission for T_e=5000K and a range of electron densities (normalized to the H102α point); in both cases τ<<1 and ΔV_L=200 km s^{-1}. (Reprinted from Shaver, Churchwell and Walmsley, 1978, Astron. Astrophys., 64, 1.)

Using the Effelsberg & ARO data, Bell & Seaquist (1978) derived the parameters of a simple model for M82 (Table I) which assumed an HII

TABLE I

Parameters Derived for Uniform H II Region from Peak Line Strengths

T_e	5000K
n_e	150 cm^{-3}
Ω	8.4 × 10^{-10}sr
E_c	2.3 × $10^6 cm^{-6}$pc
$S_N{}^0$	1.8 Jy
$T_b{}^\dagger$	246K
Thermal flux	197 mJy

†Of nonthermal background at 8.1 GHz.

region (characterised by the kinetic Temp T_e, electron density n_e, angular area Ω and emission measure E_c) in the form of a uniform cube of ionized gas between the observer and the nonthermal source. Figure 3 shows the model curve fitted to the data. The radio recombination line data for M82 thus demonstrates that the lines, produced in ionized gas between the observer and the nucleus, arise primarily through stimulation by the background continuum source. The data suggests the lines are formed in gas with a temperature of about 5×10^3K and emission

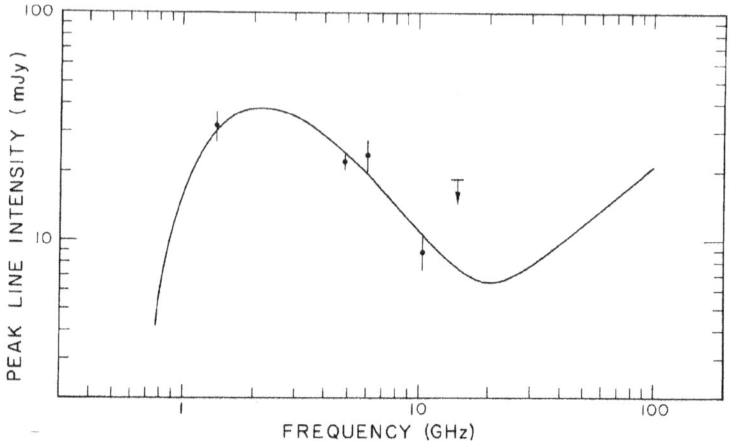

Fig. 3. Theoretical curve computed for M82 using the model parameters in Table I. (Reprinted from Bell and Seaquist, 1978, Ap. J., 223, 378.)

measure of $2{\times}10^6 cm^{-6}$ pc and these values are in good agreement with estimates by other observers from independent evidence. In particular, the value for the emission measure listed in Table I is within a factor of about 3 of the value found by Kronberg and Wilkinson (1975) for the region near M82 II, 10" west of M82 I. Their estimate is based on the assumption that the low-frequency turnover in the spectrum of the compact radio source nearly coincident with this H II region is produced by free-free absorption. Similar figures were derived for the emission measure of M82 I and II by Recillas-Cruz and Peimbert (1970) from the H_α intensities.

In May 1977, one year after the first detection of a recombination line from M82, the H102α line was detected in NGC253 at ARO (Seaquist & Bell, 1977). The H112α line was later detected at Bonn (Mebold et al.) and the combined results to date, shown plotted against frequency in Fig. 4 suggest, that as in M82, the recombination lines are strongly enhanced through stimulation by the nonthermal nuclear continuum source. The curves in figure 4 were computed by Mebold et al. (1979) from a model in which an HII region with T_e = 5000 K, $E_M = 10^6 cm^{-6}$ pc and angular diameter 10 arcsec lies directly in front of the 10" nonthermal nuclear source which has an intrinsic spectral index of -1.1.

The results suggest a value of n_e of $\sim 2{\times}10^3 cm^{-3}$ and other parameters are similar to those obtained for M82. The 85 GHz thermal continuum flux predicted by this model is 0.15 Jy which is close to the value of ∿ 0.20 Jy expected from the NeII line intensity (Beck et al., 1979) and half the peak value of 0.30 ± .07 Jy measured with a 72" beam by Kellermann (Mebold et al. 1979). It is therefore unlikely that

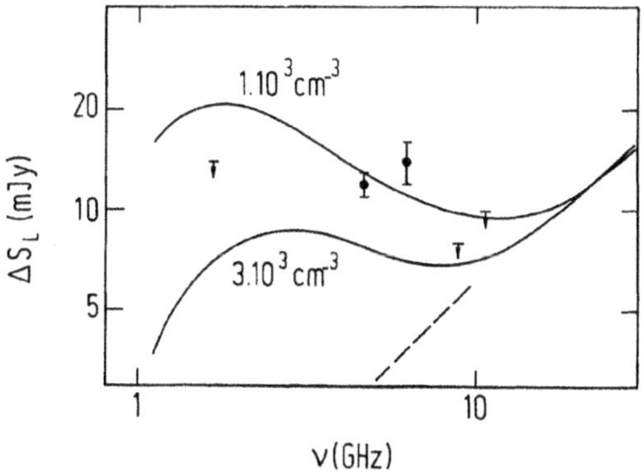

Fig. 4. Variation of peak line flux (ΔS_L) with frequency for NGC253. The smooth curves were computed for the model described in the text with electron densities as indicated. The dashed line shows the spontaneous emission predicted by the model. (Reprinted from Mebold et al. 1979, Astron. Astrophys. 77.)

the line emitting region is much more extended than the 10" size used in the model.

3. SEARCHES FOR RECOMBINATION LINES FROM GALAXIES MORE DISTANT THAN M82 & NGC253

Two independent searches for recombination lines from galaxies more distant than M82 & NGC253 were carried out by Bell and Seaquist (1978) at ARO and by Churchwell & Shaver (1979) at Aricebo. In 1977 Bell & Seaquist observed M83, M51, IC342 and Maffei II at the H102α line frequency. None of the spectra showed evidence for observable line emission to 2σ limits of ∿ 7 mJy. Similarily, Churchwell and Shaver (1979) made a sensitive search of the galaxies NGC315, NGC4278, NGC4631, 3C293, 4C12.50 and 4C31.04 at 1.4 GHz. Except for a possible detection of the H166α line (∿ 4 mJy) toward the nucleus of the edge-on galaxy NGC4631, all results were negative to levels of $\Delta S_L < 10$ mJy (5σ) and $\Delta S_L/S_c \lesssim 10^{-3}$. Since the lines observed in M82 & NGC253 are nearly an order of magnitude more intense than these limits it is clear that they are outstanding objects as seen from our vantage point in space. However, these results may not be too surprising. Bell & Seaquist (1978) showed, using the thermal continuum emission for M51 and IC 342 measured by van der Kruit (1977) and Baker et al. (1977) respectively, that the spontaneous line emission expected was substantially below their 2σ detection limit. Since the nonthermal continuum flux density of the galaxies they observed is low, the line radiation produced by stimulated emission is likely also to be weak.

4. SEYFERT GALAXIES AND QUASARS

The search for recombination lines from extragalactic objects has also extended to Seyfert galaxies and quasars. Since the expected strength of the lines is related to the strength of the continuum source and the properties of the intervening medium, there is every reason to suspect that very distant objects should also show detectable lines unless saturation or other effects are important. Using the Aricebo telescope Churchwell and Shaver (1979) observed 12 quasars and 2 Seyfert galaxies (NGC1068 and 3C120) with negative results. Bell & Seaquist, using the wider observing window (90 MHz) provided by the ARO filter spectrometer, searched a similar number of quasars and several Seyferts, chosen because they were strong, compact, showed a low frequency continuum turnover and had little or no observable polarization at cm wavelengths.

Two Seyfert galaxies (3C84 and OQ208) and one quasar (DA193) were found at ARO to exhibit possible spectral features at frequencies corresponding to their optical emission line redshifts. Although the Seyfert galaxy features have not been confirmed elsewhere both the H99α and H83α lines have now been detected in OQ208 (Figure 5). Since the low frequency turnover in the continuum spectrum of this object has previously been interpreted as being due to intervening ionized gas, (Kojoian et al., 1978; Sanamyan et al., 1978) it seems likely the radio emission features are real. If so it would make this object, with a redshift of 0.0767, (Osterbrock and Cohen, 1979) the most distant source in which a hydrogen recombination line has so far been detected.

In December 1977 a broad absorption feature was observed in the spectrum of the quasar DA193 at ARO which corresponded to the H68α line at a redshift of Z = 2.365. The optical redshift of DA193, listed as tentative (from one line assumed to be Lα) by Wills & Wills (1978) has now been confirmed (D. Wills, private communication).

Since the December 1977 observation, DA193 has been observed eight times at ARO. The broad ($\sim$ 500 km s^{-1}) absorption feature is evident in seven spectra and at 3 different line frequencies (H57α, H67α and H68α). Examples of the ARO spectra are shown in Fig. 6. However attempts to detect this feature by Shaver and Mebold at Effelsberg have so far been unsuccessful. Although this may not be too surprising at 8 GHz, taking into account the width of the observing window (40 MHz), their negative results at 4.5 GHz are more difficult to explain. However a different observing technique was used at the Algonquin Observatory which enabled that instrument to detect very weak, broad lines more reliably and if the feature in DA193 is real, this new technique, along with the wider observing window, would be the most likely explanations for its detection there. In fact, one of the most important results of the search for extragalactic lines at ARO has been the discovery of an observing technique which provides a baseline stability with the wide (90 MHz) observing window comparable to that obtained with much narrower windows. The technique is a simple adaptation to line work of the "wagging" technique used successfully at most observatories to make flux measurements of weak continuum sources. In practice it is similar to the conventional Dicke mode of operation with the exception that the "off-source" spectrum is obtained with the source in the reference beam instead of returning the telescope to the "on-source" starting hour angle. The amplitude of a spectral feature

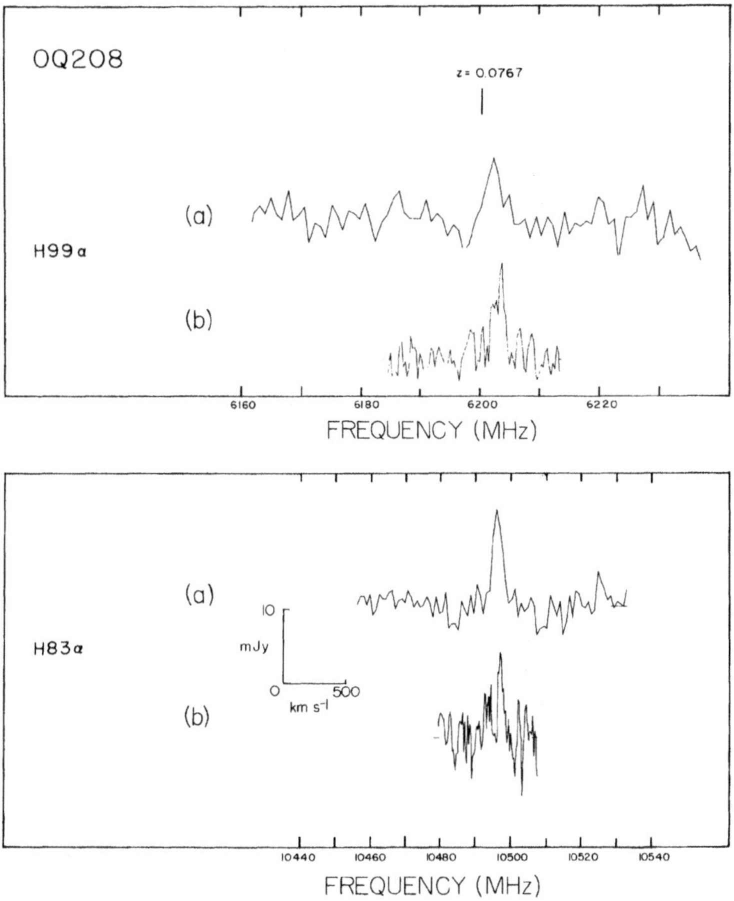

Fig. 5. H99α and H83α spectra for OQ208 observed at the Algonquin Radio Observatory with (a) 900 kHz and (b) 300 kHz resolution. The arrow indicates the LSR velocity derived from the optical emission lines.

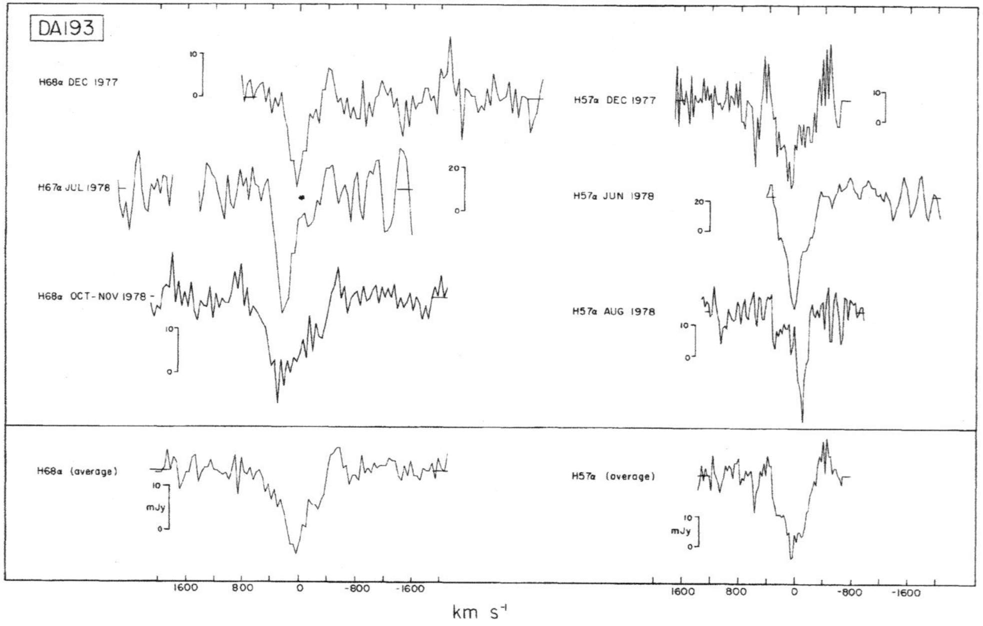

Fig. 6. Examples of DA193 spectra observed at the Algonquin Radio Observatory with 900 kHz resolution. The velocity is LSR and the zero velocity is for a redshift of $z = 2.367$.

in the resulting difference spectrum ("on-source" - "off-source") is thereby increased by a factor of two over that obtained using the conventional Dicke mode. This technique thus combines the Dicke stability with the total power sensitivity. It should be kept in mind however, that the technique can only be used for sources whose angular size is less than the separation between the two beams. For strong sources, the increased stability enables the source produced ripple components which as in other observing techniques are not cancelled in the "on-source" - "off-source" subtraction, to be removed by comparison with a scaled reference spectrum. Even using this technique, however, the results obtained at ARO on DA193 must be viewed with caution until independent confirmation can be obtained.

SUMMARY

The results obtained in the last four years indicate that hydrogen recombination lines can be detected in extragalactic objects and that they are produced primarily by stimulated emission in ionized gas located in front of a nonthermal continuum source. The most likely candidates for detection appear to be nearby edge-on normal galaxies or strong Seyferts which exhibit a low frequency turnover in their continuum spectrum and negligible polarization at centimeter wavelengths. To date no conclusive evidence for recombination lines has been seen in quasars. Whether this is due to maser saturation by intense radiation from the central continuum source, a small "covering factor" as discussed by P. Shaver in the previous paper, or whether the lines like their optical counterparts are wider and thus more difficult to detect, remains uncertain. If saturation is not a problem, there appears to be no reason why the line producing mechanism observed in M82 & NGC253 cannot also occur in quasars and with minor improvements in instrumentation and baseline stability they should be detectable.

REFERENCES

Baker, J.R., Haslam, C.G.T., Jones, B.B., and Wielebinski, R., 1977, Astron. Astrophys., 59, 261.
Beck, S.C., Lacy, J.H. and Geballe, T.R., 1979, Astron. J., 231, 28.
Bell, M.B., and Seaquist, E.R., 1977, Astron. Astrophys., 56, 461.
Bell, M.B., and Seaquist, E.R., 1978, Ap. J., 223, 378.
Chaisson, E.J., and Rodriguez, L.F., 1977, Ap. J., 214, L111.
Churchwell, E., and Shaver, P.A., 1979, Astron. Astrophys., (in press).
Kojoian, G., Dickinson, D.F., Tovmassian, H.M. and Dinger, A.S., 1978, Astrofizika, 14, No. 3, 455.
Kronberg, P.P. and Wilkinson, P.N., 1975, Ap. J., 200, 430.
Mebold, U., Shaver, P.A., Bell, M.B., and Seaquist, E.R., 1979, Astron. Astrophys. (in press).
Osterbrock, D.E., and Cohen, R., 1979, Mon. Not. R. Soc., 187, 61P.
Recillas-Cruz, E., and Peimbert, M., 1970, Bol. Obs. Tonantzintla y Tacubaya, No. 35, p. 247.
Sanamyan, V.A., and Kandalyam, R.A., 1978, Astrofizika, 14, No. 4, 687.

Seaquist, E.R., and Bell, M.B., 1977, Astron. Astrophys., 60, L1.
Shaver, P., Churchwell, E., and Rots, A.H., 1977, Astron. Astrophys., 55, 435.
Shaver, P.A., Churchwell, E. and Walmsley, C.M., 1978, Astron. Astrophys., 64, 1.
van der Kruit, P.C., 1977, Astron. Astrophys., 59, 359.
Wills, D., and Wills, B.J., 1976, Ap. J. Suppl. 31, 143.

CLOSING REMARKS

John S. Mathis
Washburn Observatory
University of Wisconsin-Madison
Madison, Wisconsin, USA

The great potential of radio recombination lines (RL) for providing diagnostics of the physical conditions and abundances in many astronomical objects has been clearly established for several years now. This workshop has shown their possible applications to a still greater variety of sources, and I expect some very exciting science to be learned from them in the future. The big questions are, how reliable are they as diagnostics -- can they provide unambigious information about Nature? Do we really know how they are produced in a given plasma? Are they so subject to the vagaries of fluctuating conditions along the line of sight that they are of very limited quantitative use? A great deal of the workshop has been concerned with these questions. I will address them before giving a short and subjective discussion of applications to various objects. Since often the same point was raised by several speakers, I will not mention individual contributors in most instances.

Let us first consider the basic theory of RL production. Professors Seaton's and Kleppner's talks were very convincing that here there is little uncertainty -- plasma with a given temperature and density will produce and amplify RL's in a known way. Very encouraging! Contrast the situation in even as old a field as stellar interiors, where fundamental quantities (e.g., opacities, to say nothing of convection!) are still unknown or hotly debated. Professor Kleppner showed beautiful experimental results on Stark broadening of high-n atoms. We conclude that atoms really do believe in quantum mechanics, although there are some surprises like the ionization rates involving mixing of different levels. In an astrophysical plasma, with its low densities, line broadening is from impacts but is still well understood. A nice simplification is that for the radio RL collisions mix the various ℓ states with the same n. However, we were warned that in the millimeter and submillimeter range this simplification is no longer true, and it is a very bad approximation in the optical region.

P.A. Shaver (ed.), Radio Recombination Lines, 269–272.

If the theory, then, is so well understood, why is there any debate (e.g., Brown, Lockman, and Knapp 1978) about the interpretation of the observations? Several speakers addressed this question, and I believe that a rough consensus emerged. While one has to be quite careful about simplistic interpretations of middle-frequency ($\sim$ 5 GHz) and certainly low-frequency RL's, the high-frequency lines (say, H66α; 23 GHz) are little affected by the masering and broadening effects which muddy the interpretation of the others. Of course, with observations of many RLs and continuum intensities at various frequencies, including the lower ones, one can extract information about the filling factor and density distribution as well as the temperature. Many speakers emphasized that one is observing only a part of the gas at low frequencies. So far, the Orion Nebula is the only object for which elaborate models seem justified, and their uniqueness has not been properly investigated. Several speakers referred to the "conspiracy" between masering (which boosts the line strength at line center) and collisions (which decrease the strength). For realistic emission measures and densities, these effects become important at about the same n. As a result, one obtains about the right temperature for Orion by making the simplest (and certainly wrong!) assumption -- that the lines are formed with the levels in thermodynamic equilibrium. Lines of widely varying frequency give about the same T_e^*, which fact can be used to derive the filling factor of the nebula to be a few hundreths, as Malcolm Walmsley showed in his talk. It is most interesting that T_e^* does not vary much across the face of the nebula -- an effect which seems to be another conspiracy, since broadening and masering are small at the edges but important at the center. However, the fact that in Orion one finds $T_e \approx T_e^*$ for say, H109α does not imply that it is true in general; one is safe only by observing at high frequencies, say H66α (in which case neither masering nor collisional broadening can be important). The study of lower-frequency lines provides constraints on the geometry of and clumping of the gas in the object.

There is one general problem which is persistent and troublesome -- baseline removal, which has been mentioned several times. In fact, there were suggestions at one point that a workshop on Creative Baseline Removal would be a good idea.

There were many really interesting applications of RL's to various objects. One is the question of gradients of both nebular temperatures and of He abundances in the Galaxy (Churchwell and Mezger's talks). The temperature gradient, expressed by an increase of T_e^* (the "LTE temperature") with D_G (galactic distance). For a given line there is a large spread of T_e^* at a given D_G, but the T_e^* - D_G correlation is present for lines spanning a large range of frequencies. It is true that there is a correlation of nebular excitation parameters, U (proportional to the cube root of the number of recombinations per second), with distance from the sun, so there may be selection effects of the sort suggested by Brown, Lockman, and Knapp (1978). However, there seems to be no correlation of T_e^* with U (Churchwell's paper). I feel that the best interpretation of the T_e^* observations is in terms of a galactic composition gradient -- very reasonable, in view of the gradients observed optically in other galaxies and over a limited range of D_G within our own. The observed gradient in He^+/H^+ as found from RL's (reviewed by Mezger) is also explained by the same heavy element gradient. Panagia showed that a larger Z nearer the center reduces He^+/H^+ by increasing line blanketing, decreasing T_{eff} for a given mass by adding opacity in the interior (thusly increasing the stellar radius), and, most importantly, decreasing the high-mass cutoff in stellar formation by increasing the dust opacity. This explanation seems much more likely to me than having dust absorb He-ionizing photons more efficiently than H-ionizing, which was a reasonable suggestion but has quantative difficulties when confronted with the predicted absorption characteristics of candidates for grain materials.

This closing talk is not the place to summarize the many other interesting talks we have heard. I will just mention some that I find especially exciting or puzzling. The aperture synthesis observations at H109α (van Gorkom), showing tight knots, with widely varying line/continuum ratios, were fascinating. Very narrow hydrogen RLs were mentioned several times. In H II regions, how can ionized hydrogen be so cold and quiescent? Why can't one predict where C^+ RLs will be strong in dark clouds? What ionizes the very narrow hydrogen RLs there? The use of RLs as diagnostics in both the C^+ regions and in planetary nebulae is very exciting. The RLs from the galactic plane, and the survey reported by Lockman, showing the gas is quite cool and not very well correlated kinematically with star formation regions, will put constraints on the conditions in the diffuse ISM throughout the Galaxy. Of course, the galactic center has long been a source of great interest and debate. It is clear that RLs have played a key role in distinguishing between thermal and nonthermal sources. The huge RL widths, supplemented by the [Ne II] measurements, are consistent with Keplerian motion around a very compact mass. The He^+/H^+ observations seem to show geometrical effects of some sort, but the 76α line in Sgr B2 still seems to have significantly different He^+/H^+ ratios as seen from Effelsberg (Mezger) and NRAO (Brown) -- an unusual parallax effect, no doubt. The strong masering of low-frequency lines as discussed by Rod Davis will provide an interesting constraint on conditions in the ISM, although the geometry is a major problem. Extragalactic RL studies are really just beginning. QSO's and active galactic nuclei are promising, and the detection of RLs in M82 in 1976 provided great encouragement. However, they are not strong, and the field of extragalactic RLs has not been investigated very completely. The prospect of seeing them in QSO's is intriguing, although we must realize that the quenching of the forbidden lines in these objects indicates very high densities. Seyferts are also good candidates for future observations.

One aspect of RLs which we did not hear about was solar lines, which are formed in very different conditions than those in nebulae. Berger and Simon (1972) failed to detect high-frequency ($\sim$ 90 GHz)

RLs in heavy ions (O VI, etc.), and Greve (1977) showed that severe broadening makes them very weak, but perhaps still detectable.

I think the Workshop has shown that RLs are potentially powerful tools for studying ionized gas in a large variety of objects and situations. Progress in understanding their limitations seems rapid now, and we can all look forward to learning a great deal about Nature from them.

REFERENCES

Berger, P.S., and Simon, M. 1972, Astrophys. J., **171**, 191
Brown, R.L., Lockman, F.J., and Knapp, G.R. 1978, Ann. Rev. Astron. & Astrophys., **16**, 445
Greve, A., Solar Physics, **52**, 423.

NAME INDEX

Underlined numbers refer to the first page of papers in the present volume contributed by the author in question

SUBJECT INDEX

ASTROPHYSICS AND SPACE SCIENCE LIBRARY

Edited by

J. E. Blamont, R. L. F. Boyd, L. Goldberg, C. de Jager, Z. Kopal, G. H. Ludwig, R. Lüst, B. M. McCormac, H. E. Newell, L. I. Sedov, Z. Švestka, and W. de Graaff

1. C. de Jager (ed.), *The Solar Spectrum, Proceedings of the Symposium held at the University of Utrecht, 26–31 August, 1963.* 1965, XIV + 417 pp.
2. J. Orthner and H. Maseland (eds.), *Introduction to Solar Terrestrial Relations, Proceedings of the Summer School in Space Physics held in Alpbach, Austria, July 15–August 10, 1963 and Organized by the European Preparatory Commission for Space Research.* 1965, IX + 506 pp.
3. C. C. Chang and S. S. Huang (eds.), *Proceedings of the Plasma Space Science Symposium, held at the Catholic University of America, Washington, D.C., June 11–14, 1963.* 1965, IX + 377 pp.
4. Zdeněk Kopal, *An Introduction to the Study of the Moon.* 1966, XII + 464 pp.
5. B. M. McCormac (ed.), *Radiation Trapped in the Earth's Magnetic Field. Proceedings of the Advanced Study Institute, held at the Chr. Michelsen Institute, Bergen, Norway, August 16–September 3, 1965.* 1966, XII + 901 pp.
6. A. B. Underhill, *The Early Type Stars.* 1966, XII + 282 pp.
7. Jean Kovalevsky, *Introduction to Celestial Mechanics.* 1967, VIII + 427 pp.
8. Zdeněk Kopal and Constantine L. Goudas (eds.), *Measure of the Moon. Proceedings of the 2nd International Conference on Selenodesy and Lunar Topography, held in the University of Manchester, England, May 30–June 4, 1966.* 1967, XVIII + 479 pp.
9. J. G. Emming (ed.), *Electromagnetic Radiation in Space. Proceedings of the 3rd ESRO Summer School in Space Physics, held in Alpbach, Austria, from 19 July to 13 August, 1965.* 1968, VIII + 307 pp.
10. R. L. Carovillano, John F. McClay, and Henry R. Radoski (eds.), *Physics of the Magnetosphere, Based upon the Proceedings of the Conference held at Boston College, June 19–28, 1967.* 1968, X + 686 pp.
11. Syun-Ichi Akasofu, *Polar and Magnetospheric Substorms.* 1968, XVIII + 280 pp.
12. Peter M. Millman (ed.), *Meteorite Research. Proceedings of a Symposium on Meteorite Research, held in Vienna, Austria, 7–13 August, 1968.* 1969, XV + 941 pp.
13. Margherita Hack (ed.), *Mass Loss from Stars. Proceedings of the 2nd Trieste Colloquium on Astrophysics, 12–17 September, 1968.* 1969, XII + 345 pp.
14. N. D'Angelo (ed.), *Low-Frequency Waves and Irregularities in the Ionosphere. Proceedings of the 2nd ESRIN-ESLAB Symposium, held in Frascati, Italy, 23–27 September, 1968.* 1969, VII + 218 pp.
15. G. A. Partel (ed.), *Space Engineering. Proceedings of the 2nd International Conference on Space Engineering, held at the Fondazione Giorgio Cini, Isola di San Giorgio, Venice, Italy, May 7–10, 1969.* 1970, XI + 728 pp.
16. S. Fred Singer (ed.), *Manned Laboratories in Space. Second International Orbital Laboratory Symposium.* 1969, XIII + 133 pp.
17. B. M. McCormac (ed.), *Particles and Fields in the Magnetosphere. Symposium Organized by the Summer Advanced Study Institute, held at the University of California, Santa Barbara, Calif., August 4–15, 1969.* 1970, XI + 450 pp.
18. Jean-Claude Pecker, *Experimental Astronomy.* 1970, X + 105 pp.
19. V. Manno and D. E. Page (eds.), *Intercorrelated Satellite Observations related to Solar Events. Proceedings of the 3rd ESLAB/ESRIN Symposium held in Noordwijk, The Netherlands, September 16–19, 1969.* 1970, XVI + 627 pp.
20. L. Mansinha, D. E. Smylie, and A. E. Beck, *Earthquake Displacement Fields and the Rotation of the Earth, A NATO Advanced Study Institute Conference Organized by the Department of Geophysics, University of Western Ontario, London, Canada, June 22–28, 1969.* 1970, XI + 308 pp.
21. Jean-Claude Pecker, *Space Observatories.* 1970, XI + 120 pp.
22. L. N. Mavridis (ed.), *Structure and Evolution of the Galaxy. Proceedings of the NATO Advanced Study Institute, held in Athens, September 8–19, 1969.* 1971, VII + 312 pp.

23. A. Muller (ed.), *The Magellanic Clouds. A European Southern Observatory Presentation: Principal Prospects, Current Observational and Theoretical Approaches, and Prospects for Future Research, Based on the Symposium on the Magellanic Clouds, held in Santiago de Chile, March 1969, on the Occasion of the Dedication of the European Southern Observatory.* 1971, XII + 189 pp.
24. B. M. McCormac (ed.), *The Radiating Atmosphere. Proceedings of a Symposium Organized by the Summer Advanced Study Institute, held at Queen's University, Kingston, Ontario, August 3–14, 1970.* 1971, XI + 455 pp.
25. G. Fiocco (ed.), *Mesospheric Models and Related Experiments. Proceedings of the 4th ESRIN-ESLAB Symposium, held at Frascati, Italy, July 6–10, 1970.* 1971, VIII + 298 pp.
26. I. Atanasijević, *Selected Exercises in Galactic Astronomy.* 1971, XII + 144 pp.
27. C. J. Macris (ed.), *Physics of the Solar Corona. Proceedings of the NATO Advanced Study Institute on Physics of the Solar Corona, held at Cavouri-Vouliagmeni, Athens, Greece, 6–17 September 1970.* 1971, XII + 345 pp.
28. F. Delobeau, *The Environment of the Earth.* 1971, IX + 113 pp.
29. E. R. Dyer (general ed.), *Solar-Terrestrial Physics/1970. Proceedings of the International Symposium on Solar-Terrestrial Physics, held in Leningrad, U.S.S.R., 12–19 May 1970.* 1972, VIII + 938 pp.
30. V. Manno and J. Ring (eds.), *Infrared Detection Techniques for Space Research. Proceedings of the 5th ESLAB-ESRIN Symposium, held in Noordwijk, The Netherlands, June 8–11, 1971.* 1972, XII + 344 pp.
31. M. Lecar (ed.), *Gravitational N-Body Problem. Proceedings of IAU Colloquium No. 10, held in Cambridge, England, August 12–15, 1970.* 1972, XI + 441 pp.
32. B. M. McCormac (ed.), *Earth's Magnetospheric Processes. Proceedings of a Symposium Organized by the Summer Advanced Study Institute and Ninth ESRO Summer School, held in Cortina, Italy, August 30–September 10, 1971.* 1972, VIII + 417 pp.
33. Antonin Rükl, *Maps of Lunar Hemispheres.* 1972, V + 24 pp.
34. V. Kourganoff, *Introduction to the Physics of Stellar Interiors.* 1973, XI + 115 pp.
35. B. M. McCormac (ed.), *Physics and Chemistry of Upper Atmospheres. Proceedings of a Symposium Organized by the Summer Advanced Study Institute, held at the University of Orléans, France, July 31–August 11, 1972.* 1973, VIII + 389 pp.
36. J. D. Fernie (ed.), *Variable Stars in Globular Clusters and in Related Systems. Proceedings of the IAU Colloquium No. 21, held at the University of Toronto, Toronto, Canada, August 29–31, 1972.* 1973, IX + 234 pp.
37. R. J. L. Grard (ed.), *Photon and Particle Interaction with Surfaces in Space. Proceedings of the 6th ESLAB Symposium, held at Noordwijk, The Netherlands, 26–29 September, 1972.* 1973, XV + 577 pp.
38. Werner Israel (ed.), *Relativity, Astrophysics and Cosmology. Proceedings of the Summer School, held 14–26 August, 1972, at the BANFF Centre, BANFF, Alberta, Canada.* 1973, IX + 323 pp.
39. B. D. Tapley and V. Szebehely (eds.), *Recent Advances in Dynamical Astronomy. Proceedings of the NATO Advanced Study Institute in Dynamical Astronomy, held in Cortina d'Ampezzo, Italy, August 9–12, 1972.* 1973, XIII + 468 pp.
40. A. G. W. Cameron (ed.), *Cosmochemistry. Proceedings of the Symposium on Cosmochemistry, held at the Smithsonian Astrophysical Observatory, Cambridge, Mass., August 14–16, 1972.* 1973, X + 173 pp.
41. M. Golay, *Introduction to Astronomical Photometry.* 1974, IX + 364 pp.
42. D. E. Page (ed.), *Correlated Interplanetary and Magnetospheric Observations. Proceedings of the 7th ESLAB Symposium, held at Saulgau, W. Germany, 22–25 May, 1973.* 1974, XIV + 662 pp.
43. Riccardo Giacconi and Herbert Gursky (eds.), *X-Ray Astronomy.* 1974, X + 450 pp.
44. B. M. McCormac (ed.), *Magnetospheric Physics. Proceedings of the Advanced Summer Institute, held in Sheffield, U.K., August 1973.* 1974, VII + 399 pp.
45. C. B. Cosmovici (ed.), *Supernovae and Supernova Remnants. Proceedings of the International Conference on Supernovae, held in Lecce, Italy, May 7–11, 1973.* 1974, XVII + 387 pp.
46. A. P. Mitra, *Ionospheric Effects of Solar Flares.* 1974, XI + 294 pp.
47. S.-I. Akasofu, *Physics of Magnetospheric Substorms.* 1977, XVIII + 599 pp.

48. H. Gursky and R. Ruffini (eds.), *Neutron Stars, Black Holes and Binary X-Ray Sources.* 1975, XII + 441 pp.
49. Z. Švestka and P. Simon (eds.), *Catalog of Solar Particle Events 1955–1969. Prepared under the Auspices of Working Group 2 of the Inter-Union Commission on Solar-Terrestrial Physics.* 1975, IX + 428 pp.
50. Zdeněk Kopal and Robert W. Carder, *Mapping of the Moon.* 1974, VIII + 237 pp.
51. B. M. McCormac (ed.), *Atmospheres of Earth and the Planets. Proceedings of the Summer Advanced Study Institute, held at the University of Liège, Belgium, July 29–August 8, 1974.* 1975, VII + 454 pp.
52. V. Formisano (ed.), *The Magnetospheres of the Earth and Jupiter. Proceedings of the Neil Brice Memorial Symposium, held in Frascati, May 28–June 1, 1974.* 1975, XI + 485 pp.
53. R. Grant Athay, *The Solar Chromosphere and Corona: Quiet Sun.* 1976, XI + 504 pp.
54. C. de Jager and H. Nieuwenhuijzen (eds.), *Image Processing Techniques in Astronomy. Proceedings of a Conference, held in Utrecht on March 25–27, 1975.* XI + 418 pp.
55. N. C. Wickramasinghe and D. J. Morgan (eds.), *Solid State Astrophysics. Proceedings of a Symposium, held at the University College, Cardiff, Wales, 9–12 July 1974.* 1976, XII + 314 pp.
56. John Meaburn, *Detection and Spectrometry of Faint Light.* 1976, IX + 270 pp.
57. K. Knott and B. Battrick (eds.), *The Scientific Satellite Programme during the International Magnetospheric Study. Proceedings of the 10th ESLAB Symposium, held at Vienna, Austria, 10–13 June 1975.* 1976, XV + 464 pp.
58. B. M. McCormac (ed.), *Magnetospheric Particles and Fields. Proceedings of the Summer Advanced Study School, held in Graz, Austria, August 4–15, 1975.* 1976, VII + 331 pp.
59. B. S. P. Shen and M. Merker (eds.), *Spallation Nuclear Reactions and Their Applications.* 1976, VIII + 235 pp.
60. Walter S. Fitch (ed.), *Multiple Periodic Variable Stars. Proceedings of the International Astronomical Union Colloquium No. 29, held at Budapest, Hungary, 1–5 September 1976.* 1976, XIV + 348 pp.
61. J. J. Burger, A. Pedersen, and B. Battrick (eds.), *Atmospheric Physics from Spacelab. Proceedings of the 11th ESLAB Symposium, Organized by the Space Science Department of the European Space Agency, held at Frascati, Italy, 11–14 May 1976.* 1976, XX + 409 pp.
62. J. Derral Mulholland (ed.), *Scientific Applications of Lunar Laser Ranging. Proceedings of a Symposium held in Austin, Tex., U.S.A., 8–10 June, 1976.* 1977, XVII + 302 pp.
63. Giovanni G. Fazio (ed.), *Infrared and Submillimeter Astronomy. Proceedings of a Symposium held in Philadelphia, Penn., U.S.A., 8–10 June, 1976.* 1977, X + 226 pp.
64. C. Jaschek and G. A. Wilkins (eds.), *Compilation, Critical Evaluation and Distribution of Stellar Data. Proceedings of the International Astronomical Union Colloquium No. 35, held at Strasbourg, France, 19–21 August, 1976.* 1977, XIV + 316 pp.
65. M. Friedjung (ed.), *Novae and Related Stars. Proceedings of an International Conference held by the Institut d'Astrophysique, Paris, France, 7–9 September, 1976.* 1977, XIV + 228 pp.
66. David N. Schramm (ed.), *Supernovae. Proceedings of a Special IAU-Session on Supernovae held in Grenoble, France, 1 September, 1976.* 1977, X + 192 pp.
67. Jean Audouze (ed.), *CNO Isotopes in Astrophysics. Proceedings of a Special IAU Session held in Grenoble, France, 30 August, 1976.* 1977, XIII + 195 pp.
68. Z. Kopal, *Dynamics of Close Binary Systems,* XIII + 510 pp.
69. A. Bruzek and C. J. Durrant (eds.), *Illustrated Glossary for Solar and Solar-Terrestrial Physics.* 1977, XVIII + 204 pp.
70. H. van Woerden (ed.), *Topics in Interstellar Matter.* 1977, VIII + 295 pp.
71. M. A. Shea, D. F. Smart, and T. S. Wu (eds.), *Study of Travelling Interplanetary Phenomena.* 1977, XII + 439 pp.
72. V. Szebehely (ed.), *Dynamics of Planets and Satellites and Theories of Their Motion. Proceedings of IAU Colloquium No. 41, held in Cambridge, England, 17–19 August 1976.* 1978, XII + 375 pp.
73. James R. Wertz (ed.), *Spacecraft Attitude Determination and Control.* 1978, XVI + 858 pp.

74. Peter J. Palmadesso and K. Papadopoulos (eds.), *Wave Instabilities in Space Plasmas. Proceedings of a Symposium Organized Within the XIX URSI General Assembly held in Helsinki, Finland, July 31–August 8, 1978.* 1979, VII + 309 pp.
75. Bengt E. Westerlund (ed.), *Stars and Star Systems. Proceedings of the Fourth European Regional Meeting in Astronomy held in Uppsala, Sweden, 7–12 August, 1978.* 1979, XVIII + 264 pp.
76. Cornelis van Schooneveld (ed.), *Image Formation from Coherence Functions in Astronomy. Proceedings of IAU Colloquium No. 49 on the Formation of Images from Spatial Coherence Functions in Astronomy, held at Groningen, The Netherlands, 10–12 August 1978.* 1979, XII + 338 pp.

GPSR Compliance

The European Union's (EU) General Product Safety Regulation (GPSR) is a set of rules that requires consumer products to be safe and our obligations to ensure this.

If you have any concerns about our products, you can contact us on ProductSafety@springernature.com

In case Publisher is established outside the EU, the EU authorized representative is:

Springer Nature Customer Service Center GmbH
Europaplatz 3
69115 Heidelberg, Germany

Batch number: 10761599

Printed by Printforce, the Netherlands